Annals of Mathematics Studies

Number 79

DISCONTINUOUS GROUPS AND RIEMANN SURFACES

Proceedings of the 1973 Conference at the University of Maryland

EDITED BY

LEON GREENBERG

PRINCETON UNIVERSITY PRESS
AND
UNIVERSITY OF TOKYO PRESS

PRINCETON, NEW JERSEY
1974

L.C. Card: 73–16783

ISBN: 0–691–08138–7

Published in Japan exclusively by
University of Tokyo Press;
in other parts of the world by
Princeton University Press

Printed in the United States of America

Library of Congress Cataloging in Publication Data
will be found on the last printed page of this book.

PREFACE

This volume contains almost all of the papers that were read at the Conference on Discontinuous Groups and Riemann Surfaces, which was held at the University of Maryland during May 21-25, 1973. The conference was the third of a sequence of conferences on this subject in recent years. The other two were held at Tulane University (1965) and SUNY at Stony Brook (1969). The Proceedings of the Stony Brook Conference appeared as Advances in the Theory of Riemann Surfaces (Annals Study 66). Thanks are due the National Science Foundation for supporting the conference; the University of Maryland for making its facilities available; the Princeton University Press for assuring rapid publication; and of course the participants of the conference for creating the mathematics contained herein.

L. GREENBERG

CONTENTS

DISCONTINUOUS GROUPS AND RIEMANN SURFACES

CONSTRUCTABILITY AND BERS STABILITY OF KLEINIAN GROUPS

William Abikoff*

The study of small deformations of Kleinian groups was initiated by Poincaré, who claimed without proof that a small deformation of a Fuchsian group results in a discontinuous group. The deformations of surfaces of finite Poincaré area was studied by Teichmüller and he gave this deformation space a real analytic structure. Later, Ahlfors and Bers used the results of Teichmüller, to parametrize the homotopy equivalence classes of finite Riemann surfaces. It was, however, the beautiful result of Teichmüller that there is an intimate tie between the deformations of a surface and the bounded quadratic differentials that live on that surface. Ahlfors finiteness theorem yields the fact that the quotient of the set of discontinuity $\Omega(G)$ of a finitely generated Kleinian group by the group G is a finite union of conformally finite Riemann surfaces. It therefore becomes relevant to consider whether each small deformation of a finitely generated Kleinian group may be induced by deformations of the quotient $\Omega(G)/G$, or, more specifically, infinitesimally by the bounded quadratic differentials on that quotient. These deformations are the quasi-conformal deformations of G whose dilatation is supported on the ordinary set of G. Bers [4], Kra [6], and Maskit [10] have studied these deformations and have woven an elegant theory of those quasi-conformal deformations of Kleinian groups. It is however not clear when all small type-preserving deformations of a finitely generated Kleinian group G are induced by quasi-conformal deformations whose dilatation is supported on its ordinary set. Such groups are called *Bers stable*.

* Research partially supported by the National Science Foundation.

Bers [3] first raised questions of stability in the study of B-groups, and proved that a finitely generated quasi-Fuchsian group is Bers stable. Gardiner and Kra [5] have derived a cohomological condition for Bers stability. Marden [8] showed that a finitely generated Kleinian group without torsion is Bers stable if it is geometrically finite, i.e., has a finite sided Dirichlet region in the upper half-space. We will examine here the relationship between Bers stability and the construction algorithms developed by Maskit and used by him to discover some of the deepest results in Kleinian group theory. Some of the results presented here have been announced in [2]. The combination theorems as stated there have not appeared in print but will appear shortly in Maskit's book.

We will call a group *Maskit-constructible* if it is finitely generated and may be constructed from cyclic groups using the Maskit algorithms. The principal results are:

THEOREM 1. *A Maskit constructible finitely generated Kleinian group is Bers stable.*

COROLLARY. *A totally degenerate group is not Maskit constructible from cyclic groups.*

Since constructibility is a topological property Theorem 1 has a stronger form, conjectured by Bers [4], namely

THEOREM 2. *Any quasi-conformation deformation of a Maskit constructible group is Bers stable, i.e., is uniformly stable.*

These theorems are direct consequences of Theorem 3 and 4. In Section 3, we show

THEOREM 3. *If* G *is a finitely generated Kleinian group, formed by application of Maskit Combination* I *to its subgroups* G_1 *and* G_2, *and both* G_1 *and* G_2 *are Bers stable, then* G *is Bers stable.*

Section 4 is devoted to the proof of

THEOREM 4. *If* G *is a finitely generated Kleinian group which is generated by a subgroup* G_1 *and an element* γ *by application of Maskit Combination* II *and* G_1 *is Bers stable, then* G *is Bers stable.*

We note that the statements of the combination theorems given in Section 2 are not the most general given by Maskit. If H in Combination I or H_1 and H_2 in Combination II are non-elementary quasi-Fuchsian of the second kind, the theorems remain valid. While it may be conjectured that Bers stability is preserved in the above cases, the techniques given here do not yield a proof. No Kleinian groups known to this author are constructible using the most general combination theorems and not constructible using the combination theorems as stated here.

§1. *Kleinian groups and stability*

A *Kleinian group* is a discrete subgroup of the Möbius group

$$\mathfrak{M} = \mathfrak{M}_{\mathbf{C}} = \left\{ z \mapsto \frac{az+b}{cz+d} \mid a, b, c, d, \in \mathbf{C},\ ad\text{-}bc = 1,\ z \in \hat{\mathbf{C}} \right\}$$

which acts discontinuously at some point of $\hat{\mathbf{C}}$ and which is neither trivial nor a finite extension of a cyclic group. If $\Omega(G)$ is the subset of $\hat{\mathbf{C}}$ on which the Kleinian group G acts discontinuously, $\Omega(G)$ is called the *set of discontinuity* of G. $\Lambda(G)$, the *limit set* of G, is $\hat{\mathbf{C}} - \Omega(G)$. G is *elementary* if $\Lambda(G)$ has less than three points. We define the *projection map*

$$\pi = \pi_G : \Omega(G) \to {}^{\Omega(G)}/_G \,.$$

All groups considered here have *finite quotient*, i.e., ${}^{\Omega(G)}/_G$ is a finite union of conformally finite Riemann surfaces.

A Kleinian group G is called *quasi-Fuchsian* if there is a Jordan domain B so that $\gamma(B) = B$ for all $\gamma \in G$. It is of the *first kind* if $\Lambda(G) = \partial B$ and of the *second kind* otherwise. We say that Ω_1 is a component of G if Ω_1 is a component of $\Omega(G)$. If $A \subset \hat{\mathbf{C}}$, then the *stability subgroup* of A, denoted G_A is the set of those $\gamma \in G$ so that $\gamma(A) = A$. If Ω_1 is a component of G, and Ω_{11} is a component of $\mathbf{C} - \bar{\Omega}_1$, then $J = \partial\Omega_1 \cap \partial\Omega_{11}$ is a quasi-circle and is called a *separator* for G (see Abikoff [1]).

A subset F of $\Omega(G)$ is called a *fundamental set* for G if,

$$G\bar{F} = \bigcup_{\gamma \in G} \gamma(\bar{F}) \supset \Omega(G)$$

and $F \cap \gamma F = \phi$ for each $\gamma \in G - \{id\}$.

If $f : \hat{\mathbf{C}} \to \hat{\mathbf{C}}$ is a quasi-conformal homeomorphism, and $G_1 = fGf^{-1} < \mathfrak{M}_{\mathbf{C}}$ for $G < \mathfrak{M}_{\mathbf{C}}$, then G_1 is called a *quasi-conformal deformation* of G.

We will discuss those stability notions relevant to the following discussion. Detailed arguments may be found in Bers [3] and [4]. Let $\chi : G \to \mathfrak{M}_{\mathbf{C}}$ be a homomorphism. χ is called *type-preserving* if $\mathrm{trace}^2(\chi(\gamma)) = \mathrm{trace}^2(\gamma)$ for each parabolic or elliptic element of G. If $\gamma_1, \ldots, \gamma_r$ generate G, $\chi(G)$ is called a *small deformation* of G if $\chi(\gamma_i)\gamma_i^{-1}$ is close to the identity. G is called *quasi-conformally stable* if each deformation close to the identity is quasi-conformal and *Bers stable* if it is quasi-conformally stable and the map f can be taken conformal on $\Lambda(G)$.

The deformation space, $\chi(G)$, is the space of type-preserving homomorphisms of G into $\mathfrak{M}_{\mathbf{C}}$. It has a natural structure as an affine algebraic variety. There is a canonical injection of $B_2(G)$ into the tangent space to $\chi(G)$ at the identity. If $\beta = \dim_{\mathbf{C}} B_2(G) = \tau(G) - 3$, where $\tau(G)$ is the dimension of $\chi(G)$ at the identity, and if $\chi(G)$ is locally irreducible at the identity, then G is Bers stable. Locally irreducibility is preserved by the processes given here, hence the argument reduces to a dimension count.

Let G be a Kleinian group and H be a subgroup of G. A topological disc B in $\hat{\mathbf{C}}$ is called *precisely invariant* under H if $\gamma(B) = B$ for $\gamma \in H$ and $\gamma(B) \cap B = \phi$ for $\gamma \in G - H$. If G has finite quotient then we claim H is trivial, cyclic or a finitely generated quasi-Fuchsian group of the first or second kind. Since $B \cap \Lambda(G) = \phi$, it follows that $J = \partial B$ is either a subset of $\Lambda(G)$ or $J \cap \Omega(G)$ is contained in a single component of $\Omega(G)$. In the first case J is a separator and in the latter case, the image of $J \cap \Omega(G)$ in $\Omega(G)/G$ is a finite set of disjoint simple loops. $\pi(B)$ is a topologically finite (possibly ramified) surface. The two cases now coalesce. If J is a separator then H is finitely generated as is shown in [1]. Otherwise H is the cover group of a planar cover of a finite surface, hence is finitely generated. It must be either elementary or quasi-Fuchsian of the second kind by Marden [7] and Maskit [9].

§2. *The Maskit combination theorems*

The combination theorems are techniques for building a Kleinian group from simpler groups. Combination I gives a geometric condition that a group G be describable as a free product of two groups G_1 and G_2 with amalgamated common subgroup H. Combination II gives such a condition for adjoining a transformation γ to a group G_0 which conjugates a subgroup H_1 of G_0 into a subgroup H_2. We proceed to state the theorems.

2.1 COMBINATION THEOREM I. *Let* G_1 *and* G_2 *be Kleinian groups with a common subgroup* H, *and* B_1 *and* B_2 *precisely invariant discs under* H *with* $J = \partial B_1 = \partial B_2$. *We further assume:*

(I-i) H *is trivial, cyclic, or quasi-Fuchsian of the first kind.*

(I-ii) $\gamma_i(B_i) \cap B_i = \phi$ *for* $\gamma_i \in G_i - H$, $i = 1, 2$, *and*

(I-iii) *if* F_i *is a fundamental set for* G_i, *then*

$$F'_i = F_i \cap G_i(B_i) \subset B_i, \quad i = 1, 2 .$$

(I-iv) $F'_i \cap F_j$ *has interior.*

We then have the following conclusions:

(I-v) *the group* G *generated by* G_1 *and* G_2 *is the free product of* G_1 *and* G_2 *with amalgamated subgroup* H.

(I-vi) $\Omega(G)/_G$ *is the union of* $(\Omega(G_1) - G_1(B_1))/_G$ *and* $(\Omega(G_2) - G_2(B_2))/_{G_2}$ *with* $(J \cap \Omega(H))/_H$ *identified in each.*

2.2. COMBINATION THEOREM II. *Let* G_0 *be a finitely generated Kleinian group with subgroups* H_1 *and* H_2 *and* B_i *a topological disc precisely invariant under* H_i, $i = 1, 2$. *We assume:*

(II-i) H_1 *and* H_2 *are either trivial, cyclic or finitely generated quasi-Fuchsian of the first kind.*

(II-ii) *There is a transformation*

$$\hat{\gamma} \in \mathfrak{M}_C - G_0 \text{ so that } \hat{\gamma} H_1 \hat{\gamma}^{-1} = H_2 .$$

(II-iii) $\gamma(B_1) \cap B_2 = \phi$ *for each* $\gamma \in G_0$.

(II-iv) $\hat{\gamma}(\partial B_1) = \partial B_2$.

(II-v) $\hat{\gamma}(B_1) \cup B_2 = \phi$.

G_0 *has a fundamental set* F *with the following properties,*

(II-vi) $F \cap B_i$ *is a fundamental set for the action of* H_i *on* B_i.

(II-vii) $F \cap (C - B_1 \cup B_2)$ *has non-void interior.*

We then have the following conclusions,

(II-viii) *The group* G *generated by* G_0 *and* γ *is Kleinian.*

(II-ix) $\Omega(G)/_G$ *is exactly*

$$(\Omega(G_0) - G_0(B_1 \cup B_2))/_{G_0}$$

with the identification, induced by

$$\hat{\gamma}, \text{ of } \pi(\partial B_1) \text{ with } \pi(\partial B_2) .$$

§3. *Proof of Theorem 3*

We maintain the notion of 2.1 and set $G_3 = H$, $\beta_i = \dim_C B_2(G)$, with the following conventions, for $i = 1, 2, 3$:

$\beta_i = 0$ if G_i is trivial
$\beta_i = -2$ if G_i is cyclic parabolic or elliptic
$\beta_i = -3$ if G_i is cyclic loxodromic.

Since we assume G_1 and G_2 are Bers stable, we have $\tau(G_i) = \beta_i + 3$, $i = 1, 2$, and we wish to show $\tau(G) = \beta + 3$.

It is quite simple to compute β using property (I-vi). If H is quasi-Fuchsian of the first kind,

$$\beta = \beta_1 + \beta_2 - \beta_3 \tag{3.1}$$

since $\Omega(G)/_G$ differs from $\left(\Omega(G_1)/_{G_1}\right) \cup \left(\Omega(G_2)/_{G_2}\right)$ only, topologically, in the loss of $\Omega(H)/_H$.

We now assume $J \cap \Omega(G_1)$ is contained in the component Ω_i of G_i, $i = 1, 2$.

If H_i is not quasi-Fuchsian of the first kind,

$$\Omega(G)/_G = \left((\Omega(G_1) - G_1\Omega_1)/_{G_1}\right) \cup \left((\Omega(G_2) - G_2\Omega_2)/_{G_2}\right) \cup \Delta \tag{3.2}$$

where Δ is $G_1(\overline{\Omega_1 - B_1})/_{G_1} \cup G_2(\overline{\Omega_2 - B_2})/_{G_2}$ with $G_1(\partial B_1 \cap \Omega(G_1))/_{G_1}$ identified pointwise with $G_2(\partial B_2 \cap \Omega(G_2))/_{G_2}$. $\Omega(G)/_G$ has exactly one less connected component than $\left(\Omega(G_1)/_{G_1}\right) \cup \left(\Omega(G_2)/_{G_2}\right)$ and

$$\beta = \beta_1 + \beta_2 - \beta_3 + 3\ . \tag{3.3}$$

We now compute $\sigma(G)$, for the two above cases. If $\hat{G}$ is a small deformation of G, i.e., $\hat{G} = \chi(G)$, then we obtain $\hat{H} = \chi(H)$ must satisfy a compatibility condition. More precisely $\hat{H}$ considered as a subgroup of either $\widehat{G_1}$ or $\widehat{G_2}$ must be the same group. This enforces a restriction on those small deformations of G_1 and G_2 which are restrictions of deformations of G. The condition that $\hat{H}$ must satisfy to be a common subgroup is that it have the identical conformal structure and position. In the case that H is non-elementary, the number of additional complex conditions is $\beta_3 + 3_0$. If H is trivial it is three. If H is parabolic or elliptic cyclic, it is two and, if H is loxodromic cyclic, it is three. That these new conditions are independent amongst themselves follows from the Bers stability of finitely generated quasi-Fuchsian groups and cyclic groups. Were we to show that these conditions are independent at the identity in $\chi(G)$, then the implicit function theorem guarantees that the natural injection of

$$X(G) \to X(G_1) \times X(G_2)$$

defined by restriction would have maximal rank, at the identity, equal to $\beta + 3$ and G would be Bers stable. Therefore it remains to show that the conditions implicit in the statement $\chi(H) = \hat{H}$ are not consequences of the relations defining $X(G_1) \times X(G_2)$. If H is cyclic or trivial, so is the statement, since the relations defining $X(G_i)$ cannot specify the location of fixed points or, infintesimally, the multipliers of loxodromic transformations. If H is a non-elementary quasi-Fuchsian group, then the conformal structure of ${}^{B_i}/_{H_i}$ is not specified on $X(G_i)$ and we obtain $\beta_{3/2}$ independent conditions for $i = 1, 2$. Likewise, by conjugation by Mobius transformations near the identity, we obtain three additional independent conditions, and we obtain the required number. The proof of Theorem 3 is complete.

§4. *Proof of Theorem* 4

The proof of Theorem 4 differs only technically from that of Theorem 3. We use the notation of 2.2. First using the properties of G_0, H_1, and γ, we compute $\beta = \dim_C B_2(G)$. Let $\beta_0 = \dim_C B_2(G_0)$. We consider the various possible cases.

If H_1 is quasi-Fuchsian of the first kind, let $\beta_1 = \dim_C B_2(H_1)$. It then follows that

$$\beta = \beta_0 - \beta_1 \tag{4.1}$$

since, topologically,

$$\Omega(G)/_G = \left(\Omega(G_0)/_{G_0}\right) - \left(B_1/_H \cup B_2/_{H_2}\right).$$

If H_1 is trivial,

$$\beta = \beta_0 + 3 \ . \tag{4.2}$$

If H_1 is parabolic or elliptic cyclic,

$$\beta = \beta_0 + 1 \ . \tag{4.3}$$

If H_1 is loxodromic cyclic, one obtains $\Omega(G)/_G$ from $\Omega(G_0)/_{G_0}$ by opening $\Omega(G_0)/_{G_0}$ along two homotopically non-trivial, non-intersecting simple loops and identifying the border components in different ways. There are several cases, but all yield

$$\beta = \beta_0 \ . \tag{4.4}$$

We compute $\sigma(G)$ when H_1 is trivial, cyclic and quasi-Fuchsian of the first kind, and show that in these cases, G is Bers stable. Again, the implicit function theorem on the natural injection $i : X(G) \hookrightarrow X(G_0) \times X(<\hat{\gamma}>)$.

If H_1 is trivial, i has maximal rank $\tau(G) \leq \tau(G_0) + 3$. G_0 is Bers stable hence $\tau(G_0) = \beta_0 + 3$ and

$$\beta + 3 \leq \tau(G) \leq \beta + 3 \ . \tag{4.5}$$

Thus G is Bers stable.

If H_1 is the cyclic group $<\theta_1>$ and $H_2 = <\theta_2>$, we have the conjugacy relation $\hat{\gamma}\theta_1\hat{\theta}^{-1} = \theta_2$ on $X(G)$ which does not exist on $X(G_0)$ by 2.2 assumptions (II-iv) and (II-v). We next compute the local dimension at the identity of $p_1 \circ i$, where p_1 is the projection

$$p_1 : X(G_0) \times X(<\hat{\gamma}>) \to X(G_0) \ .$$

If θ is parabolic or elliptic, the conjugacy relation gives two analytic conditions on $X(G_0)$, hence

$$\tau(G) \leq \tau(G_0) + 1 .$$

This gives, together with (4.3), the stability of G.

If θ is loxodromic, the conjugacy relation gives three analytic conditions on $X(G_0)$ and

$$\tau(G) \leq \tau(G_0) .$$

Equation (4.4) gives the stability of G.

COLUMBIA UNIVERSITY

REFERENCES

[1]. Abikoff, W., Residual Limit sets of Kleinian groups, *Acta Math.* 130 (1973), pp. 127-144.

[2]. ———, On the decomposition and deformation of Kleinian groups. To appear.

[3]. Bers, L., On boundaries of Teichmüller spaces and on Kleinian groups, I, *Ann. of Math.* 91 (1970), pp. 570-600.

[4]. ———, Spaces of Kleinian groups, *Lecture Notes in Math.*, 155 (1970), Springer, Berlin, pp. 9-34.

[5]. Gardiner, F., and Kra, I., Stability of Kleinian groups. To appear.

[6]. Kra, I., Deformations of Fuchsian groups, *Duke Math. J.* 36 (1969), pp. 537-546.

[7]. Marden, A., Geometry of finitely generated Kleinian groups. To appear.

[8]. ———, On finitely-generated Fuchsian groups, *Comment. Math. Helv.* 42 (1967), pp. 81-85.

[9]. Maskit, B., On boundaries of Teichmüller spaces and on Kleinian groups, II, *Ann. of Math.* 91 (1970), pp. 607-639.

[10]. ———, Self-mappings on Kleinian groups, *Amer. J. of Math.* 93 (1971), pp. 840-856.

ALGEBRAIC CURVES AND HALF-CANONICAL LINEAR SERIES

Robert D. M. Accola

1). *Introduction*

Let W_p be a Riemann surface of genus $p \geq 2$, and let $(\pi iE,B)^{p\times 2p}$ be a period matrix associated with a particular canonical homology basis of W_p. Torelli's theorem states that the conformal structure of W_p is determined by (the symplectic equivalence class of) the $p \times p$ matrix B. Consequently, one can hope to extract from the B matrix information about W_p, in particular whether or not W_p has some property not shared by all surfaces of genus p. Recently surfaces admitting certain groups of automorphisms have been characterized by properties of the B matrix. These results have often concerned vanishing properties of the theta function, $\theta(u; B)$, associated with the Jacobian of W_p, $J(W_p)$. It turns out that similar kinds of vanishing properties for $\theta(u; B)$ occur when the Riemann surface in question admits certain plane models.

Let $g(r, n)$ denote a linear series of dimension r and degree n. On W_p a $g(r, p-1)$ will be called *half-canonical* if the double of every divisor in $g(r, p-1)$ is canonical. A half-canonical $g(r, p-1)$ will be denoted $g^*(r)$. By Riemann's solution to the Jacobi inversion problem it follows that there is a one-to-one correspondence between the complete $g^*(r)$'s on W_p and the half-periods of $J(W)$ where $\theta(u; B)$ vanishes. In fact, if the half-period $e(\varepsilon J(W))$ corresponds to $g^*(r)$, then $\theta(u; B)$ vanishes at e to order $r + 1$. Thus extraordinary vanishing properties of θ at half-periods correspond to the existence of extraordinary half-canonical linear series. Therefore, we shall make statements in the language of half-canonical linear series in describing problems and results realizing that such statements can be readily translated into the language of vanishing properties of $\theta(u; B)$ at half-periods.

We now give some pertinent facts concerning plane algebraic curves. These facts are well known. If $F_n(x, y, z)$ is an irreducible homogeneous polynomial of degree n over $\mathbf{C}$ and the algebraic curve $F_n = 0$ (which we will simply denote by F_n) in $\mathbf{P}^2$ has only s ordinary singularities of multiplicities $k_1, k_2, \ldots, k_s$, then the genus of the corresponding Riemann surface, W_p, is given by

$$p = (n-1)(n-2)/2 - d \tag{1}$$

where $d = \sum_1^s k_j(k_j-1)/2$. Moreover it follows immediately:

$$2p - 2 = n(n-3) - 2d\ . \tag{2}$$

(If $k_j = 2$, the singularity is called a node and if all $k_j = 2$ then d is simply the number of nodes. d is zero if and only if F_n is non-singular.) An *adjoint* to F_n is a curve f_m (possibly reducible) with singularities where those of F_n occur, and where F_n has an ordinary k_j-fold singularity f_m has a singularity of multiplicity at least $k_j - 1$. (If all $k_j = 2$ then an adjoint is required to pass through the singular points of F_n. If $d = 0$ any f_m is an adjoint.) A *special adjoint*, f_{n-3}, is an adjoint of degree $n-3$. In addition to the $\sum k_j(k_j-1)$ $(=2d)$ intersections of a special adjoint with F_n occuring at the singularities, there are $2p-2$ other "free" points, by formula (2). In fact, the divisors of "free points" cut out on F_n by the special adjoints form the complete canonical series. By a *double adjoint* we shall mean a curve of degree $2n-6$ having at each k_j-point of F_n a singularity of multiplicity $2k_j - 2$. The "free points" of the double adjoints cut out the complete bicanonical series on F_n.

We shall refer to the following result of Severi in Section 2) [5]. Fix positive integers n and d and consider the family of plane curves of degree n having d nodes among its singularities. Each component of the family which contains a curve with precisely d nodes has dimension $n(n+3)/2 - d$.

We shall also refer to *simple* $g(1,n)$'s. Since a $g(1,n)$ is composite in the usual terminology we shall call a $g(1,n)$ simple if the map from W_p onto P^1 given by this $g(1,n)$ admits no proper factorization. Otherwise the $g(1,n)$ will be called composite.

Finally, let T_p denote Teichmüller space for genus $p \geq 2$. (T_p could just as well stand for Torelli space or the Jacobian locus of the Siegel upper half space.) The (complex) dimension of T_p is $3p-3$. Let H_p be the locus in T_p corresponding to hyperelliptic surfaces. Then the dimension of H_p is $2p-1$; that is, the codimension of H_p in T_p is $p-2$.

2). *Riemann surface admitting a* $g^*(2)$

We first consider a Riemann surface, W_p, admitting a plane model F_{p-1} where the $g(2,p-1)$ cut out by the lines of P^2 is half-canonical. By formula (1) $d = (p-1)(p-6)/2$ since $n = p-1$. It follows that the $g(2,p-1)$ is half-canonical if and only if there is an adjoint of degree $p-6$, f_{p-6} [4]. Thus, if L is a line then $f_{p-6}L^2$ cuts out a canonical divisor. We give three examples.

(i) $p = 6$; $g^*(2,5)$. F_5 is a non-singular plane quintic.

(ii) $p = 7$, $g^*(2,6)$. F_6 has three collinear nodes. For example, $F_4F_1^2 - F_3^2$ has nodes where F_1 and F_3 intersect.

(iii) $p = 9$, $g^*(2,8)$. F_8 has twelve nodes lying on a cubic. For example, $F_2F_3^2 - F_4^2$. But for this example the given $g^*(2,8)$ is not complete.

Let $G^*(r)$ be the locus in T_p of surfaces admitting a complete $g^*(r)$. We will now try to compute the dimensions of the various components of $G^*(2)$. Let $G_s^*(2)$ be the components of $G^*(2)$ containing the simple $g^*(2)$'s. As a guess we give an argument that indicates that each component of $G_s^*(2)$ may have codimension three in T_p. This result may seem surprising since $G^*(2)$ is the locus in T_p where $\frac{\partial\theta}{\partial u_1}(e;B) = \dots = \frac{\partial\theta}{\partial u_p}(e;B) = 0$ for some half period e where θ has odd order. Thus the codimension, three, is considerably less than p, the number of equations defining the locus in terms of the theta function.

If we first consider F_{p-1}'s with $(p-1)(p-6)/2$ (=d) nodes, but with no restriction on the position of the nodes, then Severi's result gives the dimension of this family of curves to be $(p-1)(p+2)/2-d$. Since we wish the nodes to lie on a f_{p-6}, the number of degrees of freedom for the nodes is reduced by $d-(p-6)(p-3)/2$; for the first $(p-6)(p-3)/2$ nodes determine f_{p-6} and the remaining nodes have but one degree of freedom since they must lie on the f_{p-6} so determined. Thus the family of F_{p-1}'s with d nodes on a f_{p-6} has dimension $(p-1)(p+2)/2-d-(d-(p-6)(p-3)/2)=3p+2$. Since every Riemann surface has at most a finite number of $g^*(2)$'s, the dimension of $G_s^*(2)$ will be $3p+2$ minus the dimension, eight, of the groups of collincations of P^2. Thus we guess that $G_s^*(2)$ has dimension $3p-6$ (= dim T_p-3).

The method of arriving at the guess furnishes a proof only for $p=6$, 7 and 8. For $p=9$ the twelve nodes may not be chosen arbitrarily on a f_3; only eleven can be chosen arbitrarily. The guess has been verified for $p=6, 7, 8, 9$ and 10.

The guess does not really deserve the status of a conjecture since there is some counter evidence. For $p=10$ there are two components of $G_s^*(3)$ of different dimensions. Also for $p=6$ there is a question of the dimension of all components of $G^*(2)$. For a W_6 is hyperelliptic if and only if (i) $W_p \in G^*(2)$ and (ii) every $g^*(2)$ is composite. Thus $H_6 \subset G^*(2)$. But it is not clear, at least to the author, whether or not each component of H_6 is a component of $G^*(2)$ or whether H_6 is in $G_s^*(2)$. If the former case held, then those components of $G^*(2)-G_s^*(2)$ would have codimension 4, the known codimension of H_6 in T_6.

Of course, for $p=5$, $G^*(2)=H_5$, so all components of $G^*(2)$ have codimension three in T_5. Also every $g^*(2)$ on a W_5 is composite.

3). *Syzygetic families of half-canonical linear series*

DEFINITION. Let G be a subgroup of the half-periods of $J(W_p)$ of order $2^V (0 \leq V \leq p)$. A coset, H, of G in the half-periods of $J(W_p)$ will be called a *syzygetic coset* if the orders of vanishing of $\theta(u;B)$ at the points of H all have the same parity (odd or even). If there are 2^V

complete $g^*(r_i)$'s on W_p, $i = 1, 2, \ldots, 2^V$; $r_i \geq 0$; $r_i \equiv r_j \pmod 2$, corresponding to the points of H then the $g^*(r_i)$'s will be called a *syzygetic family*.

For example, any complete $g^*(r)$ will be considered a syzygetic family of order one ($V = 0$). If $g^*(r_1)$ and $g^*(r_2)$ are distinct, complete, and $r_1 \equiv r_2 \pmod 2$ then they form a syzygetic family of order two ($V = 1$). $g^*(r_i)$, $i = 1, 2, 3, 4$, will form a syzygetic family of order four ($V = 2$) if (i) they are distinct and complete, (ii) $r_1 \equiv r_2 \equiv r_3 \equiv r_4 \pmod 2$, and (iii) if D_i is an arbitrary divisor from $g^*(r_i)$ we have that the divisor $D_1 + D_2 + D_3 + D_4$ is bicanonical.

Riemann surfaces admitting automorphisms of period two (involutions) usually admit extraordinary syzygetic families. Our purpose in this section and the next is to exhibit Riemann surfaces as plane curves which admit syzygetic families but which do not admit obvious involutions.

3.0. $V = 0$; $g^*(p/3)$'s: A theorem due essentially to Castelnuovo [3] asserts that $g^*(r, p-1)$ is composite for $r \geq 2$ if $p < 3r$. If $3|p$ a problem arises of exhibiting Riemann surfaces admitting a simple $g^*(p/3)$. For $p = 6$ and $p = 9$ the examples given in Section 2) suffice. A simple $g^*(4)$ on a W_{12} can be exhibited by considering a F_7 with a single ordinary triple point. Conics through the triple points cut out a $g^*(4)$ since special adjoints are quartics with double points at the triple point of F_7. Similarly, arbitrary conics cut out a $g^*(5)$ on a non-singular F_7 ($p = 15$). In a similar manner examples can be constructed for higher genera p where $3|p$.

Now we ask whether the existence of a simple $g^*(p/3)$ on W_p implies the existence of such a plane model. For $p = 6$ the existence of a simple $g^*(2)$ implies, by definition, a non-singular plane quintic for a model. For $p = 9$ the existence of a simple $g^*(3)$ implies a plane model whose equation is $F_3^2 F_2 - F_4^2$. The method of Castelnuovo's theorem shows that the $g^*(3)$ gives a space model for W_9 which is the complete intersection of a quadric and quartic surface. Then eliminating one variable (projecting from a point not on the quadric) gives an equation of a plane curve of the indicated form.

3.1. $V = 1$; two $g^*((p-2)/3)$'s: By an extension of the theorem of Castelnuovo cited above in subsection 3.0 a Riemann surface of genus p cannot admit two simple $g^*(r)$'s for $p-2 < 3r$ ($p=4$ is excluded). Examples of plane curves of genus $p = 2+3r$ admitting two simple $g^*((p-2)/3)$'s are now given.

If A and B are points of P^2 let L_{AB} be the line connecting them. Let A, B, C, and D be four points of P^2 no three of which are collinear. Let P be the point where L_{AB} and L_{CD} meet and let Q be the point where L_{AC} and L_{BD} meet.

For $p = 5$ consider a curve F_7 with nodes at A, B, C, and D and with ordinary triple points at P and Q (formula (1)). The pencil of lines through Q cuts out a $g(1,4)$ on F_7. Since $L_{AB} L_{CD} L_Q^2$ is a special adjoint if L_Q is a line through Q we see that this $g(1,4)$ is half-canonical. Similarly, the lines through P cut out a $g^*(1)$. It can be shown that every Riemann surface of genus five without involutions admitting two $g^*(1)$'s admits such a plane model [2].

For $p = 8$ consider F_9 with nodes at A and D, ordinary triple points at B and C and ordinary four points at P and Q. Consider conics, F_2, through P, Q, and B. Then $L_{AC} L_{CD} F_2^2$ is a special adjoint and so the F_2's cut out a $g^*(2)$ on F_7. Similarly for conics through P, Q and C.

3.2. $V = 2$; four $g^*((p-3)/3)$'s: By an extension of the theorem of Castelnuovo cited above in subsection 3.0, a surface of genus p cannot admit a syzygetic family of four simple $g^*(r)$'s for $3r > p-3$. Examples are now given to show this bound is sharp.

Using the notation of 3.1 let R be the point where L_{AD} and L_{BC} meet, that is, the third diagonal point for the quadrangle A, B, C, D.

Consider F_8 with ordinary triple points at A, B, C, and D and nodes at P, Q, and R. Then $p = 6$ (formula (1)). If L_A is a line passing through A then $L_{BC} L_{CD} L_{BD} L_A^2$ is a special adjoint and so the pencil of such L_A's cut out a $g^*(1)$ on F_8. Using analogous notation for B, C and D we see that

$$L_A L_B L_C L_D L_{AB} L_{AC} L_{AD} L_{BC} L_{BD} L_{CD}$$

is a double adjoint and so the sum of the four linear series is bicanonical. It can be shown that a W_6 without involutions admitting a syzygetic family of four $g^*(1)$'s admits such a plane model.

Consider a F_{10} with ordinary four points at A, B, C and D and nodes at P, Q, and R. Then $p = 9$. Let F_A be a conic through B, C and D. Then $F_A^2 L_{AB} L_{AC} L_{AD}$ is a special adjoint so the F_A's cut out a $g^*(2)$ on F_{10}. Since

$$F_A F_B F_C F_D L_{AB} L_{AC} L_{AD} L_{BC} L_{BD} L_{CD}$$

is a double adjoint, the sum of the four $g^*(2)$'s is bicanonical.

4). *Construction of more general syzygetic families*

We now give a more general procedure for producing plane curves which have fairly obvious syzygetic families. All the previous examples for $V = 1$ and 2 are special cases of this procedure.

Let k and t be positive integers with $k \geq t$. Let F_{k+4} be an irreducible plane curve with t tacnodes at points $P_1, P_2, \ldots, P_t$ no three of which are collinear. Let $L_1, \ldots, L_t$ be the t tacnodal tangents and suppose they are distinct and all meet in a single point Q. Suppose F_{k+4} has an ordinary k-fold singularity at Q. Since each tacnode contributes two to d of formulas (1) and (2) we have

(3) $$d = 2t + k(k-1)/2 \text{ and so}$$

(4) $$2p - 2 = (k+4)(k+1) - 4t - k(k-1) = 6k - 4t + 4 .$$

Special adjoints have degree $k+1$ and must be tangent to L_j at P_j in addition to the usual condition at Q. In the following computations we must consider separately the cases k even and odd although the final results are the same for both cases.

Suppose k is odd. Pick an integer s so that $0 \leq 2s \leq t$. In one of the $\binom{t}{2s}$ possible ways, choose $2s$ of the lines $L_1, \ldots, L_t$, say $L_1, L_2, \ldots, L_{2s}$. Then $L_1 L_2 \ldots L_{2s} f_m^2$ will have degree $k+1$ if $m = (k+1-2s)/2$. Suppose $f_{(k+1-2s)/2}$ passes through $P_{2s+1}, \ldots, P_t$ and has a $(k-1-2s)/2$-fold singularity at Q. Then $L_1 L_2 \ldots L_s f_m^2$ will be a special adjoint. The "free points" of such f_m's cut out on F_{k+4} a $g(R, N)$ where

$$N = (k+4)(k+1-2s)/2 - 2(t-2s) - k(k-1-2s)/2 = p-1$$

and

$$R \geq (k+1-2s)(k+7-2s)/8 - (t-2s) - (k-1-2s)(k+1-2s)/8$$

or

$$R \geq k-t+1 \ .$$

Thus such $f_{(k+1-2s)/2}$'s cut out $\binom{t}{2s}$ $g^*(R)$'s. Considering all choices of $2s$ of the lines we get $2^{t-1}\left(= \sum \binom{t}{2s}\right)$ $g^*(R)$'s which can be shown to form a syzygetic family. It is to be expected that in general $R = k-t+1$ for all the $g^*(R)$'s.

For k even and s chosen so that $0 \leq 2s+1 \leq t$, choose $2s+1$ lines, say $L_1, L_2, \ldots, L_{2s+1}$. Then $L_1 L_2 \ldots L_{2s+1} f_m^2$ will be a special adjoint if $m = (k-2s)/2$, f_m passes through $P_{2s+2}, \ldots, P_t$ and f_m has a $(k-2s-2)/2$-fold singularity at Q. The "free points" of such f_m's again cut out on F_{k+4} a $g^*(R)$, $R \geq k-t+1$, and so we again obtain a syzygetic family of order $2^{t-1}\left(= \sum \binom{t}{2s+1}\right)$.

There remains, of course, the problem of the actual existence of irreducible algebraic curves F_{k+4} of genus $p = 3k-2t+3$ having the prescribed singularities. Call a pair of integers (t, k), $k \geq t > 0$, *admissible* if for any $t+1$ points $P_1, \ldots, P_t, Q$, no three of which are collinear, there exists an irreducible algebraic plane curve F_{k+4} of genus $p = 3k - 2t + 3$ having the prescribed singularities. By using Bertini's theorem (concerning irreducible curves in a linear family of curves) and other standard methods we can obtain the following results about admissible pairs:

(1) $(t, t+w)$ is admissible if $t \leq 13 + 5w$ $(w \geq 0)$

(2) If (t, k) is admissible so is $(t, k+1)$.

Now set $V = t - 1$ for the following theorem.

THEOREM 1. *For integers* V, k *so that* $(V+1, k)$ *is admissible, there exists a Riemann surface of genus* $p = 3k - 2V + 1$ *admitting a syzygetic family of* 2^V $g^*(R)$'s *where each* $R \geq k - V$.

The cases considered in subsection 3.1 occur when $V = 1$ and $k = 2$ or 3. One transforms the curves derived in Theorem 1 by a standard quadratic transformation with fundamental points at the two tacnodes and at any third non-singular point of F_{k+4}. The cases considered in subsection 3.2 occur when $V = 2$ and $k = 3$ or 4. One transforms F_{k+4} by a standard quadratic transformation with the fundamental points at the three tacnodes.

The methods used for the above constructions can be considerably extended by considering singularities of F where several tacnodes occur on each line L and/or more complicated tacnodes occur on each L. The restriction on admissible pairs (t, k) given in (1) immediately preceding the statement of Theorem 1 seems inherent in the method of construction but hardly seems inherent in the nature of things. Nevertheless, the above constructions give a sharpness for the following generalization of Castelnuovo's theorem which subsumes all the special instances cited in Section 3). We omit the proof [1].

THEOREM 2. *For* $V = 1, 2,$ *or* 3 *and any* $k > V$, *a surface of genus* $p < 3k - 2V + 1$ *admitting a syzygetic family of* $g^*(k-V)$'s *of order* 2^V *must admit an involution and either no* $g^*(k-V)$ *of the syzygetic family is simple or* W_p *is hyperelliptic.*

Castelnuovo's theorem and its extension can also be used to show that all R's of Theorem 1 are equal to $k - V$ for low values of V, namely $V = 1, 2, 3,$ and 4.

UNIVERSITY OF MANCHESTER AND BROWN UNIVERSITY

REFERENCES

[1]. Accola, Robert D. M., Vanishing properties of theta functions for abelian covers of closed Riemann surfaces. Part III (to appear).

[2]. Accola, Robert D. M., Some loci of Teichmüller space for genus five defined by vanishing theta-nulls (to appear).

[3]. Castelnuovo, G., Sur multipli du una serie lineare di gruppi di punti. *Rendiconti del Circolo Matematico di Palermo* Vol. VIII (1893), pp. 89-110.

[4]. Kraus, Note über ausgewohnlicke Specialgruppen auf algebraischen Curven. *Mathematischen Annalen* Vol. XVI (1880), p. 310.

[5]. Severi-Löffler, *Vorlesunger über algebraische Geometrie*, Leipzig 1921 (Chelsea).

SUFFICIENT CONDITIONS FOR QUASI-CONFORMAL EXTENSION

Lars V. Ahlfors*

1. A locally homeomorphic mapping of a sphere into itself is automatically a homeomorphism. This simple topological fact can be used to prove that a function on a subregion is schlicht. Indeed, it will be schlicht if it can be extended to a locally homeomorphic mapping of the whole sphere. The easiest way to show that a mapping is locally homeomorphic is to show that the Jacobian is $\neq 0$. This will be so if the mapping f satisfies $f_z \neq 0$ and

$$|f_{\bar{z}}| \leqq k|f_z| \tag{1}$$

for some $k < 1$. Hence the importance of quasi-conformal extensions.

To be specific, suppose that f satisfies (1) together with $f_z \neq 0$ in $\Delta = \{|z| < 1\}$. If $k \leqq k' < 1$, under what additional conditions can f be extended to a k'-q.c. mapping of the Riemann sphere? The question may be regarded as consisting of two parts:

1) When does there exist a k'-q.c. locally homeomorphic mapping f^* of $\Delta^* = \{|z| > 1\}$ such that f and f^* have equal continuous extensions to $|z| = 1$.

2) If f and f^* have this property, when do they together form a locally homeomorphic mapping of the sphere, and hence a k'-q.c. homeomorphism of the sphere on itself.

There are some trivial answers to the second question. For instance, if f and f^* extend to locally homeomorphic mappings of the closed disks $\bar{\Delta}$ and $\bar{\Delta}^*$, then they determine a global homeomorphism. More

---

* Research partially supported under NSF Grant GP-38886.

refined results would be interesting, but seem to be difficult to come by. At present the best method is to approximate f by a sequence of functions f_n with good boundary behavior that allow k_n-q.c. extensions f_n^*. If $k_n \to k'$ it is possible to extract a subsequence which converges to a k'–q.c. extension of the original f.

2. Not much generality is lost by assuming that $k = 0$ or, in other words, that f is conformal. Indeed, by the general existence theorem for quasi-conformal mapping there exists a q.c. homeomorphism h of the sphere on itself that preserves symmetry with respect to the unit circle and satisfies $h_{\bar{z}}/h_z = f_{\bar{z}}/f_z$ in Δ. This means that $f \circ h^{-1}$ is conformal in Δ, and if f has a quasi-conformal extension so does $f \circ h^{-1}$, and vice versa.

Of course, since h is not explicitly known, the general case remains less tractable. It is therefore less by choice than of necessity that we limit ourselves to the discussion of holomorphic f. If a holomorphic f has a k-q.c. extension to the sphere we shall say that f is k-schlicht.

3. Two separate sufficient conditions are known. In 1962 Ahlfors and Weill [1] proved that f is k-schlicht if its Schwarzian derivative satisfies

$$|\{f,z\}|\ (1-|z|^2)^2 \leqq 2k \ . \tag{2}$$

This result has been important for the theory of Teichmüller spaces and Kleinian groups. In 1972 Becker [2] proved the sufficiency of the condition

$$|z f''/f'|\ (1-|z|^2) \leqq k \ . \tag{3}$$

This was an improvement on a result by Duren, Shapiro, and Shields [3].

Actually, Becker proves much more. Under the condition (2) or (3) he shows that f is the initial link in a chain of subordinate schlicht functions. This leads to a Löwner-type differential equation and to more intrinsic proofs than by the original method. In this paper we do not pursue this interesting aspect, but we prove the following generalizations of conditions (2) and (3).

THEOREM. *In order that* f *be* k-*schlicht it is sufficient that either*

(A) $$\left| z \frac{f''}{f'} (1-|z|^2) + c|z|^2 \right| \leqq k$$

for some complex constant c *with* $|c| \leqq k$, *or that*

(B) $$\left| \frac{1}{2} \{f, z\} (1-|z|^2)^2 - c(1-c)\bar{z}^2 \right| \leqq k|c|$$

with $|c-1| \leqq k$.

Note that (2) is (B) with $c = 1$, and (3) is (A) with $c = 0$.

4. We proceed directly to the proof, but under the preliminary condition that f is holomorphic on the closed disk $\bar{\Delta}$. It is convenient to define the extension in the form

$$F(z) = \begin{cases} f(z) & \text{for } |z| \leqq 1 \\ g(1/\bar{z}) & \text{for } |z| \geqq 1 \end{cases}$$

where g is sense reversing and k-q.c. in Δ with $g = f$ on $\{|z| = 1\}$.

If we set $g = f + u$ the boundarv condition is $u = 0$, and the quasi-conformality is expressed by

(3) $$|f' + u_z| \leqq k|u_{\bar{z}}| \;.$$

In addition we require that $u_{\bar{z}} \neq 0$ in $\bar{\Delta}$ and that u is sufficiently regular, for instance real-analytic on $|z| = 1$. In these circumstances it is evident that F will be a k-q.c. homeomorphic extension of f.

Nothing prevents us from making u dependent on f. If we choose

$$u = f'/\sigma$$

condition (3) becomes

(4) $$\left|\sigma \frac{f''}{f'} + \sigma^2 - \sigma_z\right| \leqq k|\sigma_{\bar{z}}| \;,$$

and $1/\sigma$ must be zero on $|z| = 1$. It is also required that $\sigma_{\bar{z}}/\sigma^2 \neq 0$ in the closed unit disk.

As a second step we substitute

$$\sigma = v - \frac{1}{2} f''/f'$$

in (4). This introduces the Schwarzian derivative

$$\phi(z) = \{f, z\} = f'''/f' - \frac{3}{2} (f''/f')^2 ,$$

for the condition is now

$$\left|\frac{1}{2}\phi + v^2 - v_z\right| \leqq k\,|v_{\bar{z}}| \tag{5}$$

together with $v \to \infty$ for $|z| \to 1$ and $v_{\bar{z}}/v^2 \neq 0$.

The reader is asked to verify that condition (A) is (4) with $\sigma = (c+1)\bar{z}\,(1-|z|^2)^{-1}$, and that (B) is (5) with $v = c\bar{z}\,(1-|z|^2)^{-1}$. The boundary condition is fulfilled, and the gradient does not vanish provided that $c \neq -1$ in the first case and $c \neq 0$ in the second. The theorem has now been proved for functions that remain regular on $|z| = 1$.

5. The passage to arbitrary f is quite easy due to the special form of (A) and (B). We shall apply the part of the theorem that has already been proved to the functions f_r defined by $f_r(z) = f(rz)$, $0 < r < 1$. For this purpose we have to show that if f satisfies (A) or (B), so does f_r, provided that c conforms to the conditions in the theorem.

This is a simple verification. If f satisfies (A), then

$$\left| rz\,\frac{f''}{f'}(rz) + \frac{cr^2|z|^2}{1-r^2|z|^2} \right| \leqq \frac{k}{1-r^2|z|^2} .$$

We want to deduce that

$$\left| rz\,\frac{f''}{f'}(rz) + \frac{c|z|^2}{1-|z|^2} \right| \leqq \frac{k}{1-|z|^2}$$

if $|c| \leqq k$. This conclusion is trivial in view of the identity

$$\frac{|z|^2}{1-|z|^2} - \frac{r^2|z|^2}{1-r^2|z|^2} = \frac{1}{1-|z|^2} - \frac{1}{1-r^2|z|^2} .$$

Similarly, if f satisfies (B), then

$$\left|\frac{1}{2} r^2\phi(rz)(1-r^2|z|^2)^2 - c(1-c)r^4\bar{z}^2\right| \leqq k|c|r^2 ,$$

while we need to know that

$$\left|\frac{1}{2} r^2\phi(rz)(1-|z|^2)^2 - c(1-c)\bar{z}^2\right| \leqq k|c| .$$

If $|c-1| \leqq k$ the second inequality follows from the first in view of the fact that

$$\left|\frac{\bar{z}^2}{(1-|z|^2)^2} - \frac{r^4\bar{z}^2}{(1-r^2|z|^2)^2}\right| \leqq \frac{1}{(1-|z|^2)^2} - \frac{r^2}{(1-r^2|z|^2)^2} .$$

We conclude that f_r has a k-q.c. extension, and on letting r tend to 1 through a subsequence we find that f itself has a k-q.c. extension, thereby completing the proof of the theorem.

6. It is of some interest to examine condition (B) a little more closely. The condition restricts the value of $\{f,z\}$ to a disk whose center and radius vary with z. There is some advantage in passing to the quantity

$$\alpha(f,z) = \frac{1}{2} z^2\{f,z\}(1-|z|^2)^2 ,$$

for with this notation (B) becomes

$$\text{(B}') \qquad |\alpha(f,z) - c(1-c)|z|^4| \leqq k|c|\,|z|^2 ,$$

which is a radially symmetric constraint. This is a more natural form of the condition, for the full region of variability of $\alpha(f,z)$ is a function of $|z|$ alone.

The theorem asserts that f is k-schlicht if for some fixed c with $|c-1| \leqq k$ the value of $\alpha(f,z)$ lies in a variable disk $K(c,|z|)$. Observe first that c and $1-c$ correspond to disks with the same center, but that c gives the larger radius if $\operatorname{Re} c < \frac{1}{2}$. For this reason we consider only values c with $|c-1| \leqq k$, $\operatorname{Re} c \leqq \frac{1}{2}$.

We set $c = 1 - re^{i\theta}$ and keep $r \leqq k$ fixed. For $r \leqq \frac{1}{2}$ the center $c(1-c)$ describes a smooth closed curve, but for $r > \frac{1}{2}$ the curve is heart shaped. The radius of $K(c, |z|)$ has a maximum of $k(1+r)|z|^2$ when the center is at the extreme left, and a minimum $k(1-r)|z|^2$ when the center is on the right. This lack of rotational symmetry is suggestive, but in the absence of asymmetric necessary conditions no further conclusions can be made.

For k and $|z|$ close to 1 and $c = 1+k$ the disk $K(c, |z|)$ contains points at distance nearly 4 from the origin. On the other hand, if f is schlicht we know by Nehari's theorem that $|\alpha(f, z)| \leqq 3$. This leads to a curious consequence that has nothing to do with schlicht functions.

COROLLARY. *If* ϕ *is holomorphic and satisfies*

$$\left|\frac{1}{2} z^2\phi(z)(1-|z|^2)^2 - c(1-c)|z|^4\right| \leqq |c|\,|z|^2$$

in the unit disk for some constant c *with* $|c-1| < 1$, *then it also satisfies*

$$|\phi(z)|(1-|z|^2)^2 \leqq 6\ ,$$

and this is not an arithmetic triviality when, for instance, $\sqrt{3} \leqq c \leqq 2$.

7. Condition (B) is the result of arbitrary choices and differs only by its relative simplicity from many similar conditions. Clearly, we can get a whole family of such conditions if we replace f by $f \circ L$ where L is a linear transformation. An interesting formula is obtained by having f defined in the upper half-plane $\{y > 0\}$ and using a linear transformation that maps the half-plane on the unit disk with it mapped into 0. It is found that f has a k-q.c. extension if

$$|2y^2\{f,z\} - c(c-1)\left(\frac{\overline{z}+it}{z+it}\right)^2| \leqq k|c|$$

with $|c-1| \leqq k$ and $t > 0$. This seems too complicated to be useful, but if allowed to let t tend to ∞ one would arrive at the condition

$$|2y^2\{f,z\} - c(c-1)| \leqq k|c|$$

where the disk is no longer variable. Without additional assumptions the writer has not been able to show that this is a sufficient condition.

HARVARD UNIVERSITY

REFERENCES

[1]. Ahlfors, L. V., and Weil, G., A uniqueness theorem for Beltrami equations. Proc. Am. Math. Soc. 13 (1962), pp. 975-978.

[2]. Becker, J., Löwnersche Differentialgleichung und quasi-konform fortsetzbare schlichte Funktionen, Journal f. reine u. angewandte Math., 255 (1972), pp. 23-43.

[3]. Duren, P. L., Shapiro, H. S., and Shields, H. L., Singular measures and domains not of Smirnov type. Duke Math. J. 33 (1966), pp. 247-254.

FUNDAMENTAL DOMAINS FOR KLEINIAN GROUPS

A. F. Beardon

§1. *Introduction*

For simplicity we shall consider Fuchsian groups only until further notice. We suppose then that G is a Fuchsian group which acts on the unit disc Δ or the upper half-plane Σ. A *fundamental domain* D for G acting in Δ is a subset of Δ such that

A1: D is open and connected and

A2: each point of Δ is G-equivalent to exactly one point in D or at least one point in $\bar{D}$ (the closure of D in Δ).

The definition when G acts on Σ is clear and in this case too we shall use $\bar{D}$ to denote the closure of D in Σ.

This is the usual definition yet with it one can prove very little. Many writers have considered problems concerning fundamental domains and have found it necessary to impose additional requirements (we mention, for example [1], [2], [3], [4]). We shall study the relationships between certain common additional requirements: in particular we show that the almost universal requirement that D is locally finite (see Section 2) is usually redundant.

§2. *Fundamental domains*

There is one additional property which we shall immediately impose: this is for convenience only. If D is a fundamental domain and if $z \in D$ then $D-\{z\}$ is also a fundamental domain. To avoid the subsequent complications of this we shall henceforth require that a fundamental domain shall satisfy

A3: $D = \text{Int}\,\bar{D}$.

If D_1 satisfies A1 and A2 then, as is easily seen, $D = \text{Int}\, \overline{D}_1$ satisfies A1, A2 and A3 and is the unique maximal fundamental domain which contains D_1.

The following result is now clear: *if* D *is a fundamental domain then* D *is simply connected.* Indeed if not we find (using A3) that D separates some V(D) from $\{|z| = 1\}$. This implies that D separates V(D), $V^2(D), \ldots$ from $\{|z| = 1\}$ which is false.

We come now to the additional requirements which are usually imposed on D. These fall into two categories: (i) those concerned with the G-equivalence of the boundary ∂D of D in Δ and (ii) those concerned with the topology or geometry of D.

We say that a fundamental domain D is *locally finite* if and only if each point in Δ has a neighborhood meeting only finitely many images of D.

Next, we say that ∂D is *accessible* if every point of ∂D (in Δ) is accessible from within D. A stronger assumption is that ∂D consists of Jordan arcs: stronger still is the assumption of convexity.

Finally, for each z in Δ we denote by n(z) the number (possibly $+\infty$) of V in G such that $z \in V(\overline{D})$. The condition that D has sides which are paired implies that $n(z) \geq 2$ on ∂D: if D is locally finite then $2 \leq n(z) < +\infty$ on ∂D.

§3. *Locally finite domains*

We begin by proving

THEOREM 1. *Let* D *be a fundamental domain such that* (i) ∂D *is accessible and* (ii) $n(z) \geq 2$ *on* ∂D. *Then* D *is locally finite. Further, neither* (i) *nor* (ii) *alone is sufficient to imply that* D *is locally finite.*

Proof. We shall call a point z in Δ *exceptional* if every neighborhood of z meets infinitely many distinct images of D and shall denote the set of exceptional points by E. We shall show that (i) and (ii) imply that E is countable whereas if D is not locally finite, then E is uncountable.

It is clear that exceptional points can only occur on the boundary of some image of D: as (ii) holds they occur only in sets of the form $U(\partial D) \cap V(\partial D)$ with U and V in G. If z_i and z_j are two points in $U(\partial D) \cap V(\partial D)$ then by (i) there is a Jordan curve $\sigma_{i,j}$ in $U(D) \cup V(D) \cup \{z_i, z_j\}$. Now take three distinct points z_1, z_2 and z_3 in $U(\partial D) \cap V(\partial D)$. We find that (apart from a relabelling) z_3 is separated from $\{|z| = 1\}$ by $\sigma_{1,2}$ and thus z_3 is not exceptional. We conclude that each of the countable sets $U(\partial D) \cap V(\partial D)$, $U, V \in G$, contains at most two exceptional points and so E is countable.

We now suppose that D is not locally finite. We may assume (by conjugation) that $0 \in \partial D$ and that there are distinct elements V_n of G and points z_n in $V_n(D)$ such that $z_n \to 0$ as $n \to \infty$.

As D is not locally finite it cannot be bounded in Δ, thus we can find points w_n in $V_n(D)$ with $|w_n| \to 1$ as $n \to \infty$. We can now construct a curve σ_n which joins z_n and w_n in $V_n(D)$: If $0 < t < 1$, then apart from a finite set of n, σ_n meets $\{|z| = t\}$ in t_n and these distinct points accumulate at some point z_t. Clearly $|z_t| = t$, $z_t \in E$ and so E is uncountable.

We come now to the two examples to show that neither (i) nor (ii) alone suffices.

EXAMPLE 1. We define V_1 and V_2 by

$$V_1(z) = 2z, \; V_2(z) = (3z+8)/(z+3) .$$

Then the group G generated by V_1 and V_2 acts on Σ and has a (locally finite) fundamental region given by

$$D = \{z \in \Sigma : 2 < |z| < 4, \; |z-3| > 1, \; |z+3| > 1\} .$$

Now write D_1 for the intersection of D with the open quadrant $\{x > 0, y > 0\}$ and D_2 the intersection of D with the quadrant $\{x \leq 0, y > 0\}$. We first replace D_2 by $V_2(D_2)$ which adjoins D along the side $|z-3| = 1$ of D. Next, it is a matter of elementary geometry to show that D_1 may be replaced by

$$D_1{}^* = \{x+iy = z : 2 < x < 4,\ y > 0,\ |z-3| > 1\}$$

and so $V(D_2) \cup D_1{}^*$ is now a convex fundamental region for G which is not locally finite. Precisely, the images of this region under V_1^{-n} (n = 1, 2, ...) accumulate along the positive y axis. [In order to see that D_1 may be replaced by $D_1{}^*$ it is only necessary to use powers of V_1 and to consider the parts of D_1 above and below the tangent to $|z-3| = 1$ from the origin.]

We conclude that the modified fundamental domain is convex (and so (i) holds) but is not locally finite. We note for future use that Δ/G is a punctured torus and that there is a hyperbolic fix-point in $\bar{D}$, the closure of D in the extended plane.

EXAMPLE 2. We consider the same group G and domain D as in Example 1. For $n = 1, 2, \ldots$ define $z_n = 2 \exp(\pi i/4n)$. Our aim is to again modify D to produce a new fundamental region D_0. First, D_0 shall contain the quadilateral in D with vertices $-4, -2, z_1, 2z_1$. Next, we replace the quadilateral with vertices $z_1, 2z_1, 2z_3, z_3$ by the non-convex polygon with vertices $z_1, 2z_1, 4z_2, 2z_3, z_3, 2z_2, z_1$. The next stage will be to make a similar modification to the quadilateral with vertices $z_3, 2z_3, 2z_7, z_7$. This is replaced by the polygon with vertices

$$z_3,\ 2^{-1}z_4,\ 2^{-2}z_5,\ 2^{-1}z_6,\ z_7,\ 2z_7,\ z_6,\ 2^{-1}z_5,\ z_4,\ 2z_3,\ z_3$$

and this gives a displacement of greater amplitude and in the opposite direction to the previous one. Continuing in the obvious way we get a new fundamental domain D_0 for G which is not locally finite. In fact every point on the geodesic joining +1 to +3 is on the boundary of $V_1{}^n(D_0)$ for every n. Indeed the points w on this geodesic have the property that their cycle on $\bar{D}_0$ is infinite yet they have a neighborhood which meets only those images of $\overline{D}_0$ which contain w on their boundary. In any event, (ii) holds but D_0 is not locally finite.

§4. *Quotient spaces*

The group G induces an equivalence relation on Δ and another on $\overline{D}$: we shall use π and π_D to denote the natural maps

$$\pi : \Delta \to \Delta/G\,,\ \ \pi_D : \overline{D} \to \overline{D}/G$$

where $\overline{D}$, $\overline{D}/G$ and Δ/G have the natural topologies.

Next, there is a well defined bijection $\theta : \overline{D}/G \to \Delta/G$ defined by

$$\theta(\pi_D(z)) = \pi(z), \qquad (z \in \overline{D})\ .$$

Observe that if A is an open subset of Δ/G, then

$$\pi_D^{-1}(\theta^{-1}(A)) = \overline{D} \cap \pi^{-1}(A)$$

and, as π and π_D are continuous, $\theta^{-1}(A)$ is open in $\overline{D}/G$. We conclude that θ is continuous.

We thus have the situation

$$\begin{array}{ccc} \overline{D} & \xrightarrow{\ \ \iota\ \ } & \Delta \\ {\scriptstyle \pi_D}\downarrow & & \downarrow{\scriptstyle \pi} \\ \overline{D}/G & \xrightarrow{\ \ \theta\ \ } & \Delta/G \end{array}$$

where (i) ι is the inclusion map (ii) all maps are continuous and (iii) $\pi_D \circ \theta = \pi \circ \iota$.

The following result is simple.

THEOREM 2. D *is locally finite if and only if* θ *is a homeomorphism.*

Proof. We suppose first that D is locally finite: we need only show that θ is an open map. Thus we consider an open subset A of $\bar{D}/G$ and suppose that $\zeta \in A$.

As D is locally finite we may write

$$\pi_D^{-1}\{\zeta\} = \{\zeta_0, \zeta_1, \ldots, \zeta_s\}$$

where $\zeta_j = V_j(\zeta_0)$, $V_j \in G$ and V_0 is the identity. Because of this and the fact that $\pi_D^{-1}(A)$ is open in $\bar{D}$ there is a disc N with center ζ_0 such that

(i) $V_j(N) \cap D \subset \pi_D^{-1}(A)$, $j = 0, 1, \ldots, s$, and

(ii) $V^{-1}(\bar{D}) \cap N \neq \phi$ if and only if V is some V_j.

Now observe that by (i) and (ii)

$$\begin{aligned}\pi(N) &= \pi(N \cap \bigcup_j V_j^{-1}(\bar{D})) \\ &= \bigcup_j \pi(\bar{D} \cap V_j(N)) \\ &= \bigcup_j \theta\pi_D(\bar{D} \cap V_j(N)) \\ &\subset \theta(A) .\end{aligned}$$

We deduce that $\theta(\zeta)(= \pi(\zeta_0))$ is in $\pi(N)$ which is an open subset (as π is open) of $\theta(A)$. It follows that $\theta(A)$ is open.

The reverse implication is even easier. If D is not locally finite then there are points z_n in D and distinct V_n in G with $V_n(z_n)$ converging to some point ζ on the boundary of D. The set $K = \{z_n\}$ is closed in D for if N is a compact neighborhood of ζ, $z_n \in V_n^{-1}(N)$ and so $|z_n| \to 1$ as $n \to \infty$.

Observe also that $\pi_D^{-1}\pi_D(K) = K$ and, as K is closed in D, $\pi_D(K)$ is closed in $\bar{D}/G$. Note however that

$$\pi(z_n) = \pi(V_n(z_n)) \to \pi(\zeta)$$

and so $\pi(K)$ is not closed in Δ/G. We conclude that θ is not a homeomorphism as $\pi(K) = \theta\pi_D(K)$.

As G is a Fuchsian group we can adjoin to Δ the parabolic fix-points of G to form the space Δ^*: the corresponding quotient space (with the appropriate topology) is denoted by Δ^*/G and we extend the action of π to Δ^*. In a similar way we consider $\bar{D}^*$ and $\bar{D}^*/G$.

THEOREM 3. D *is locally finite if and only if for every convergent sequence* ζ_n *on* Δ^*/G *there is a convergent sequence* z_{n_j} *on* $\bar{D}^*$ *with* $\pi(z_n) = \zeta_n$.

Proof. If D is not locally finite we can find points z_n in D and a point z^* in $\bar{D}$ such that $V_n(z_n) \to z^*$, V_n distinct elements of G. Obviously $\zeta_n = \pi(z_n)$ converges on Δ^*/G: we need only prove that z_n has no convergent subsequence in $\bar{D}^*$. First, we must have $|z_n| \to 1$ as $n \to \infty$ and so a subsequence z_{n_j} can only converge (if at all) to a parabolic fix-point p. However, as is well known, we can find a G-invariant system of horocycles at the orbit of p which do not meet some neighborhood of z^*. We conclude that the z_n cannot lie in any of these horocycles and so do not converge in $\bar{D}^*$.

We now assume that D is locally finite. Our first task is to show that the local finiteness of D extends to neighborhoods of parabolic fix-points.

LEMMA 1. *Let* D *be locally finite and let* p *be any parabolic fix-point with a horocycle* $\bar{H}$ *at* p. *Then* $\bar{D}^*$ *meets a positive finite number of images of* $\bar{H}$.

With this result available we proceed as follows. If ζ_n is a convergent sequence on Δ^*/G, say to ζ, we can find a point z in $\bar{D}^*$ with $\pi(z) = \zeta$, points z_n in $\bar{D}^*$ with $\pi(z_n) = \zeta_n$ and a neighborhood N of z such that

$$z_n \in \Big[\bigcup_{V \in G} V(N)\Big] \cap \bar{D}^* .$$

(If z is a parabolic fix-point, N is a horocycle.) If $z \in \bar{D}$, then only finitely many copies of $\bar{D}^*$ meet N; thus only finitely many copies of N meet $\bar{D}^*$ and so z_n contains a convergent subsequence. A slight modification is required when z is a parabolic fix-point: this is justified by Lemma 1.

Proof of Lemma 1. It is convenient to conjugate so that G acts on Σ, H is $\{x+iy : y > t\}$ (so $p = \infty$) and the stabilizer J of ∞ is generated by $P : z \to 1 + z$.

As D is connected, $V(\bar{D}) \subset H$, $V(\bar{D})$ meets the line $L = \{y=t\}$ or $V(\bar{D})$ lies "beneath" L. If $U(\bar{D})$ and $V(\bar{D})$ are subsets of H, then $UV^{-1}(H)$ meets H and so $JU = JV$. If $V(\bar{D})$ meets L, then for some n, $P^n V(\bar{D})$ meets the compact set $\{y=t, 0 \leq x \leq 1\}$ and this can only happen for a finite set of elements of G.

We conclude that $V(\bar{D})$ meets $\bar{H}$ if and only if

$$V \in JV_1 \cup \ldots \cup JV_s$$

for some finite subset $\{V_1 \ldots V_s\}$ of G. This is equivalent to the conclusion of the Lemma.

For the moment, we restrict our attention to finitely generated groups of the first kind. We shall briefly discuss results (which are well-known) of the nature "D is finite-sided." Without referring to "sides" of D we are interested in the number of V in G such that $\bar{D} \cap V(\bar{D}) \neq \emptyset$. If this holds for an infinite sequence V_n we select z_n in $\bar{D} \cap V_n(\bar{D})$ and observe that if D is locally finite the Finiteness Theorem together with Theorem 3 forces the z_n to converge (on a subsequence) to a parabolic fix-point p in the topology of $\bar{D}^*$: this is the crucial step. In view of Lemma 1 we see that for some V in G, $V_n = P^{k_n}V$ where P generates the stabilizer of p (again on a subsequence) and where the k_n are distinct integers. It is now clear that this cannot be so if D is reasonably well-behaved in the horocyclic neighborhoods. The problem is thus reduced solely to the behavior of D in these neighborhoods regardless of the nature of D except that it be locally finite.

§5. *Remarks on Fuchsian groups*

The assumption that D is locally finite leads to other results which we only sketch here.

An *elliptic fix-point* must be represented on $\bar{D}$. The cycle may be infinite: it is finite if D is locally finite.

A *hyperbolic fix-point* can occur on the boundary of a convex domain (Example 1) or on the boundary of a locally finite domain (such a situation is easily constructed using a cyclic group). If, however, D is convex and locally finite then there are no hyperbolic fix-points on the boundary of D [1].

Example 3 (below) shows that *parabolic fix-points* need not appear on the boundary of D. If D is locally finite, then every parabolic fix-point is represented on the boundary of D (Lemma 1).

EXAMPLE 3. We construct a sequence of disjoint semicircles Q_n in Σ the closed diameters of which are disjoint, lie in the interval $0 < x < \frac{1}{2}$ and converge to the origin. We let Q_n' be the image of Q_n under reflection in the line $x = \frac{1}{2}$. We construct hyperbolic elements H_n which preserve Σ and map Q_n to Q_n', we define P by $P(z) = z + 1$ and let G be the group generated by P and the H_n. This has the region outside the Q_n and Q_n' and inside the strip $\{x+iy : y < 0,\ 0 < x < 1\}$ as a locally finite fundamental domain. Observe that the only limit points on the boundary of this region are ∞ (a parabolic fix-point) and 0 and 1 (which are not parabolic fix-points).

We may suppose the Q_n are labelled in the obvious way so that x_n separates the diameters of Q_n and Q_{n+1} where

$$x_0 = \frac{1}{2} > x_1 > x_2 > \ldots, \quad x_n \to 0 .$$

The closed region D_n bounded by the *Euclidean* polygon with vertices $x_n, x_{n+1}, x_{n+1} + i/x_{n+1}, 1 + i/x_{n+1}, 1 + i/x_n, x_n + i/x_n, x_n$ may be replaced by its image $H_n(D_n)$. If this is done for all n we obtain a fundamental domain for G which has $z = 1$ as the only limit point on its boundary and this is not a parabolic fix-point.

Finally we remark that, as we have seen, D may be convex but not locally finite. Example 1 shows that this is so and in this case Δ^*/G has genus one. The author can prove that in the case of the Modular group (and certain other groups) any convex fundamental domain is necessarily locally finite. It remains an open question, however, as to whether or not there exists a group G and a fundamental domain D such that (i) Δ^*/G has genus zero and (ii) D is convex but not locally finite.

§6. *Kleinian groups*

Let G be a Kleinian group with set of discontinuity Ω. Let Δ_j be a maximal inequivalent set of components of Ω, let G_j be the stability subgroup of Δ_j and D_j a connected fundamental domain for G_j in Δ_j.

A parabolic fix-point p is a *cusp* if it corresponds to a puncture on Ω/G: then there are horocyclic neighborhoods of p lying in some image of Δ_j.

It is evident that the preceding results are true in much greater generality than stated and that the proofs can easily be modified where necessary. We add (as usual) the cusps to the Δ_j and the three theorems are true for D_j in Δ_j.

The same arguments apply to finitely generated Kleinian groups. We suppose that ∞ is a cusp whose stabilizer contains $P: z \to z+1$ and that $H = \{y > t\}$ lies in Δ_j: we suppose also that $\bar{D}_j \cap H$ can only meet a finite number of sets of the form $H \cap P^n V(D_j)$ (for a fixed V and varying n). With this supposition, if G is finitely generated and D_j is locally finite, then D_j is "finite sided."

REFERENCES

[1]. Beardon, A. F., and Maskit, B., "Limit points of Kleinian groups and finite sided fundamental polyhedra," (unpublished).

[2]. Greenberg, L., "Fundamental polygons for Fuchsian groups," *J. d'Anal. Math.*, 18 (1967), pp. 99-105.

[3]. Heins, M., "Fundamental polygons of Fuchsian and Fuchsoid groups," *Ann. Acad. Sci. Fenn.*, 1964, p. 30.

[4]. Marden, A., "On finitely generated Fuchsian groups," *Comment. Math. Helv.*, 42 (1967), pp. 81-85.

SPACES OF DEGENERATING RIEMANN SURFACES*

Lipman Bers

This is a preliminary report on work in progress, prepared for the Maryland conference.** Proofs will appear elsewhere. Since the manuscript is not completed, I prefer to label all statements of results conjectures. The aim is to construct analogues of Teichmüller spaces for algebraic curves which have or acquire ordinary double points.

§1. RIEMANN SURFACES WITH NODES

A *Riemann surface with nodes*, S, is a connected complex space such that every $P \in S$ has arbitrarily small neighborhoods isomorphic either to the set $|z| < 1$ in $\mathbf{C}$ or to the set $|z| < 1$, $|w| < 1$, $zw = 0$ in $\mathbf{C}^2$. In the second case, P is called a node. Every component of the complement of the nodes is called a *part* of S, it is an ordinary (non-singular) Riemann surface. S is called *exceptional* if a part of S is a sphere with 0, 1 or 2 punctures, or a torus with 0 or 1 punctures. If S is non-exceptional, every part carries a canonical *Poincaré metric*. If S is also compact, the Poincaré area of every part is a multiple of 2π, and the Poincaré area A of S is a multiple of 4π. The *genus* p of a compact non-exceptional S is defined by the relation $A = 4\pi(p-1)$; the number k of nodes satisfies the inequality $0 \leq k \leq 3p - 3$. If S has $3p - 3$ nodes, it is called *terminal*. There are, up to isomorphisms, only finitely many terminal S, for a given p.

From now on the letter S, with or without subscripts, denotes a compact non-exceptional Riemann surface with nodes, of a *fixed genus* p. We denote by M_p the *space of isomorphism classes* of all such S.

---

* Work partially supported by the National Science Foundation.

** International Conference on Riemann Surfaces and Discontinuous Groups, University of Maryland, College Park, Maryland, May 21-25, 1973.

Let $q > 0$ be an integer. A *regular q-differential* on S is a holomorphic form of type $(q, 0)$ on each part of S, which has poles of order at most q at the punctures, the "residues" at any two punctures joined in a node being equal (if q is even) or opposite (if q is odd). [The residue of a holomorphic form w of type $(q, 0)$ at a point P of a non-singular Riemann surface is the residue, at P, of the Abelian differential $wz^{q-1}dz$, where z is a local parameter with $z = 0$ at P.]

The number of linearly independent regular q-differentials is p for $q = 1$ and $(2q-1)(p-1)$ for $q > 1$. Every basis of regular q-differentials defines a holomorphic (q-*canonical*) embedding of S into $\mathbf{P}_{(2q-1)(p-1)-1}$ and exhibits S as a "stable" (in the sense of Mayer and Mumford) algebraic curve with nodes, of arithmetic genus p.

A continuous surjection $f: S_1 \to S$ will be called a *deformation* if the image of a node is a node, the inverse image of a node is a node or a Jordan curve avoiding all nodes, and the restriction of f to the complement of the inverse images of the nodes is a sense preserving homeomorphism.

We choose *once and for all* an integer $\nu > 3$; it will remain fixed throughout this paper. Two deformations $f_1 : S_1 \to S$ and $f_2 : S_2 \to S$ will be called *equivalent* if there exist a homeomorphism g of S_1 onto S_2 such that $f_1 = f_2 \circ g$, and g is homotopic to a product of ν-th powers of Dehn twists about Jordan curves mapped by f_1 into nodes, followed by an isomorphism.

Note that a deformation which is not a homeomorphism *increases* the number of nodes. A terminal S admits no deformations except for isomorphisms.

§2. RIEMANN SURFACES WITH NODES AND BOUNDARIES OF TEICHMÜLLER SPACES

Before describing our present approach, we report briefly on the connection between Riemann surfaces with nodes and boundaries of Teichmüller spaces. In this section, which may be omitted without loss of continuity, S denotes a *non-singular* Riemann surface, $\overline{S}$ its *mirror image*, and π_p its *fundamental group*. By a *sense reversing deformation* we mean a deformation followed or preceded by the canonical mapping of a Riemann surface onto its mirror image.

By the theorem on simultaneous uniformization, equivalence classes of sense reversing homeomorphic deformations $\overline{S} \to S_1$ can be represented by conjugacy classes (in the Möbius group) of *quasi-Fuchsian groups* isomorphic to π_p, and by points of the *Teichmüller space* T(S). It is known that T(S) is a complex manifold which admits a *standard embedding* as a bounded domain in $\mathbb{C}^{3p-3}$, viewed as the space of regular 2-forms on $\overline{S}$, and that every *boundary point* $\psi \in \partial T(S)$, relative to this embedding, represents a Kleinian group G^{ψ}, isomorphic to π_p. While almost all of these groups are totally or partially "degenerated," some may be non-degenerate or, as one calls them now, regular b-groups.

A Kleinian group G isomorphic to π_p (we need not consider other groups) is called a *regular* b-*group* if its region of discontinuity $\Omega(G)$ has precisely one invariant component Δ_0 and $r > 0$ non-invariant non-conjugate components $\Delta_1, \ldots, \Delta_r$ with stability subgroups $G_1, \ldots, G_r$, such that Δ_0/G is a non-singular Riemann surface $\overline{S}$ and $\Delta_1/G_1, \ldots, \Delta_r/G_r$ are the parts of a Riemann surface S_1 with $k > 0$ nodes, two punctures joined into a node corresponding to conjugate parabolic subgroups of G. The group G defines an equivalence class of sense reversing deformations $\overline{S} \to S_1$.

The construction of regular b-groups representing a *given* sense reversing deformation is due to Maskit. Maskit also showed that $\partial T(S)$ contains regular b-groups representing deformations of $\overline{S}$ onto terminal surfaces, and that such groups yield new, *non-standard*, embeddings of T(S) into $\mathbb{C}^{3p-3}$. Important recent results are due to Marden (who uses three-dimensional topology), Earle and Marden, Abikoff (who relies on two-dimensional and function-theoretical methods) and Harvey (who operates mainly within the framework of Fuchsian groups). In view of their investigations, which are to appear, one knows that *every* regular b-group is indeed a boundary group, that is, occurs on the boundary of a Teichmüller space, and that such a group is determined, up to a conjugacy in the Möbius group, by the equivalence class of the corresponding deformation. One can extend the action of the modular group from $T(S)$ to the part of $\partial T(S)$ consisting of regular b-groups.

§3. SPACES OF RIEMANN SURFACES WITH NODES

We proceed to construct a complex domain parametrizing all Riemann surfaces with nodes which can be deformed into a given surface S, surfaces with nodes occurring not on the boundary but in the interior of this domain. The construction depends on various choices, which will later turn out to have been irrelevant, as will the complex structure of S, though not its topological structure.

Let S have r parts $\Sigma_1, \ldots, \Sigma_r$ and k nodes $P_1, \ldots, P_k$. Let Σ_j have genus p_j and n_j punctures. Then

$$2k = n_1 + \ldots + n_r, \quad p = p_1 + \ldots + p_r + k + 1 - r .$$

We choose r Fuchsian groups $G_1, \ldots, G_r$ acting on discs $\Delta_1, \ldots, \Delta_r$ with disjoint closures, such that G_j has n_j non-conjugate maximal elliptic subgroups, each of the same fixed order $\nu > 3$, the Riemann surface Δ_j/G_j, with the images of all elliptic vertices removed, is conformally equivalent to Σ_j, and $G_1, \ldots, G_r$ generate a Kleinian group G which is their free product and has an invariant component Δ_0 such that Δ_0/G is a compact Riemann surface Σ of genus $p_1 + \ldots + p_r$, the mapping $\Delta_0 \to \Delta_0/G$ being ramified over $2k$ points. (Our conditions can be satisfied, in view of the limit circle theorem of Klein and Poincaré and of Klein's combination theorem.)

Now we assign to each node P_i of S two non-conjugate maximal elliptic subgroups Γ'_i, Γ''_i of G, so that if P_i joins Σ_j to Σ_ℓ, then $\Gamma'_i \subset G_j$ and $\Gamma''_i \subset G_\ell$. Two elliptic vertices, not in Δ_0, will be called *related* if they are fixed under elliptic subgroups conjugate to either Γ'_i

or to Γ''_i. The Γ_i are chosen so that the union of the Δ_j/G_j, with the images of any two related elliptic vertices identified, is isomorphic to S.

If $s_i \in \mathbf{C}$, $|s_i| > 0$ and small, then there exists a unique loxodromic Möbius transformation g_{i,s_i} which conjugates the group Γ'_i into the group Γ''_i, has multiplier s_i, and has fixed points in Δ_j and Δ_ℓ (j and ℓ being as before). Set $s = (s_1, \ldots, s_k)$. If $|s| = \max |s_i|$ is small, let $G_{o,s}$ denote the group generated by G and all elements g_{i,s_i}, with $s_i \neq 0$. By Maskit's second combination theorem, $G_{o,s}$ is a Kleinian group.

Let s be as before, and let W be a quasi-conformal automorphism of $\mathbf{C}$ such that W leaves $0, 1, \infty$ fixed, $WG_{o,s}W^{-1}$ is a Kleinian group, and $W|\Delta_0$ is conformal. Then $W|\Delta_j$, $j = 1, \ldots, r$, defines an element τ_j of the Teichmüller space $T(G_j)$, which we represent as a bounded domain in $\mathbf{C}^{3p_j-3+n_j}$. If $s_i \neq 0$, set $t_i = a_i - \hat{a}_i$ where a_i is the repelling fixed point of $W \circ g_{i,s_i} \circ W^{-1}$ and $\hat{a}_i$ the fixed point of $W\Gamma'_i W^{-1}$ in Δ_j. If $s_i = 0$, set $t_i = 0$. The point

$$(\tau, t) = (\tau_1, \ldots, \tau_r, t_1, \ldots, t_k) \in \mathbf{C}^{3p-3}$$

determines the group $WG_{o,s}W^{-1}$ completely; we denote this group by $G_{\tau,t}$, and we denote the set of all (τ, t) for which a group $G_{\tau,t}$ is defined by $X_\alpha(S)$. Here α represents the choices made: the groups G_j and the subgroups Γ'_i, Γ''_i.

I. $X_\alpha(S)$ *is a bounded domain in* $\mathbf{C}^{3p-3}$.

We denote by $\Omega_\alpha(\tau, t)$ the part of the region of discontinuity of $G_{\tau,t}$ which does not map onto Σ, and by $\Omega^0_\alpha(\tau, t)$ the complement in $\Omega_\alpha(\tau, t)$ of the set of elliptic vertices. The images of the pair Γ'_i, Γ''_i under the conjugation $G_{o,s} \to WG_{o,s}W^{-1}$ will be denoted by $\Gamma'_i(\tau, t), \Gamma''_i(\tau, t)$. Two elliptic vertices in $\Omega_\alpha(\tau, t)$ are called *related* if they are fixed under elliptic subgroups conjugate to either $\Gamma'_i(\tau, t)$ or to $\Gamma''_i(\tau, t)$. The quotient $\Omega_\alpha(\tau, t)/G(\tau, t)$, with images of related elliptic vertices identified, is a Riemann surface with nodes $S_{\tau,t}$. The genus of $S_{\tau,t}$ is p, and

$S_{\tau,t}$ has as many nodes as there are zeros in $(t_1, \ldots, t_r)$. Furthermore, each $S_{\tau,t}$ is equipped with a deformation $f_{\tau,t} : S_{\tau,t} \to S$, determined up to equivalence.

II. $X_\alpha(S)$ *is in a natural one-to-one correspondence with the set of equivalence classes of deformations* $S' \to S$.

The fiber space $Y_\alpha(S)$ over $X_\alpha(S)$ is the set of points (τ, t, z) with $(\tau, t) \in X_\alpha(S)$, $z \in \Omega_\alpha^0(\tau, t)$.

III. $Y_\alpha(S)$ *is a domain in* $\mathbf{C}^{3p-2}$.

Let $\lambda(\tau, t, z)\,|dz|$ denote the *Poincaré metric* in $\Omega_\alpha^0(\tau, t)$. It projects onto the Poincaré metric on $S_{\tau,t}$. Set $\partial = \partial/\partial z$, $\overline{\partial} = \partial/\partial \overline{z}$.

IV. *All functions* $\partial^m \overline{\partial}^n \lambda$, $m \geq 0$, $n \geq 0$, *are continuous in* $Y_\alpha(S)$.

The proof depends on standard potential theoretic estimates.

If S is non-singular, $X_\alpha(S)$ may be identified with the Teichmüller space $T(S)$, and $Y_\alpha(S)$ with the standard fiber space of Jordan domains over $T(S)$.

§4. AUTOMORPHIC FORMS

One would like to attach to each point (τ, t) of $X_\alpha(S)$ a basis of regular q-forms on $S_{\tau,t}$, which depends holomorphically on (τ, t). At present we can do this only after removing from $X_\alpha(S)$ a "small" set (except when S is non-singular or $q = 1$).

V. *For every* q, $1 < q \leq \nu$, *and every* $(\tau^0, t^0) \in X_\alpha(S)$, *there is a closed set* $\sigma \subset X_\alpha(S) - \{(\tau^0, t^0)\}$ *which is either empty or of real codimension* 2, *and* $d = (2q-1)(p-1)$ *holomorphic functions* $\phi_j(\tau, t; z)$ *defined for* $(\tau, t) \in X_\alpha(S) - \sigma$, $z \in \Omega_\alpha^0(\tau, t)$, *which form a basis of regular* q-*differentials on* $S_{\tau,t}$. (*If* $t^0 = 0$, σ *is an analytic subvariety. If* S *is non-singular,* σ *may be chosen as empty.*)

We shall sketch the construction of these functions for $k > 0$. If $k = 0$, the result is known from Teichmüller space theory.

If $t^0 = 0$, the first d−k functions ϕ_j are Poincaré series

$$\phi_j(\tau, t; z) = \Sigma\, \Phi_j(g(z))\, g'(z)^q, \quad g \in G_{\tau,t} ,$$

where the Φ_j are appropriately chosen polynomials. The next k functions ϕ_j are defined as follows. Let $a_i(\tau, t)$, $b_i(\tau, t)$ be the fixed points of the loxodromic transformation $g_{i,\tau,t}$ in the group $G_{\tau,t}$ which conjugates one of the two groups $\Gamma'_i(\tau, t)$, $\Gamma''_i(\tau, t)$ onto the other. These are holomorphic functions which, for $t_i \to 0$, become the two fixed points of $\Gamma'_i(\tau, t)$, $\Gamma''_i(\tau, t)$ lying in $\Omega_\alpha(\tau, t)$. Set

$$\Psi_i(\tau, t; z) = [z - a_i(\tau, t)]^{-q}\, [z - b_i(\tau, t)]^{-q}$$

and

$$\phi_{i+d-k} = \sum_{\ell=0}^{\infty} \Psi_i(\tau, t; \gamma_\ell(z))\, \gamma'_\ell(z)^q ,$$

where $\gamma_1, \gamma_2, \ldots$ runs through all $G_{\tau,t}$ if $t_i = 0$, and through a complete list of right coset representatives of $G_{\tau,t}$ modulo the subgroup generated by $g_{i,\tau,t}$ if $t_i \neq 0$. In the case just considered, the functions ϕ_i are defined in the whole domain $Y_\alpha(S)$, and σ is the divisor of the Wronskian of the functions $\phi_1, \ldots, \phi_d$.

We say that a point $(\tau, t) \in X_\alpha(S)$ lies *not below* $(\tau^0, t^0) \in X_\alpha(S)$ if $t_i = 0$ implies that $t_i^0 = 0$, $i = 1, \ldots, r$.

If $t^0 \neq 0$, let k_0 be the number of zeros among $t_1^0, \ldots, t_k^0$. The first $d - k_0$ functions ϕ are Poincaré series formed with $d - k_0$ functions

$$\Phi_j(\tau, t; z) = \sum_{\ell=1}^{N} a_\ell^j [z - A(\tau, t)]^{-1}\, [z - B(\tau, t)]^{-1}\, [z - C_\ell(\tau, t)]^{-1}$$

where a_ℓ^j are appropriately chosen constants and $A(\tau, t)$, $B(\tau, t)$, $C_1(\tau, t), \ldots, C_N(\tau, t)$ appropriately chosen holomorphic functions defined for (τ, t) not below (τ^0, t^0), and representing $N + 2$ distinct loxodromic fixed points of $G_{\tau,t}$. The remaining k_0 functions ϕ_j are the ϕ_{i+d-k} defined above, with $t_i^0 = 0$.

We need now a modification of the Petersson scalar product.

Let δ be a geodesic Jordan curve or a node of S. A *collar* about δ, of area 2a, in the union of two doubly connected regions on S, each of Poincaré area a, each bounded by δ and by a Jordan curve δ' freely homotopic to δ and orthogonal to the pencil of geodesic arcs entering δ if δ is a point, orthogonal to δ if δ is a curve.

We choose, once and for all, sufficiently small positive numbers a_0 and ε_0, and a non-increasing function $a(t)$, $0 \leq t < +\infty$ with $a(0) = a_0$ and $a(\varepsilon_0) = 0$. Let $S^{\#}$ denote the Riemann surface S from which we removed every collar of area $2a_0$ about a node and every collar of area $2a(\varepsilon)$ about a geodesic Jordan curve of length $\varepsilon < \varepsilon_0$. (Cf. the paper by L. Keen in this volume.)

Let $\phi(t)\,dt^q$ and $\psi(t)\,dt^q$ be two regular q-forms on S, and let $\lambda(t)\,|dt|$ be the Poincaré metric on S; here t is a local parameter. We set

$$<\phi,\psi> = i \iint_{S^{\#}} \phi(t)\,\overline{\psi(t)}\,\lambda(t)^{2-2q}\,dt \wedge d\bar{t}\,.$$

A basis of regular q-forms on S is called *orthonormal* if it is so with respect to this scalar product. Such a basis is determined by the conformal structure of S, except for a unitary transformation.

Using IV and V and the compactness of the unitary group one establishes

VI. *If* $S_{\tau',t'}$ *is not isomorphic to* $S_{\tau'',t''}$, *then the points* (τ',t') *and* (τ'',t'') *have neighborhoods* N' *and* N'' *such that no* $S_{\tau,t}$ *with* $(\tau,t) \in N'$ *is isomorphic to a* $S_{\sigma,s}$ *with* $(\sigma,s) \in N''$.

Using V one can also prove

VII. *There are* p *holomorphic functions* $\phi_j(\tau,t;z)$ *defined in* $Y_a(S)$ *which form a basis of regular* 1*-differentials on* $S_{\tau,t}$.

The proof involves normalizing regular 1-differentials on $S_{\tau,t}$ by the condition:

$$\int_{\beta_j} \phi_k(t)\,dt = \delta_{jk}$$

where $\beta_1,\ldots,\beta_p$ are simple closed curves which either define non-vanishing homology classes or are homotopic to nodes.

One may use this result to study the behavior of a period matrix under degenerations.

§5. FENCHEL-NIELSEN COORDINATES

A *partition* of S is a set of $3p-3-k$ disjoint *geodesic* Jordan curves on S, where k is the number of nodes. The Jordan curves and the nodes are called the *boundary elements* of the partition. Every boundary element has two *banks*. A partition is *ordered* by ordering the $6p-6$ banks, with the provision that every bank of a boundary element which does not separate S has a lower number than any bank of an element which does. An ordering of the partition induces an ordering of the boundary elements, and induces a direction on every boundary element which is not a node. Following the celebrated manuscript by Fenchel and Nielsen, we can describe the conformal structure of a Riemann surface, endowed with an ordered partition, by $3p-3$ complex numbers $\zeta_1,\ldots,\zeta_{3p-3}$, such that $|\zeta_j|$ is the length of the j-th boundary element, and for $\zeta_j \neq 0$ we have $\zeta_j = |\zeta_j| e^{i\theta_j}$ where the *twist parameter* θ_j is determined by the position of endpoints of certain geodesic arcs joining and orthogonal to distinct boundary elements.

Let S_0 be a terminal Riemann surface; it is partitioned by its nodes. If we order this partition, then a deformation $g: S \to S_0$ induces an ordered partition on S, obtained by deforming every Jordan curve on S, mapped into a node of S_0, into a geodesic. For $(\tau, t) \in X_\alpha(S)$, one obtains a mapping $\hat{g}: (\tau, t) \mapsto (\zeta_1,\ldots,\zeta_{3p-3})$ where $(\zeta_1,\ldots,\zeta_{3p-3})$ are the Fenchel-Nielsen coordinates of $S_{\tau,t}$ with respect to the ordered partition induced by $g \circ f_{\tau,t}$. We call $\hat{g}$ a Fenchel-Nielsen mapping. It is never holomorphic.

VIII. *Let* $\hat{g}: X_\alpha(S) \to C^{3p-3}$ *be a Fenchel-Nielsen mapping. If* S *has* $k \leq 3p-3$ *nodes,* $\hat{g}(X_\alpha(S))$ *is the complement of* $3p-3-k$ *coordinate hyperplanes, and* $\hat{g}$ *is a covering ramified over* k *coordinate hyperplanes, a universal covering, if* $k = 0$, *a* ν^{3p-3}*-sheeted one if* $k = 3p-3$.

§6. RELATIONS BETWEEN THE SPACES $X_\alpha(S)$

Statement VIII has several important consequences.

IX. *A homeomorphic deformation* $f: S_1 \to S_2$ *induces an isomorphism (of complex manifolds)* $f_*: X_\alpha(S_1) \to X_\beta(S_2)$.

Thus we may write X(S) instead of $X_\alpha(S)$. Note that X(S) has k *distinguished analytic hypersurfaces*, the sets $t_i = 0$ in $X_\alpha(S)$.

X. *A deformation* $f: S_1 \to S_2$, *where* S_2 *has* ℓ *more nodes than* S_1, *induces a holomorphic mapping* $f_*: X(S_1) \to X(S_2)$ *such that* $f_*(X(S_1))$ *is the complement of* ℓ *distinguished hypersurfaces and* f_* *is a universal covering.*

Another consequence of VIII is

XI. X(S) *is a cell.*

Question: is X(S) a domain of holomorphy?

If S_1 has no nodes, so that $X(S_1)$ can be identified with $T(S_1)$, there always is a deformation $S_1 \to S$, and therefore a universal covering $X(S_1) \to X^0(S)$, where $X^0(S)$ is the complement of all distinguished hypersurfaces of X(S). The subgroup of the Teichmüller modular group Mod (S_1), which normalizes the covering group of this covering, induces a group of holomorphic self-mappings of $X^0(S)$; this extends to a group $\Gamma(S)$ of holomorphic self-mappings of X(S).

XII. $\Gamma(S)$ *is discrete, and is the group of holomorphic automorphisms of* X(S) *induced by topological automorphisms of* S.

§7. MODULI SPACES

Let $\Pi = \Pi_S$ denote the canonical mapping $X(S) \to M_p$ which takes $(\tau, t) \in X_\alpha(S)$ into the isomorphism class of $S_{\tau,t}$. It turns out that $\Pi \circ \gamma = \Pi$ for $\gamma \in \Gamma(S)$, but that there are points in $X(S)$ which are not equivalent by $\Gamma(S)$ and have the same image under Π_S. Let $\Gamma_0(S)$ denote the (finite) *isotropy group*, in $\Gamma(S)$, of the *origin* in $X(S)$, that is, of the point represented by $\mathrm{id}: S \to S$.

XIII. *The origin in* $X(S)$ *has a neighborhood* N, *stable under* $\Gamma_0(S)$, *such that if* $X_1, X_2 \in N$, *then* $\Pi(x_1) = \Pi(x_2)$ *if and only if* $\gamma(x_1) = x_2$ *for some* $\gamma \in \Gamma_0(S)$.

Using statements IX - XIII one obtains

XIV. M_p *has a Hausdorff topology and a structure of a normal complex space (and a* V-*manifold) which makes all mappings* Π *holomorphic.*

Note that isomorphism classes of Riemann surfaces with $k > 0$ nodes form an analytic subvariety of M_p, of codimension k.

XV. *There is a number* L, *depending only on* p, *such that every* S *has a partition in which no boundary element has length exceeding* L.

This is a geometric lemma which generalizes results by Mumford and by Keen.

There are finitely many non-isomorphic terminal Riemann surfaces, $S_1, \ldots, S_m$, of genus p, and M_p is contained in the union of all $\Pi_{S_i}(X(S_i))$, $i = 1, \ldots, m$. By XV, each $X(S_i)$ has a compact subset $\hat{X}_i$ such that M_p is contained in the union of all $\Pi_{S_i}(\hat{X}_i)$, $i = 1, \ldots, m$. Thus we obtain

XVI. M_p *is compact.*

That the space of moduli of non-singular algebraic curves of genus p can be compactified by adding moduli of stable curves with nodes was announced by Mayer and Mumford 10 years ago. The desire to establish this result without using higher dimensional algebraic geometry was the original stimulus for this investigation.

APPENDIX. SCHOTTKY GROUPS

Besides the spaces constructed in Section 2, there are many other spaces of Kleinian groups representing Riemann surfaces with nodes. A particularly simple example is that of Schottky groups.

Let $0, 1, \infty, \zeta'_2, \zeta_3, \zeta'_3, \ldots, \zeta_p, \zeta'_p$ be $2p$ distinct points in the extended complex plane, and let $t_1, \ldots, t_p$ be complex numbers, $0 < |t_j| < 1$, such that the loxodromic Möbius transformations g_j with fixed points ζ_j, ζ'_j and multipliers t_j are free generators of a Schottky group (of course, $\zeta_1 = 0$, $\zeta'_1 = \infty$ and $\zeta_2 = 1$). Such $(3p-3)$-tuples

$$(*) \qquad (\zeta'_2, \zeta_3, \ldots, \zeta'_p, t_1, \ldots, t_p) \in \mathbf{C}^{3p-3}$$

form the *Schottky space*. We form the *extended Schottky space* $\mathcal{S}_p$ by adding points with some or all of the t_j being 0, provided that the g_i with $t_i \neq 0$ generate a Schottky group. A point in $\mathcal{S}_p$ represents, in an obvious way, a Riemann surface of genus p, with one part, and with as many nodes as there are zeros among $t_1, \ldots, t_p$.

Many of the results stated above have valid analogues for the spaces $\mathcal{S}_p$. We shall describe only the construction of regular q-differentials, for $q > 1$.

Let G be the group determined by the point (*); it is understood that if $t_i = 0$, there is no i-th generator. Now set

$$\Phi_1(z) = z^{-q}, \ \Phi_i(z) = (z - \zeta_i)^{-q}(z - \zeta'_i)^{-q}, \quad i = 2, \ldots, p \ ;$$

$\phi_i = \Phi_i$ if $t = (t_1, \ldots, t_p) = 0$, and

$$\phi_i(z) = \sum_{\ell=1}^{\infty} \Phi_i(\gamma_\ell(z)) \gamma'_\ell(z)^q$$

if $t \neq 0$, where $\gamma_1, \gamma_2, \ldots$ is a list of right coset representatives of G modulo the subgroup generated by g_i, if $t_i \neq 0$, a list of all elements of G, if $t_i = 0$.

Also set $F_j(z) = z^{j+1-q}[(z-1)(z-\zeta'_2)(z-\zeta_3)\ldots(z-\zeta'_p)]^{1-q}$, and

$$\phi_{p+j}(z) = \sum_{g \in G} F_j(g(z))\, g'(z)^q, \; j = 1, \ldots, (2q-1)(p-1)-p \; .$$

Then the functions $\phi_1(z), \ldots, \phi_{(2q-1)(p-1)}(z)$ are *holomorphic functions of all* $3p-2$ *variables* $\zeta'_2, \ldots, \zeta_p, \zeta'_p, t_1, \ldots, t_p, z$, define *regular* q-*differentials* on the Riemann surface represented by G, and, for $(\zeta'_2, \ldots, t_p)$ avoiding an analytic subvariety $\sigma \subset \mathcal{S}_p$, the ϕ_j are *linearly independent*. The subvariety σ contains no points with $t_i = \ldots = t_p = 0$.

COLUMBIA UNIVERSITY

MAPPING CLASS GROUPS OF SURFACES: A SURVEY

Joan S. Birman

In the following pages we will attempt to present a summary of progress during the past 10 years on investigations into the structure of mapping class groups of orientable surfaces. Our focus will be primarily on questions relating to generators and defining relations, subgroup structure, relationships between closed and punctured surfaces, connections with braid groups, and lastly the construction of finite representations (in particular representations which do not factor through $Sp(2g, Z)$).

§1. *Basic concepts*

Let S_g denote a closed, orientable surface of genus g. The homeotopy group H_g of S_g is the group of all isotopy classes of self-homeomorphisms of S_g. By a classical result of J. Nielsen [27], later simplified and reproved by W. Mangler [25], the group H_g is canonically isomorphic to Out $\pi_1 S_g$. If we restrict our attention to orientation-preserving homeomorphisms of S_g we obtain a subgroup M_g of index 2 in H_g, and this group is referred to as the mapping class group of S_g. Algebraically, if $\pi_1 S_g$ is presented by $< a_1, \ldots, a_g, b_1, \ldots, b_g; \prod_{i=1}^{g} [a_i, b_i] >$, then M_g is the group of classes (mod Inn $\pi_1 S_g$) of proper automorphisms of $\pi_1 S_g$, that is, automorphisms induced by free substitutions on the generators a_i, b_i which map $\prod_{i=1}^{g} [a_i, b_i]$ onto a conjugate of itself (but not a conjugate of its inverse).

The group M_0 is, of course, trivial. If $g = 1$ the situation is still easy. Since $\pi_1 S_1$ is free abelian of rank 2, the group M_1 is simply

Aut $\pi_1 S_1$. If a, b are a canonical basis for $\pi_1 S_1$, then an endomorphism of $\pi_1 S_1$ is specified by the images $a^{n_{11}} b^{n_{12}}$ and $a^{n_{21}} b^{n_{22}}$ of a, b respectively. This mapping determines an element of M_1 if and only if the 2×2 matrix $\|n_{ij}\|$ has determinant $+1$, hence M_1 is precisely the group of 2×2 integral matrices with determinant $+1$. This group has been studied extensively in the classical literature. Thus the first interesting case is $g = 2$.

Generators and defining relations for the group M_2 were determined by the author and H. Hilden in [6]. Their proof was based upon a demonstration that the group M_2 is closely related to the mapping class group of S_2 with 6 punctures, and also to the mapping class group of the 6-punctured sphere. We will discuss that result in Section 4 below, but first we review the related questions of (i) the determination of generators for M_g and (ii) the relationship between the mapping class groups of closed and punctured surfaces.

§2. *Generators for* M_g

Let c be a simple closed curve on a surface S_g. Let N_c be a neighborhood of c which is homeomorphic to a cylinder. Assume that N_c has cylindrical coordinates (y, θ), $-1 \leq y \leq +1$, $0 \leq \theta \leq 2\pi$. A twist t_c about c is defined by the identity map outside N_c, with $t_c(y, \theta) = (y, \theta + \pi(y+1))$. (See Figure 1.) Observe that isotopic deformations of the

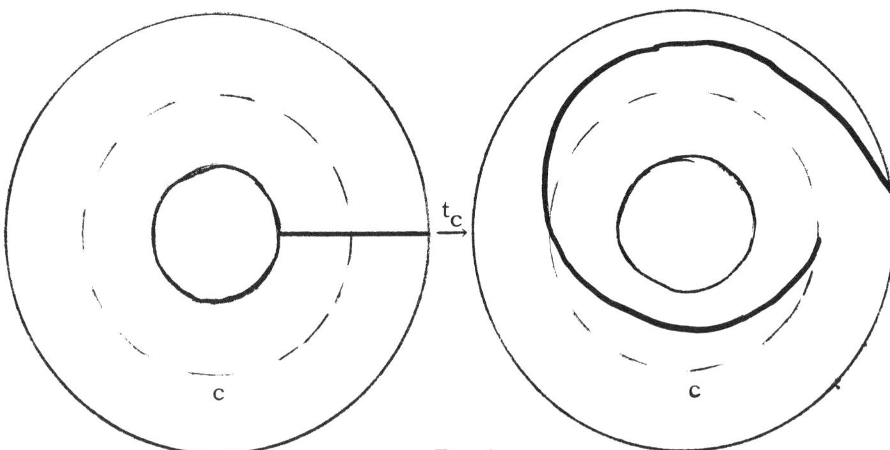

Fig. 1

Cylindrical neighborhood N_c of the curve c.

Image of N_c under the twist t_c.

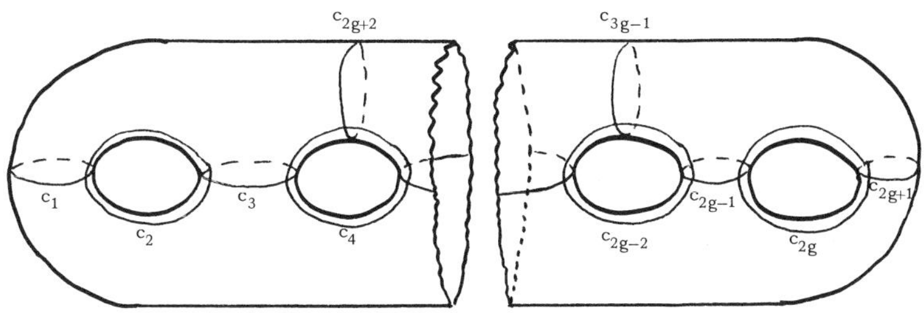

Fig. 2. Generators for M_g.

curve c will not alter the isotopy class τ_c of a twist t_c about c. The sense of τ_c depends on the choice of a positive direction for θ (but not for y) in the parametrization of N_c, and may be fixed for every $c \in S_g$ by assigning an orientation to S_g and using a rule such as "positive twists move points on a directed line segment which is approaching c to the right." Such a rule will be valid for points on either side of c. Note that the sense of τ_c does not depend on the orientation of c.

It was proved by Dehn [12] that M_g is generated by twists. Dehn's result was later reproved and simplified by Lickorish [18,19], who showed that the particular set $\tau_{c_1}, \ldots, \tau_{c_{3g-1}}$ illustrated in Figure 2 will do. It may be shown that the mapping class groups of punctured surface (where the punctures are either kept fixed as a set, or kept fixed individually) are also generated by twists [3]. This is still true if we remove discs instead of points, requiring either that the boundaries of the discs be kept fixed pointwise or that they be kept fixed setwise [30], and again it is true (with a single exception) for non-orientable surfaces [9]. It is not known whether any similar result holds for higher dimensional manifolds; one suspects not. Algebraically, there is a close relationship between the automorphisms induced by twist maps and elementary Nielsen transformations [24]. Intuitively, one may explain the universality of these particular generators for mapping class groups of surfaces by observing that non-abelian surface groups are either free or are one relator groups, and that in the latter case every automorphism is induced by a free substitution; in the abelian case, the only surface group which is not free is free abelian of rank 2, and every automorphism of a free abelian group of rank 2 is induced by a free substitution [24].

We illustrate with an example why twist maps are so pleasant to work with. As noted above, M_2 is generated by $\tau_{c_1}, \ldots, \tau_{c_5}$. Hence it must be possible to express the isotopy class τ_f of the twist t_f about the curve f in Figure 3 as a product of powers of these generators. We will now show how to find this power product explicitly. First observe that if c and d are simple closed curves on S_g, and if h is a homeomorphism of S_g which maps c onto d, then

$$t_d = h\, t_c h^{-1} . \tag{1}$$

Second, we observe that if c and d are simple closed curves on S_g which meet in precisely one point, with algebraic intersection 1, then the effect of the twist t_c on the curve d is obtained by cutting the curve d where it meets c, and inserting a copy of c. With this rule in mind, we examine the effect of the twist product $t_{c_5} t_{c_4} t_{c_3} t_{c_2}$ on the curve c_1, (see Figures 3(ii), 3(iii) and 3(iv)). From Figure 3(iv) one sees that on S_2 the curve $t_{c_5} t_{c_4} t_{c_3} t_{c_2}$ (c_1) is isotopic to f. This fact, together with equation (1), implies the "relation"

$$\tau_f = \tau_{c_5} \tau_{c_4} \tau_{c_3} \tau_{c_2} \tau_{c_1} \tau_{c_2}^{-1} \tau_{c_3}^{-1} \tau_{c_4}^{-1} \tau_{c_5}^{-1} \tag{2}$$

in M_2. With a little practice it is possible to discover very complex relations in M_g by an escalation of this simple technique. In particular, we note two general types of "local" relations which occur in all of the groups M_g, which are easily verified by the method just described. They are:

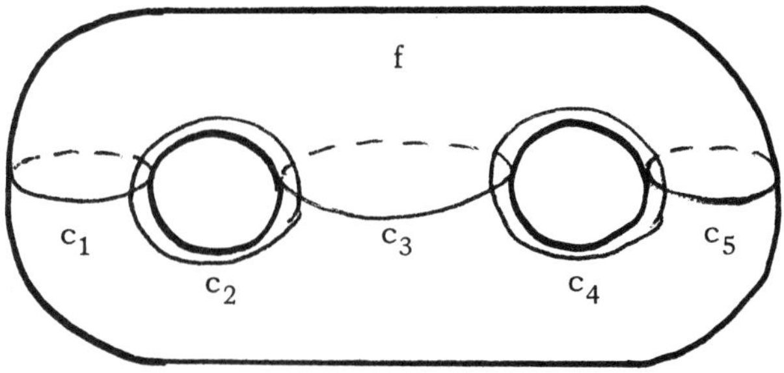

(i) The curves $c_1, c_2, c_3, c_4, c_5, f$.

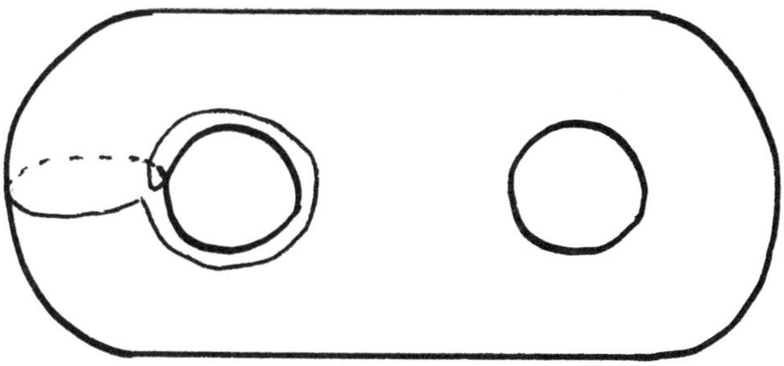

(ii) $t_{c_2}(c_1)$.

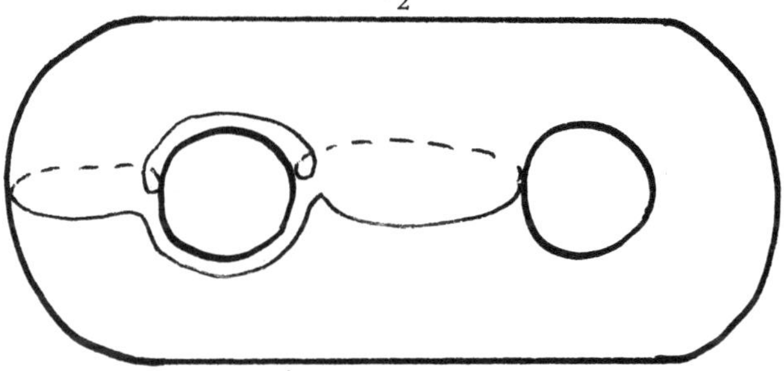

(iii) $t_{c_3} t_{c_2}(c_1)$.

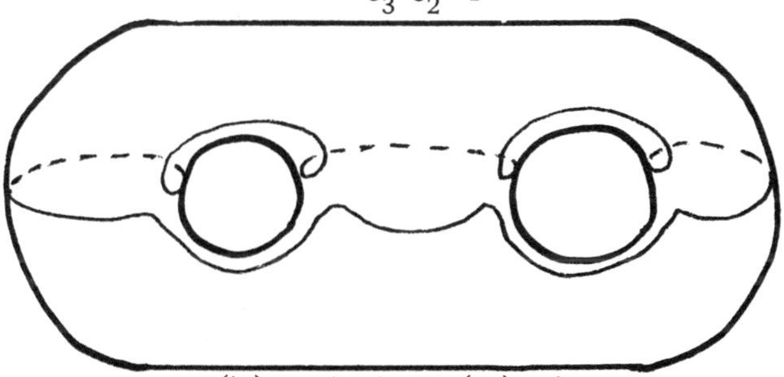

(iv) $t_{c_5} t_{c_4} t_{c_3} t_{c_2}(c_1) \sim f$.

Fig. 3. Expressing τ_f as a power product of $\tau_1, \tau_2, \tau_3, \tau_4, \tau_5$.

(3) $\tau_c \tau_d = \tau_d \tau_c$ if $c \cap d = \emptyset$

(4) $\tau_c \tau_d \tau_c = \tau_d \tau_c \tau_d$ if c and d have algebraic intersection 1 .

We observe that Artin's braid group [5,24] may be defined by a system of relations which are all of type (3) or (4). This is the first hint one has of a connection between braids and twist maps, and indeed this connection has proved to be a very important one.

A non-trivial twist map is, of course, always of infinite order. It is of interest to note that Maclachlin has found a set of elements, each of which has finite order, which also generate M_g [20], and this result has been extended by Patterson [28] to surfaces with punctures, for all except certain exceptional cases. This fact was used by Maclachlin [20] and again by Patterson [28] to prove that the space of moduli is in most cases simply-connected. These two papers are interesting examples of how knowledge about the mapping class group has been applied to Riemann surface theory.

§3. *The relationship between mapping class groups of closed surfaces and of surfaces with punctures*

Let $Q = \{q_1, \ldots, q_n\}$ denote an arbitrary but henceforth fixed set of points on the surface S_g. The group $M_{g,n}$ is defined to be the mapping class group of S_g, with admissible maps and isotopies restricted to those which keep the set Q fixed. A second group, $P_{g,n}$ the "pure" mapping class group, is defined by the condition that admissible maps and isotopies keep each point q_i fixed. It is immediate that $M_{g,n}$ is obtained from $P_{g,n}$ by extending by the full symmetric group Σ_n. We fix our attention first on $P_{g,n}$.

The relationship between the groups $P_{g,n}$ and $P_{g,m}$, $m \neq n$, was studied by the author in [3]:

THEOREM (Birman). *Let* $\Phi_{g,n}$ *be the natural homomorphism from* $P_{g,n}$ *to* $P_{g,n-1}$ *(defined by "filling in" the* n^{th} *puncture). Then* $\ker \Phi_{g,n}$ *is canonically isomorphic to* $\pi_1(S_g - q_1, \ldots, q_{n-1})/$*center.*

This result has an easy intuitive explanation. Suppose that α is a non-trivial element of $\ker \Phi_{g,n}$. Then α may be represented by a self-homeomorphism of $(S_g - q_1, \ldots, q_n)$ which is not isotopic to the identity on $(S_g - q_1, \ldots, q_n)$, but is isotopic to the identity on $(S_g - q_1, \ldots, q_{n-1})$. If α_t is the isotopy, then the orbit $\alpha_t(q_n)$ of q_n defines an element of $\pi_1(S_g - q_1, \ldots, q_{n-1})$ which we may associate with α.

Looking at this from another point of view, we will locate in $M_{g,n}$ a set of maps which generate a subgroup isomorphic to $\pi_1(S_g - q_1, \ldots, q_{n-1})$. For simplicity, first take $n = 1$, $q = 1$. Let $a_1, \ldots, a_g$, $b_1, \ldots, b_g$ be simple closed curves on S_g which form a canonical basis for $\pi_1 S_g$. Let $a_{1,1}, \ldots, a_{g,1}$, $b_{1,1}, \ldots, b_{g,1}$ be a second such system, with each a_{i1}, b_{i1} chosen to be homotopic to a_i, b_i on S_g, but separated from a_i, b_i by the puncture q_1 on $S_g - q_1$. Then $\ker \Phi_{g,1}$ is generated by the $2g$ twist products

$$\left\{\tau_{a_i}\tau_{a_{i,1}}^{-1}, \tau_{b_i}\tau_{b_{i,1}}^{-1} ;\ 1 \leq i \leq g\right\}.$$

It is not difficult to convince oneself that these $2g$ surface mappings satisfy the same relations in $P_{g,1}$ as do $\{a_i b_i;\ 1 \leq i \leq g\}$ in $\pi_1 S_g$. A similar construction holds for arbitrary n; one chooses $2g+n-1$ curves $\{a_{i1}, b_{i1}, a_{1j};\ i = 1, \ldots, g,\ j = 2, \ldots, n\}$ with a_{ik} and b_{ik} separated from a_i, b_i by the punctures $\{q_1 \cup q_2 \cup \ldots \cup q_k\}$, but homotopic to a_i on the closed surface. Then, if $n > 1, g \geq 1$, $\ker \Phi_{g,n}$ is generated freely by the $2g+n-1$ pairs

$$\left\{\tau_{a_i}\tau_{a_{i,1}}^{-1}, \tau_{b_i}\tau_{b_{i,1}}^{-1}, \tau_{a_{1,j-1}}\tau_{a_{1,j}}^{-1} ;\ 1 \leq i \leq g, 2 \leq j \leq n\right\}.$$

Knowledge of the group $\ker \Phi_{g,n}$ may be used to find a system of defining relations for $P_{g,n}$, and hence for $M_{g,n}$, by an iterative procedure, if one knows defining relations for $P_{g,0} = M_{g,0}$. (The action of $P_{g,n-1}$ on $\ker \Phi_{g,n}$ may be determined by the method described in Section 2 of this paper.) This procedure easily goes over to a procedure for finding defining relations for the groups $M_{g,n}$ since the latter may be obtained from $P_{g,n}$ by extension by Σ_n. For explicit results for $g = 1$ and arbitrary n, see [3]; for $g = 0$ and arbitrary n see [29], or [30], and also (using a different method) [21]. These results are not available for $g \geq 2$ because the group $P_{g,0} = M_{g,0}$ is not well-understood. Using recent results in [6] it should, however, be possible to determine presentations for $M_{2,n}$ for arbitrary n.

In addition to indicating a technique for finding defining relations for the groups $P_{g,n}$ or $M_{g,n}$ our theorem also yields insight into the structure of the group $P_{g,n}$, for it tells us that $P_{g,n}$ may be built up from $P_{g,0}$ by repeatedly extending $\pi_1(S_g - q_1, \ldots, q_{k-1})$ by $P_{g,k-1}$ for each $k = 1, 2, \ldots, n-1$. This yields, for example, a unique normal form in $P_{g,n}$ (modulo a normal form in $P_{g,0}$).

The n-string braid group of a surface is defined to be the fundamental group of the space of unordered n-tuples of distinct points of S. (If S is taken to be the Euclidean plane, this yields the classical braid group of Artin [24]; in the more general case, it yields a related group.) Let $\Psi_{g,n}$ denote the natural homomorphism from $M_{g,n}$ to $M_{g,0}$ defined by filling in the punctures. Then our theorem may be restated "the kernel of $\Psi_{g,n}$ is isomorphic to the n-string braid group of S_g modulo its center." The significance of this connection between braid groups of surfaces and mapping class groups lies mainly in the fact that braid groups have been studied extensively, both from a topological and from an algebraic point of view, and the possibility exists for applying some of this knowledge to Riemann surface theory. For a general introduction to the theory of braid groups, see [5] and [32]. For an explicit development of the connections between $\ker \Psi_{g,n}$ and $M_{g,n}$, see [2, 3, 29]. For a discussion of the relationship between braid groups of surfaces and the n-fold product group $\pi_1 S_g \times \ldots \times \pi_1 S_g$ see [13]. For generalizations to higher dimensions, see [11].

§4. *Lifting and projecting homeomorphisms*

In this section we discuss a method for attacking the problem of finding defining relations for the groups $M_{g,0}$. The method was used by the author and Hilden in [6] to determine a presentation for the group $M_{2,0}$, but if $g \geq 3$ the method yields, at best, a presentation for a "large" subgroup of $M_{g,0}$ (probably always of infinite index). The problem of finding defining relations for the full group $M_{g,0}$ is open if $g \geq 3$. The methods described in this section do, however, indicate relationships which exist between subgroups of $M_{g,0}$ and subgroups of $M_{g',n'}$ for suitable triplets (g, g', n').

Let $\mathcal{L}$ be a hyperelliptic involution on the surface S_g. The orbit space of S_g under the action of $\mathcal{L}$ is a sphere, and the collapsing map π exhibits S_g as a 2-fold ramified covering of S_0, with branching over the image $B = \pi(\tilde{B})$ of the $2g+2$ fixed points $\tilde{B}$ of $\mathcal{L}$. If $h : (S_0 - B) \to (S_0 - B)$ is a homeomorphism, then h always lifts to S_g, and its lift $\tilde{h}$ will be "$\mathcal{L}$-symmetric," that is:

$$\tilde{h} \mathcal{L} \tilde{h} = \mathcal{L} . \tag{5}$$

If h represents the class of the identity in $M_{0,2g+2}$, then the isotopy between h and the identity map on $S_0 - B$ also lifts, and its lift $\tilde{h}_t$ will have the property that $\tilde{h}_t$ is also $\mathcal{L}$-symmetric for each $0 \leq t \leq 1$. Let $M_g(\mathcal{L})$ denote the group of $\mathcal{L}$-symmetric mapping classes of S_g, that is admissible maps and admissible isotopies are required to commute with $\mathcal{L}$. One would not expect $M_g(\mathcal{L})$ to embed as a subgroup of $M_{g,0}$, because the requirement that two homeomorphisms $\tilde{h}, \tilde{k}$ be $\mathcal{L}$-symmetrically isotopic is a stronger requirement than isotopy. Nevertheless, it was discovered in [6] that:

(i) If $h : S_g \to S_g$ is an $\mathcal{L}$-symmetric homeomorphism which is isotopic to the identity map of S_g, then there is an $\mathcal{L}$-symmetric isotopy between h and id.

(ii) If $g = 2$, *every* element of $M_{g,0}$ has an $\mathcal{L}$-symmetric representative.

As a consequence of (i) and (ii) one obtains almost immediately:

(i)* $M_g(\mathcal{L})$ maps homomorphically onto $M_{0,2g+2}$, with kernel the subgroup of order 2 generated by the class of $\mathcal{L}$.

(ii)* $M_2(\mathcal{L})$ coincides with $M_{2,0}$.

This allows us to lift the known presentation for $M_{0,6}$ to $M_{2,0}$ and so obtain defining relations in $M_{2,0}$. Even more, it shows the close relationship between the group $M_{2,0}$ (and its appropriate generalizations to $\mathcal{L}$-symmetric subgroups of M_g) and the mapping class group $M_{0,2g+2}$ of the punctured sphere, which is in turn a close relative of Artin's braid group (cf. Section 3 above). Thus, for example, knowledge of the existence of elements of finite order in $M_{0,6}$ may be applied to describe the elements of finite order in $M_{2,0}$ [see 16].

These ideas were extended to the case where $\mathcal{L}$ is replaced by any element of finite order in $M_{g,0}$ by the author and Hilden in [7, 8], and to finite subgroups of $M_{g,0}$ (with a somewhat different emphasis) by Harvey and Maclachlan, in [16]. If G is a finite group of homeomorphisms of S_g, then one may define a symmetric mapping class group $M_g(G)$ of S_g by the requirement that admissible maps $\tilde{h}$ satisfy

$$\tilde{h}\, G\, \tilde{h}^{-1} = G\ . \tag{6}$$

Note that if $\tilde{h}, \tilde{k}$ are two representatives of an element of $M_g(G)$, then the isotopy $\tilde{h}_t$ joining $\tilde{h}$ to $\tilde{k}$ must satisfy (6) for each $0 \le t \le 1$. Harvey and Maclachlan establish in [16] that, with suitable restrictions on G, the group $M_g(G)$ embeds in $M_g = M_{g,0}$.

A third reference on the same problem is recent work of Zieschang [31], who establishes a theorem that is somewhat weaker than the results in [16], but somewhat stronger than those in [7], using purely algebraic methods. The methods of Birman and Hilden are largely topological; those of Harvey and Maclachlan use Teichmüller theory.

For a study of the *limitations* of the method of lifting and projecting homeomorphisms as a method for studying M_g, see W. Magnus, [22]. Magnus studies the situation of ramified coverings of the sphere. He establishes that if $g \ge 3$, then for every finite set of points B on the sphere, and for every element α in the automorphism group of $\pi_1(S_0-B)$, there is a Riemann surface $\tilde{S}$ which covers S_0 and is ramified over B, such that α does not lift to an automorphism of $\pi_1\tilde{S}$. The connections between these ideas and classical work of Hurwitz [17], and also recent work of V. I. Arnold [1] has also been explored by Magnus in [22], and by Magnus and Peluso in [23].

§5. *Does* $M_g/[M_g, M_g]$ *have order* 1 *or* 2, *if* $g \leq 3$?

Since M_g is generated by twists, and since these twist generators are all in a single conjugacy class, it follows immediately that $M_g/[M_g,M_g]$ is a cyclic group. Known relations between these twist maps imply that $M_g/[M_g,M_g]$ has order 2 or 1 if $g \geq 3$ [see Section 4], however it remains an unsolved problem whether $M_g/[M_g,M_g]$ does, in fact, have order 1. To prove this it would be adequate to find any relator in M_g which is a product of an odd number of twists; alternatively, to show that $M_g/[M_g,M_g]$ has order 2, it would be adequate to exhibit a single homomorphism from M_g onto a group of order 2. The problem of finding such homomorphisms will be discussed in the next section.

§6. *Finite representations of* M_g

A group G is said to be residually finite if for every $g \in G$ there is a homomorphism $\eta : G \to Q$, with Q finite, such that $\eta(g) \neq 1$. Residual finiteness is a pleasant property for an infinite group to have, because it offers the possibility that many aspects of the group G may be investigated in finite quotients of G.

One collection of finite quotients of our group M_g which has been known for a long time is the class of groups $\{S_p(2g, Z_{p^n}); \text{p prime}, n \geq 1\}$. In 1965 it was established by Mennicke [26] that this class exhausted the finite quotients of M_g which factor through $S_p(2g, Z)$, however until very recently it was not known whether M_g had any finite quotients which do not factor through $S_p(2g, Z)$. For this reason, it is particularly interesting that E. Grossman has established that M_g is residually finite [14].

R. Gilman has suggested a concrete method for constructing finite representations of M_g, and it is not difficult to arrange this construction so that the representation does not factor through $S_p(2g, Z)$. We will give a brief description of Gilman's construction, and of its geometric meaning. Let Q be an arbitrary but henceforth fixed finite group, and let $\{\eta_\lambda ; \lambda \in \Lambda\}$ be the collection of all homomorphisms from $\pi_1 S_g$ onto Q. These homomorphisms may be listed explicitly by a straightforward procedure in which one tests every possible mapping from the generators of $\pi_1 S_g$ onto Q to see whether the single defining relation in $\pi_1 S_g$ is satisfied. A homomorphism η_λ may be defined explicitly by an ordered array of 2g elements of which are the images of a standard set of generators of $\pi_1 S_g$ under η_λ. If $\eta_\lambda : \pi_1 S_g \to Q$ is one such representation, and if $t \in \text{Aut } Q$, then $t\eta_\lambda$ will be another such representation, hence we may divide our representations into equivalence classes (mod Aut Q). Let N be the collection of all such equivalence classes. Each element of N may be described geometrically as the covering space of S_g which belongs to the subgroup $\ker \eta_\lambda$ of $\pi_1 S_g$. Thus N may be regarded as the collection of all Riemann surfaces which are regular coverings of a fixed surface S_g, with group of covering transformations isomorphic to Q.

The mapping class group M_g may now be allowed to act on the set N as a group of permutations, by the rule that if $\alpha \in M_g$ is represented by $\alpha_* \in \text{Aut } \pi_1 S_g$, then the permutation π_α induced by α maps each η_λ to $\eta_\lambda \alpha_*$. It is not difficult to check that π_α is independent of the choice of representative α_*; that π_α preserves equivalence classes (mod Aut Q); and finally that $\pi_{\alpha\beta} = \pi_\alpha \pi_\beta$, $\alpha, \beta \in M_g$. Therefore one obtains in this way a finite representation of M_g as a subgroup of the group of all permutations of the finite set N.

While these representations were introduced because it was hoped that they would provide answers to certain explicit questions about M_g (such as the question which was posed in Section 5 above), they are also of considerable interest from a purely group-theoretical point of view. If the group M_g is a perfect group (i.e., $M_g/[M_g,M_g] = 1$), then all of its finite quotients will also be perfect groups, hence Gilman's construction would give an explicit method for finding infinite sequences of finite perfect groups. Two such finite representations have been studied by the author (unpublished). The first uses the group Σ_3 of permutations on 3 letters as the defining subgroup, to obtain a transitive representation of M_3 as a group of permutations of 2,530 letters. The second uses the the quaternion group of order 8, to obtain a transitive representation of M_3 as a group of permutations of 5,040 letters.

REFERENCES

[1]. Arnold, V. I., "Remarks on the branching of hyperelliptic integrals," Functional Analysis and its Applications (English translation), Vol. 2 (1968).

[2]. Birman, Joan S., "On Braid Groups," Com. Pure and App. Math., 22 (1969), pp. 41-72.

[3]. Birman, Joan S., "Mapping class groups and their relationship to braid groups," Com. Pure and Applied Math., 22 (1969), pp. 213-238.

[4]. Birman, Joan S., "Abelian quotients of the mapping class group of a 2-manifold," Bull. AMS, 76 (1970), pp. 147-150; Errata: *ibid.*, 77 (1971), pp. 479.

[5]. Birman, Joan S., "Braids, links and mapping class groups," a research monograph, to appear in Annals of Mathematics Studies Series, early 1974.

[6]. Birman, J., and Hilden, H., "Mapping class groups of closed surfaces as covering spaces," Annals of Math. Studies #66, "Advances on the theory of Riemann Surfaces," ed. Ahlfors et al., Princeton University Press, 1972.

[7]. Birman, J., and Hilden, H., "Isotopies of homeomorphisms of Riemann surfaces and a theorem about Artin's braid group," Bull. AMS, 78, No. 6, Nov. 1972, pp. 1002-1004.

[8]. Birman, J., and Hilden, H., "Isotopies of homeomorphisms of Riemann surfaces," Annals of Math., to appear.

[9]. Birman, Joan S., and Chillingworth, D. R. J., "On the homeotopy group of a non-orientable surface," Proc. Camb. Phil. Soc., 71 (1972), pp. 437-448.

[10]. Cohen, David, "The Hurwitz Monodromy Group," Ph.D. thesis, 1962.

[11]. Dahm, David, "A generalization of braid theory," Princeton Univ. Ph.D. thesis, 1962.

[12]. Dehn, M., "Die Gruppe der Abbildungsklassen," Acta Math., 69 (1938), pp. 135-206.

[13]. Goldberg, C., "An exact sequence of braid groups," to appear in Math. Scand.

[14]. Grossman, Edna K., "On the residual finiteness of certain mapping class groups," to appear.

[15]. Harvey, W., "Cyclic groups of automorphisms of a compact Riemann surface," Quart. Jnl. of Math., 17, No. 66 (1969), pp. 86-97.

[16]. Harvey, W., and Maclachlan, C., "On mapping-class groups and Teichmüller spaces," to appear.

[17]. Hurwitz, A., "Über Riemannsche Flachen mit gege benen Verzweigungs punkten," Math Annalen 39 (1891), pp. 1-61.

[18]. Lickorish, W. B. R., "A representation of orientable, combinatorial 3-manifolds," Annals of Math. 76 (1962), pp. 531-540.

[19]. Lickorish, W. B. R., "A finite set of generators for the homeotopy group of a 2-manifold," Proc. Camb. Phil. Soc., 60 (1964), pp. 769-778. Also, Corrigendum, 62 (1966), pp. 679-681.

[20]. Maclachlan, C., "Modulus space is simply-connected," Proc. AMS, 29 (1971), pp. 85-86.

[21]. Magnus, W., "Über Automorphismen von Fundamental gruppen Berandeter Flächen," Math Annalen 109 (1934), pp. 617-646.

[22]. Magnus, W., "Braids and Riemann surfaces," Comm. on Pure and Applied Math., 25 (1972), pp. 151-161.

[23]. Magnus, W., and Pelluso, A., "On a theorem of V. I. Arnold," Comm. Pure and Applied Math., 22 (1969), pp. 683-692.

[24]. Magnus, Karass and Solitar, "Combinatorial Group Theory," Interscience, John Wiley and Sons, 1966.

[25]. Mangler, "Die klassen von topologischen Abbildungen einer geschlossenen Flache auf sich," Math Zeitsch, 44 (1939), pp. 541-554.

[26]. Mennicke, "Zur Theorie der Siegelschen Modulgruppe," Math Annalen 159 (1965), pp. 115-129.

[27]. Nielsen, J., "Untersuchen zur Topologie der geschlossenen Zweiseitigen Flachen I," Acta Math. 50 (1927), pp. 184-358, also III, Acta Math 58 (1932), pp. 87-167.

[28]. Patterson, D., "The fundamental group of the moduli space," Mich. Jnl. of Math., to appear.

[29]. Scott, G. P., "Braid groups and the group of homeomorphisms of a surface," Proc. Camb. Phil. Soc. 68 (1970), pp. 605-617.

[30]. Sprows, David, "Isotopy classification of homeomorphism of multiply-punctured compact 2-manifolds," thesis, Univ. of Penna., 1972.

[31]. Zieschang, H., "On the homeotopy groups of surfaces," Math Annalen, to appear.

[32]. Magnus, W., "Braid groups; a survey," to appear.

A NOTE ON $L_2(\Gamma\backslash G)$

Su-shing Chen

Let G be a locally compact topological group and Γ be a discrete subgroup of G such that $\Gamma\backslash G$ is compact. The regular representation T of G on $L_2(\Gamma\backslash G)$ is unitary and $L_2(\Gamma\backslash G)$ splits into the direct sum of a countable number of invariant irreducible subspaces with finite multiplicities.

Consider the collection of all subgroups $\Gamma_\alpha (\alpha \in A)$ of finite index in Γ. Then each $\Gamma_\alpha\backslash G$ is compact. If $\Gamma_\alpha \subset \Gamma_\beta$, then we have a natural embedding $L_2(\Gamma_\beta\backslash G) \subset L_2(\Gamma_\alpha\backslash G)$. The direct limit of the spaces $L_2(\Gamma_\alpha\backslash G)(\alpha \in A)$ is denoted by H. This space can be described as follows. We set $\Gamma'_0 = \Gamma$, $\Gamma'_\alpha = \Gamma_1 \cap \ldots \cap \Gamma_\alpha$. Then $\Gamma'_\alpha \subset \Gamma'_{\alpha-1}$ and $L_2(\Gamma'_{\alpha-1}\backslash G) \subset L_2(\Gamma'_\alpha\backslash G)$. Denote by H_α the orthogonal complement of the subspace $L_2(\Gamma'_{\alpha-1}\backslash G)$ in $L_2(\Gamma'_\alpha\backslash G)$. Then H is the direct sum of the Hilbert spaces $L_2(\Gamma\backslash G)$ and $H_\alpha (\alpha \in A)$. There is a natural representation of G on H. Furthermore, the representation on H splits into a countable direct sum of irreducible unitary representations.

The above account is given on pp. 19-20 of [1], where the following question for $G = SL(2, R)$ is raised: Is the multiplicity with which an irreducible unitary representation occurs in this decomposition finite? In this note, we shall explain why the answer is negative.

It is shown on p. 78 of [1] that when Γ does not contain elliptic elements, the multiplicity $N_n^{\pm}(\Gamma)$ with which the representation $T_n^{\pm}$ of the discrete series occurs in T is given by the formula:

$$N_n^{\pm}(\Gamma) = [1 + (-1)^{n-1}]\frac{\mu(\Gamma\backslash G)}{\pi^2} n ,$$

where $\mu(\Gamma\backslash G)$ denotes the volume of $\Gamma\backslash G$.

Consider a sequence $\{\Gamma_i\}$ of subgroups of Γ such that $\Gamma_0 = \Gamma$, $\Gamma_{i+1} \triangleleft \Gamma_i$ and $1 < [\Gamma_i : \Gamma_{i+1}] = \lambda_i < \infty$, for $i = 0, 1, 2, \ldots$. The existence of such a sequence follows from the fact that the fundamental group of a surface of genus $g \geq 1$ can be mapped homomorphically onto a cylic group of any order. Since the projection $\Gamma_{i+1} \backslash G \to \Gamma_i \backslash G$ is a λ_i-sheeted covering, $\mu(\Gamma_{i+1} \backslash G) = \lambda_i \mu(\Gamma_i \backslash G)$ and $N_n^{\pm}(\Gamma_{i+1}) = \lambda_i N_n^{\pm}(\Gamma_i)$. Therefore $N_n^{\pm}(\Gamma_{i+1}) = \lambda_0 \lambda_1 \lambda_2 \ldots \lambda_i N_n^{\pm}(\Gamma)$, and consequently we have:

THEOREM. *Let* n *be a positive odd integer. Then the multiplicities with which the representations* $T_n^{\pm}$ *of the discrete series occur in the decomposition of* H *are infinite.*

UNIVERSITY OF FLORIDA
GAINESVILLE, FLORIDA 32601

REFERENCES

[1]. Gel'fand, I. M., Graev, M. E., and Pyatetskii-Shapiro, I. I., Representation theory and automorphic functions, Saunders, Philadelphia, London, and Toronto, 1969.

ON THE OUTRADIUS OF FINITE-DIMENSIONAL TEICHMÜLLER SPACES

T. Chu

The universal Teichmüller space, T(1), may be defined as the set of functions ϕ, holomorphic in $\Delta^* = \{z \in \hat{C} \mid |z| > 1\}$ which are Schwarzian derivatives of mapping functions admitting quasi-conformal extension to $\hat{C}$. T(1) has a natural embedding in the linear space $B_2(1)$ with

$$\|\phi\| = \sup_{z \in \Delta^*} (|z|^2 - 1)^2 \, |\phi(z)|$$

for $\phi \in B_2(1)$. It follows from well-known results of Nehari, Hille and Ahlfors and Weill that

$$o(1) = \sup_{\phi \in T(1)} \|\phi\| = 6$$

$$i(1) = \inf_{\phi \in B_2(1)-T(1)} \|\phi\| = 2 .$$

o(1) is the radius of the smallest ball about O containing T(1) and is called the *outradius* of T(1). i(1) is the radius of the largest ball about O contained in T(1). It is also known that the function $k(z) = \dfrac{z}{(z-1)^2}$ has a Schwarzian derivative $\phi_k = [k] = \dfrac{-6}{(z^2-1)^2}$ with norm 6. k(z) is the usual Koebe function composed on the right with the Möbius transformation $z \to z^{-1}$. $\phi_k \in \partial T(1)$ since there are holomorphic $\left(1 - \frac{1}{n}\right)$-th roots $k_n(z)$ defined on Δ^*, $[k_n(z)] \in T(1)$ and $\|[k_n] - [k]\| \to 0$.

If G is a Fuchsian group leaving the unit circle invariant, we define

$$B_2(G) = \{\phi \in B_2(1) | \, (\phi\circ\gamma)(z)\gamma'(z)^2 = \phi(z) \ ,$$
$$\text{for } z \in \Delta^* \text{ and } \gamma \in G\} \ .$$

T(G), the *Teichmüller space of* G, is the component containing 0 of $T(1) \cap B_2(G)$. T(G) is a topological cell which is finite dimensional if and only if G is finitely generated and of the first kind.

The major result presented here is a construction lemma leading to an asymptotic estimate for o(G), where

$$o(G) = \sup_{\phi \in T(G)} \|\phi\| \ .$$

The construction theorem is a generalization, suggested by Akaza, of a technique of Klein. The application to estimates of o(G) was proposed by Abikoff. The precise statement is

THEOREM. *For each* $\varepsilon > 0$, *there is a finitely generated hyperbolic Fuchsian group* G *of the first kind, for which* $o(G) > 6 - \varepsilon$.

Before proceeding, we need the following definition. Two Jordan curves J_1 and J_2 are said to have *Frechet distance* $\delta(J_1, J_2) < \varepsilon$, if there are parametrizations $\alpha_i(t)$ of J_i, $i = 1, 2$, $0 \le t \le 1$ so that $|\alpha_1(t) - \alpha_2(t)| < \varepsilon$ for each $t \in [0,1]$. We will state the two basic lemmas, show that the theorem follows easily from them, and then give their proofs.

LEMMA 1. *For each positive* ε, *there exist* $\varepsilon_1, \varepsilon_2 > 0$ *so that, if* J_1 *is a* $1 + \varepsilon_1$*-level curve of* k(z) *and* J_2 *is a Jordan curve with* $\delta(J_1, J_2) < \varepsilon_2$, *then a conformal map* f *of* Δ^* *onto* Ext J_2 *satisfies* $\|[f]\| > 6 - \varepsilon$.

LEMMA 2. *Given a rectifiable Jordan curve* J *and* $\varepsilon_3 > 0$, *there exists a finitely generated purely loxodromic quasi-Fuchsian group* G *of the first kind so that* $\delta(J, \Lambda(G)) < \varepsilon_3$.

Proof of Theorem. Each $\left(1+\frac{1}{n}\right)$-level curve J_n of k is a rectifiable Jordan curve hence we may find finitely generated purely loxodromic quasi-Fuchsian groups $G_{n,k}$ so that $\lim\limits_{k\to\infty} \delta(J_n, \Lambda(G_{n,k})) \to 0$. Let $f_{n,k}$ be the conformal map of Δ^* onto $\operatorname{Ext}\Lambda(G_{n,k})$ normalized by $f_{n,k}(\infty)=\infty$ and $f'_{n,k}(\infty)>0$. It follows from the Caratheodory convergence theorem [1] that $f_{n,k}$ converges normally to f_n, the conformal map of Δ^* onto $\operatorname{Ext} J_n$ with $f_n(\infty)=\infty$ and $f'_n(\infty)>o$. The Cauchy Integral theorem then yields that $[f_{n,k}]$ converges normally to $[f_n]$ hence

$$\lim_{k\to\infty} \|[f_{n,k}]\| \le \|[f_n]\| > 6-\varepsilon\ .$$

It is well known that $f_{n,k}^{-1} G_{n,k} f_{n,k}$ is a finitely generated purely hyperbolic Fuchsian group $H_{n,k}$ of the first kind and $[f_{n,k}] \in T(H_{n,k})$ which proves the theorem.

Proof of Lemma 1. We may assume $f(\infty)=\infty$ and $f'(\infty)>0$ since these conditions are satisfied by $\eta\circ f$, η a Möbius transformation and $[\eta\circ f] = [f]$. We now assume that the Lemma is false. It would then follow that for each ε and ε_1 and all $\varepsilon_2>0$, the conclusion is false. In particular, if $\varepsilon_n \to 0$, $J_{1,n}$ denotes the $(1+\varepsilon_n)$-level curve of $k(z)$ and $J_{2,n}$ is a Jordan curve with $\delta(J_{1,n}, J_{2,n}) < \frac{1}{n}$, then the maps f_n of Δ^* onto $\operatorname{Ext} J_{2,n}$, $f_n(\infty)=\infty$ and $f'_n(\infty)>0$, have the property that $\|[f_n]\| \le 6-\varepsilon$. Since the domains $\operatorname{Ext} J_{2,n}$ converge in the sense of Caratheodory to $k(\Delta^*)$, it follows that f_n converge normally to $k(z)$. In particular

$$(|z_0|^2-1)^2\,|[f_n](z_0)| \to (|z_0|^2-1)^2\,|[k](z_0)| = 6$$

$$\text{for } z_0 \in \mathbf{R}^+ \cap \Delta^*$$

and we have a contradiction.

Proof of Lemma 2. We will explicitly construct the group G as a subgroup of finite index of a reflection group with compact fundamental domain. We first approximate J by a piecewise linear Jordan curve J_1 with the property that $\delta(J, J_1) < \frac{\varepsilon_3}{2}$. Next we cover J_1 by a finite number of closed discs $B_1, \ldots, B_n$ of diameter less than $\frac{\varepsilon_3}{2}$. We assume for notational convenience that $B_0 = B_n$ and $B_{n+1} = B_1$, i.e., we have a cyclic ordering of the B_i. The covering may easily be assumed to have the following property

$$\bar{B}_i \cup \bar{B}_j = \phi \text{ for } 1 \leq i \leq n \text{ and } j \neq n-1, n+1$$

the circles ∂B_i and ∂B_{i+1} are orthogonal circles. The existence of such a covering follows from the continuous dependence of the angle of intersection of two circles on their radii and position of their centers.

Under the above assumptions, we wish to show that (*) the group generated by reflection in ∂B_i $i = 1, \ldots, n$ is a reflection group G_1 having as fundamental set $F = \hat{C} - \bigcap_{i=1}^{n} \operatorname{Ext} B_i$ and whose limit set is a Jordan curve $\Lambda = \Lambda(G_1)$ not intersecting $\bar{F}$.

Assuming (*), the Selberg Lemma shows that G_1 has a subgroup G of finite index of directly conformal transformations, none of which has finite order. Since G_1 has compact quotient, so does G and $\Lambda(G) = \Lambda(G_1)$. We may assume G leaves invariant each component of $\hat{C} - \Lambda(G)$ since G_1 either has this property or has a subgroup of index two with this property. $\Lambda(G)$ is a Jordan curve and $\delta(J_1, \Lambda(G)) < \frac{\varepsilon_3}{2}$. G is a quasi-Fuchsian group by Maskit's theorem (see Kra and Maskit [2]). $\Omega(G)/G$ is a finite sheeted cover of the non-orientable compact surface $\Omega(G_1)/G_1$, hence is compact. Since G has a compact fundamental set and is torsion-free, it is purely loxodromic and finitely generated.

We have thus reduced the proof of Lemma 1, to the proof of (*). We denote by F^{∞}, the component of F containing ∞ and by F_0 the other component of F. We will now show that F is a fundamental set for G_1 and F is compactly contained in $\Omega(G_1)$. If R_i denotes reflection in ∂B_i and $L_i = R_1 R_i$ for $2 \leq i \leq n$ then

(1) $$\Delta_0 = \bar{F}_0 \cup \left(\bigcup_{i=2}^{n} L_i \bar{F}_0\right) \cup \left(\bigcup_{i=1}^{n} R_i F_0\right)$$

is a neighborhood of $\overline{F}_0$, which except for images of ∂F_0 consists of disjoint images of F_0. By an induction argument we obtain $\gamma F_0 \cap \gamma^1(F_0) \neq \emptyset$ if and only if $\gamma = \gamma^1$, hence G_1 is discontinuous and F_0 is a fundamental set for $G_1\overline{F}_0$. Since $\Delta_0 \subset \Omega(G_1)$ and is a neighborhood of F_0, F_0 is compactly contained in $\Omega(G_1)$. A similar argument yields similar conclusions for F_∞. Both $G_1\overline{F}_0$ and $G_1\overline{F}_\infty$ are connected. We claim G_1 has exactly two invariant components. That it can have no more than two follows from the standard result on finitely generated Kleinian groups by passing to the directly conformal subgroup G_2 of G_1. If $\Omega(G_1)$ had one component, there would be a path α, avoiding images of vertices of F, from F_0 to F_∞ and lying entirely in $\Omega(G_1)$. α may then be covered by finitely many images of some fundamental set $\hat{F}$ for $G_1, \hat{F} \supset F$. But using (1) we see that finitely many images of $\overline{F}$ cover a neighborhood of F, hence α is covered by finitely many images of $\overline{F}$. In this case, there would be a map $\gamma \in G_1, \gamma\overline{F}_0 \cap \overline{F}_\infty \neq \emptyset$, which is impossible. Thus we have shown that G_1 has two invariant components. It follows that G_2 has two invariant components, hence $\Lambda(G_2) = \Lambda(G_1)$ is a quasi-circle by Maskit's Theorem. That F is a fundamental set for G_1 follows from a path-connectedness argument similar to the one given above.

COLUMBIA UNIVERSITY

REFERENCES

[1]. Goluzin, G. M., *Geometric Theory of Functions of a Complex Variable*, Amer. Math. Soc., Providence, 1969.

[2]. Kra, I., and Maskit, B., Involutions on Kleinian groups, *Bull. Amer. Math. Soc.*, 78 (1972), pp. 801-805.

SOME DIRECT LIMITS OF PRIMITIVE HOMOTOPY WORDS AND OF MARKOFF GEODESICS

Harvey Cohn*

§1. *Summary*

If a perforated torus (torus minus disk) is given a suitable metric, the closed geodesics will have periods corresponding uniquely to the primitive homotopy classes. These geodesics are called "Markoff geodesics" because of their relation to Markoff's binary quadratic forms. The Markoff geodesics are then in various homotopy classes all *equivalent* under automorphisms of the fundamental group of the perforated torus.

On the same manifold, the nonperiodic geodesics corresponding to real slope numbers θ, or "θ-geodesics" are studied, (generalizing the Markoff geodesics, for which θ is rational). When θ is quadratic, the corresponding geodesics lie in a (direct limit) infinite homotopy class which is *invariant* under an automorphism of the fundamental group.

BACKGROUND RESULTS

§2. *Sequences*

Let θ be a real number. We consider the (Lagrangian) continued fraction, or L-expansion

$$\theta = L(a_1, a_2, \ldots) = a_1 + 1/(a_2 + 1/(a_3 + \ldots)) \quad . \tag{2.1}$$

Under the usual conditions where $a_i \in Z$; $a_i \geq 1, (i > 1)$, the expansion is unique (allowing for rational termination). We denote successive approximants as $\theta, \theta^L, \theta^{L^2}$, etc., where

* Research supported by NSF Grant G-33910.

$$\theta = a_1 + 1/\theta^L = L(a_1, \theta^L),\ a_1 = [\theta] \ . \tag{2.2}$$

For two real numbers

$$\xi > 1,\ 0 > \eta > -1 \tag{2.3}$$

$$\xi = L(a_0, a_1, a_2, \ldots),\quad -\eta = L(0, a_{-1}, a_{-2}, \ldots) \tag{2.4}$$

we denote the pair $G[\xi, \eta]$ as a two-way infinite sequence

$$G[\xi, \eta] = LL(\ldots a_{-2}, a_{-1}, a_0, a_1, a_2, \ldots),\quad 1 \leq a_i \in Z \ . \tag{2.5}$$

This LL-*expansion* equally well denotes $G[\xi_m, \eta_m]$ where $m \in Z$,

$$\xi_m = L(a_m, a_{m+1}, \ldots),\quad -\eta_m = L(0, a_{m-1}, a_{m-2}, \ldots) \ . \tag{2.6}$$

A periodic L-expansion is denoted by a double parenthesis as $L((a_1, \ldots a_p))$, and likewise for $LL((a_1, \ldots a_p))$.

In the special case where the sequence (2.5) is a succession of double ones or double twos, usually mixed, we can write it as

$$G[\xi, \eta] = LL(\ldots 1^{2m_{-1}}, 2, 2, 1^{2m_0}, 2, 2, 1^{2m_1}, \ldots),\quad 0 \leq m_i \in Z \ . \tag{2.7}$$

Here the superscript always shows the tally and a zero shows a void. Except for certain cases (see (2.9a) below) we abbreviate (2.7) by an equivalent M-*expansion*

$$G[\xi, \eta] = M(\ldots m_{-1}, m_0, m_1, \ldots) \ . \tag{2.8}$$

Special cases of LL-expansions (2.7) are called *Markoff* sequences, namely those where

$$G[\xi, \eta] = LL(2^\infty, 1, 1, 2^\infty),\ LL(1^\infty, 2, 2, 1^\infty),\ LL((1)),\ LL((2)) \tag{2.9a}$$

$$G[\xi, \eta] = LL((1^{2m}, 2, 2)),\quad m \geq 1 \ , \tag{2.9b}$$

and also those LL-expansions for which m_i takes only two values m and $m+1$ in such a fashion that (with fixed choice of $\pm$ sign)

(2.10) $$m_k - m_{k+1} = \pm 1 \Longrightarrow \pm(m_{k+h+1} - m_{k-h}) > 0$$

where $h(>0)$ is the smallest integer for which $m_{k-h} = m_{k+h+1}$.

In cases where infinitely many ones and twos occur in both directions in (2.7) and $m_i = m$ or $m+1$, Markoff defined a *derived* sequence as follows: If (2.8) is written as

(2.11a) $$G[\xi,\eta] = M(\ldots m^{n_{-1}}, (m+1), m^{n_0}, (m+1), m^{n_1}, \ldots), \quad n_i \geq 0 ,$$

then the derived sequence is

(2.11b) $$G^M[\xi,\eta] = M(\ldots n_{-1}, n_0, n_1, \ldots) .$$

If the n_i take the two values n and $(n+1)$, then further derivation is possible. (Note $n_i = 0$ simply denotes consecutive occurrences of $(m+1)$ in (2.11a).)

THEOREM 2.12 (Markoff-Dickson). *A Markoff sequence either has form* (2.9ab) *or else it permits successive derivations possibly terminating only in such a form.*

THEOREM 2.13 (Markoff-Dickson). *The Markoff sequences are precisely those for which for all* m

(2.14) $$\xi_m - \eta_m \leq 3 .$$

Markoff proved these results for periodic sequences only [8], but Dickson [4] extended them more generally, also showing that if

(2.15) $$\sup(\xi_m - \eta_m) < 3$$

then the Markoff sequence must be periodic. When this happens, by Markoff's work, the sequence is finitely derivable to the trivial sequence LL((1)) or LL((2)).

An additional type of sequence to consider for θ real is the E-*expansion* or the EE-*expansion* ("E" for "entier")

$$(2.16) \qquad \theta = E(c_1, c_2, c_3, \ldots)$$

$$(2.17) \qquad \theta = EE(\ldots c_{-2}, c_{-1}, c_0, c_1, c_2, \ldots)$$

$$(2.18) \qquad c_n = [n\theta] - [(n-1)\theta] \ .$$

Here, for θ nonintegral, $-[-\theta] = [\theta] + 1$; $c_1 = [\theta]$, $c_0 = [\theta] + 1$. Therefore for θ irrational,

$$(2.19) \qquad c_0 = c_1 + 1,\ c_{-n} = c_{n+1}\ (n \geq 1) \ .$$

The corresponding LL-expansion is written as

$$(2.20a) \qquad G(\theta) = LL(\ldots 1^{2c_{-1}}, 2, 2, 1^{2c_0}, 2, 2, 1^{2c_1}, 2, 2, \ldots)$$

or as the M-expansion

$$(2.20b) \qquad G(\theta) = M(\ldots c_{-1}, c_0, c_1, \ldots) \ ,$$

with the usual special case $G(1/0) = LL((1))$.

THEOREM 2.21. *The periodic Markoff sequences correspond to* $G(\theta)$ *for* θ *rational or* $1/0$.

This was proved in [2].

Finally, let the symbol $-G$ denote the reverse expansion.

§3. *Groups*

Let $\Gamma = SL_2(Z)$ or $PSL_2(Z)$ according to context. Thus if $M \in \Gamma$ then M is a unimodular integral matrix or a function

$$(3.1) \qquad M = \begin{pmatrix} a & b \\ c & d \end{pmatrix} \in SL_2(Z),\ M(z) = \frac{az+b}{cz+d} \ .$$

Accordingly, $(-M)(z) = M(z)$ so the identity is either I or $(\pm I)(z)$.

Now the commutator of Γ is a free group (see [7])

$$(3.2) \qquad F_2 = [\Gamma, \Gamma] = \{A, B\} = \{V_1, V_2\}$$

with generators variously taken as (see [2])

(3.3a) $$V_1(z) = L(1,1,z),\ V_1 = \begin{pmatrix} 2 & 1 \\ 1 & 1 \end{pmatrix},$$

(3.3b) $$V_2(z) = L(2,2,z),\ V_2 = \begin{pmatrix} 5 & 2 \\ 2 & 1 \end{pmatrix};$$

or else they may be taken from A, B, C, where

(3.4) $$A = \begin{pmatrix} 1 & 1 \\ 1 & 2 \end{pmatrix},\ B = \begin{pmatrix} 1 & -1 \\ -1 & 2 \end{pmatrix},\ C = \begin{pmatrix} 3 & -1 \\ 1 & 0 \end{pmatrix},$$

(3.5) $$V_1 = B^{-1},\ V_2 = B^{-1}A^{-1}B^{-1}.$$

Note that the triple A, B, C has the property (see [1])

(3.6) $$ABC = I,\ CBA = [B^{-1}, A^{-1}] = \begin{pmatrix} -1 & -6 \\ 0 & -1 \end{pmatrix} = K.$$

§4. *Manifolds*

Let H^+ denote the upper half z-plane, $z = x+iy, y > 0$ and let Γ act on H^+. We truncate H^+ to form H^* by the removal of all points equivalent under Γ to

(4.1a) $$\operatorname{Im} z > \frac{3}{2}.$$

This consists of circular disks, ($\operatorname{Im} M(z) > \frac{3}{2}$, see (3.1)) or

(4.1b) $$(x+d/c)^2 + (y-1/(3c^2))^2 < 1/(3c^2)^2$$

for every pair $(c,d) = 1$, $c > 0$, (see Figure 1).

Clearly H^*/Γ forms a disk but it is more significant that

(4.2) $$H^*/[\Gamma,\Gamma] = T^*,$$

a perforated torus (see Figure 2 and [2]). Here the transformation

(4.3) $$du = dJ/\ J^{\frac{2}{3}}(1-J)^{\frac{1}{2}},$$

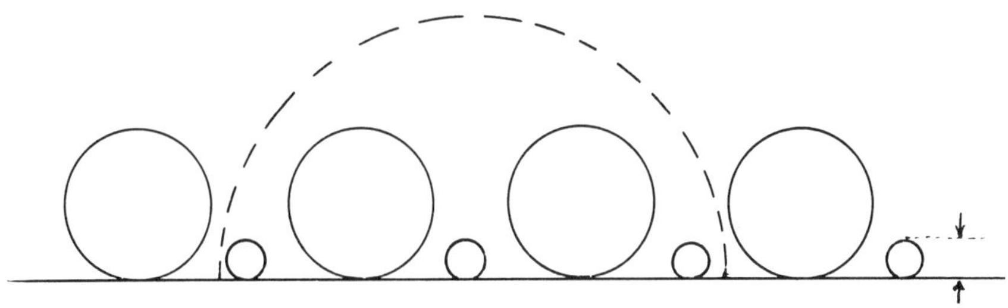

Fig. 1. Truncation of the upper half plane to form H*. Dotted curve G is geodesic remaining solely in H*.

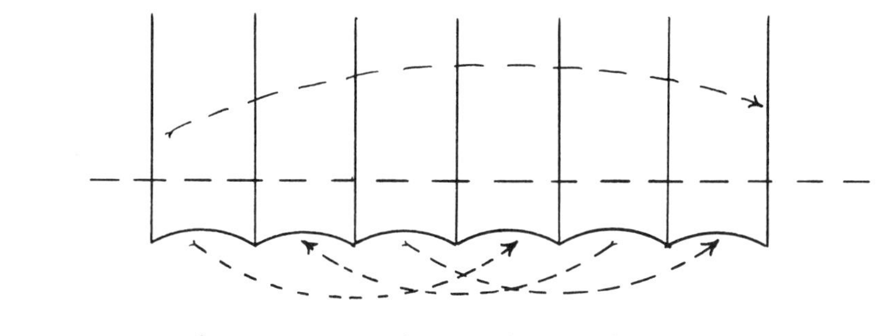

Fig. 2. Subdivision of H* by $[\Gamma,\Gamma]$ to form perforated torus T*.

(for J(z) the modular invariant), maps T^* onto the "hexagon-minus-disk" (in Figure 3), with opposite sides identified. The importance of T^* lies in the fact that its fundamental group is $F_2 = [\Gamma,\Gamma]$, (always in the presentation of Section 3).

We define H^*-geodesics, in the Poincaré metric

$$ds = |dz|/y \ , \tag{4.4}$$

as geodesics wholly in the compact manifold H^*, possibly tangent to the boundary ∂H^* but continuing in both directions. They are also "canonically" geodesics on T^*.

The H^*-geodesics (coincidentally, also denoted by G) are semicircles orthogonal to the real axis in H^* joining the real points ξ and η. By "Gaussian reduction," ξ and η can be made to satisfy (2.3) under Γ (in fact also under F_2, although this is not needed). Thus the Euclidean radius of G satisfies $\left((\xi-\eta)/2 \le \frac{3}{2}, \text{ (in fact, } (\xi_m-\eta_m)/2 \le \frac{3}{2}\right)$. Thus Theorems 2.12 and 2.13 can be interpreted (see [2]) as follows:

THEOREM 4.5. *The H^*-geodesics are equivalent under Γ (or for that matter F_2) to those where $G[\xi,\eta]$ are Markoff sequences.*

THEOREM 4.6. *The only H^*-geodesics which lie in a compact set interior to T^* are periodic.*

The periodic geodesics are called *Markoff geodesics*. The symbol G now stands for a geodesic or its expansion. The symbol $-G$ stands for a reversal of direction on G.

§5. *Automorphisms of Markoff geodesics*

It was shown in [2] that if a H^*-geodesic G is traced on T^*, the homotopy class of any segment is determined by a simple geometric construction; the terms of the LL-expansion consisting of $1,1$ correspond to V_1 and the terms $2,2$ correspond to V_2 (as traced in Figure 3, for example).

Thus, by Theorem 2.21, the Markoff geodesics correspond to $G(\theta)$ for θ rational. In the notation of (2.7) and (2.20a),

$$G(r/s) = LL((1^{2a_1}, 2, 2, 1^{2a_2}, 2, 2, \ldots, 1^{2a_s}, 2, 2)) \tag{5.1}$$

$$a_n = [nr/s] - [(n-1)r/s], \quad 1 \le n \le s \ , \tag{5.2}$$

for r/s reduced, $r \ge 0$. Symbolically, $r/s = \left(\frac{1}{0}\right)$ denotes $G\left(\frac{1}{0}\right) = LL((1))$.

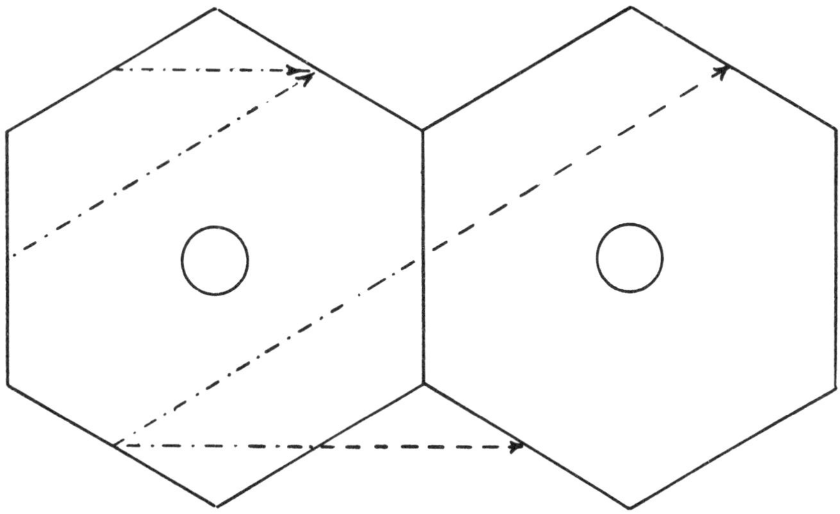

Fig. 3 Replicas of T^* in the u-plane showing how $LL(\ldots,1,1,\ldots)$ corresponds to $\ldots V_1 \ldots$ and $LL(\ldots,2,2,\ldots)$ corresponds to $\ldots V_2 \ldots$, generators of the fundamental group of the perforated torus. Note ∂T^* is the circular hole inside.

Classically, the familiar cases are given in terms of

$$\Phi(x,y) = (x-\xi y)(x-\eta y), \qquad x,y \in Z \ , \tag{5.3}$$

the *Markoff forms*. Thus

$$r/s = \frac{1}{0} : LL((1,1));\ \xi,\eta = \frac{1}{2}\left(1 \pm 5^{\frac{1}{2}}\right);\ \Phi(x,y) = x^2 - xy - y^2 \ .$$

$$r/s = \frac{0}{1} : LL((2,2));\ \xi,\eta = 1 \pm 2^{\frac{1}{2}};\ \Phi(x,y) = x^2 - 2xy - y^2 \ .$$

$$r/s = \frac{1}{1} : LL((1,1,2,2));\ \xi,\eta = \left(11 \pm 221^{\frac{1}{2}}\right)/10;\ \Phi(x,y) = 5x^2 - 11xy - 5y^2 \ .$$

Historically, Markoff's motivation was the property

$$\min |\Phi(x,y)| \,/\, (\mathrm{Disc}\ \Phi)^{\frac{1}{2}} > \frac{1}{3}\ , \tag{5.4}$$

where the minimum, of course is over $(0,0) \neq (x,y) \in Z^2$.

We find that the period (5.1) corresponds to the homotopy class $\prod_{n=1}^{s} V_1^{a_n} V_2$, or by an immediate change of basis (see (3.5)),

$$M_{r/s} = \prod_{n=1}^{s} A B^{2+a_n},\quad M_{1/0} = B\ , \tag{5.5}$$

$$a_n = [nr/s] - [(n-1)r/s]\ , \tag{5.6}$$

where r/s is reduced ≥ 0.

THEOREM 5.7. *The Markoff geodesics lie in homotopy classes equivalent under* $\mathrm{Aut}\ F_2$, *the automorphisms of the fundamental group of* T^*. *(Indeed each such geodesic is equivalent to its negative.)*

The proof is contained in [1], [2], [3]. It starts with the observation that according to the trace formulas of Fricke [5], the relations (3.6), understood for *arbitrary* matrices,

$$ABC = I,\quad CBA = K \tag{5.8}$$

lead to the following formulas for traces $3a, 3b, 3c$ of A, B, C:

$$a^2 + b^2 + c^2 = 3abc\ . \tag{5.9}$$

Meanwhile, the commutator K is preserved by a sufficiently large subgroup of $\mathrm{Aut}\ F_2$, e.g.,

$$(A,B) \mapsto (BA, A^{-1});\quad (A,B) \mapsto (A, A^{-1}B) \tag{5.10}$$

so that the generation of all Markoff's forms is possible. Indeed, by (5.10),

(5.11) $(a,b,c) \mapsto (c,a,b); \ (a,b,c) \mapsto (a,3ab-c,b)$

according to additional trace formulas discovered by Fricke [5].

The key step (see [1]) is that the minimal property (5.4) holds when ξ, η are roots of $M(x) = x$ for $M = A, B, C$.

The group generated by (5.10), (denoted by Γ_F in [3]) is isomorphic to the trefoil knot group. As a subgroup of Aut F_2 it is large enough to produce all Markoff forms by projecting onto $SL_2(Z)$ when F_2 is abelianized. We are then assured that the Markoff forms correspond precisely to the primitive words of F_2.

THEOREM 5.12. *The primitive words* $w(A,B)$ *of* F_2 *are precisely those reducible to the form* (5.5) *by inner automorphisms in* Aut F_2 *applied to one of the twelve words*:

$$w(A,B),\ w(B,B^{-1}A^{-1}),\ w(B^{-1}A^{-1},A),\ w(A^{-1},B^{-1}),\ w(B^{-1},AB),$$
$$w(AB,A^{-1}),\ w(B,A),\ w(B^{-1}A^{-1},B),\ w(A,B^{-1}A^{-1}),\ w(B^{-1},A^{-1}),$$
$$w(AB,B^{-1}),\ w(A^{-1},AB)\ .$$

The proof is in [3]. We only remark that these 12 words correspond to the dihedral group of the hexagon (e.g., in Figure 3).

DIRECT LIMITS

§6. *The* θ*-geodesics*

For an irrational $\theta > 0$ we can define a formal expansion $G(\theta)$ as in (2.20ab).

THEOREM 6.1. *The expansion* $G(\theta)$ *represents a* H^**-geodesic.*

The proof depends on the fact that the conditions (2.10) are *local*. The conditions are each covered if we approximate θ by a rational r/s sufficiently close that the interval referred to in (2.10), or $(k-h, k+h+1)$ lies within the period of $G(r/s)$. We call these $G(\theta)$ *θ-geodesics*. They are thus direct limits of Markoff geodesics.

THEOREM 6.2. *The θ-geodesics (for θ irrational) are each tangent to ∂T^* exactly once.*

Proof. Because of the symmetry in (2.19), we split (2.20a) after 1^{2c_0} and write $G(\theta) = G[\xi, \eta]$ with

(6.3) $$\xi = L(2,2,\zeta), \; -\eta = L(0,1,1,\zeta)$$

(6.4) $$\zeta = L(1^{2c_1}, 2, 2, 1^{2c_2}, 2, 2, \ldots) = L(1^{2c_0-2}, 2, 2, 1^{2c_{-1}}, 2, 2, \ldots) .$$

Now, the following identity in ζ, (known to Markoff [8]),

(6.5) $$\xi - \eta = 3$$

produces one tangency. There can not be a second tangency, either by a continued fraction argument (after Dickson [4]) or because a second tangency would produce a pair $(x,y) \in Z^2$ and not $(\pm 1, 0)$ for which

(6.6) $$\Phi(x,y) = (x - \xi y)(x - \eta y) = \pm 1 .$$

(Note Disc $\Phi = 9$, and compare (2.14) and (5.4)). This would make ζ quadratic.

§7. *Automorphisms of θ-geodesics*

Let us consider the effect of derivation on the M-expansion of $G(\theta)$ for θ irrational > 0:

(7.1) $$G(\theta) = M(\ldots a_{-2}, a_{-1}, a_0, a_1, a_2, \ldots) .$$

Since $a_n = [n\theta] - [(n-1)\theta]$ clearly $a_n = a_1$ $(= [\theta])$ or a_0 $(= 1 + [\theta])$ only. Thus a_1 starts a string of symbols denoted by $a_1^{j_1}$ for some $j_1 > 0$. More generally, for $j_n \geq 0$,

(7.2) $$G(\theta) = M(\ldots (a_1+1), a_1^{j_0}, (a_1+1), a_1^{j_1}, (a_1+1), a_1^{j_2}, \ldots) .$$

Now the symmetry (2.19) has that unequal pair a_0 and a_1. Thus

$$(7.3) \qquad j_0 = j_1 - 1, \quad j_{-n} = j_{n+1}, \quad (n>0) .$$

We can now make the derivation (see (2.11b)),

$$(7.4) \qquad G^M(\theta) = M(\ldots j_3, j_2, j_1-1, j_1, j_2, j_3, \ldots) .$$

LEMMA 7.5. *If* θ *is a real irrational*

$$(7.6) \qquad -G^M(\theta) = G(\theta^L - 1) .$$

For proof we apply the symmetry (2.19) to the following:

LEMMA 7.7. *Let* θ *have the* E-*expansion*

$$(7.8) \qquad \theta = E(a_1, a_2, \ldots) = ({a_1}^{j_1}, (a_1+1), {a_1}^{j_2}, (a_1+1), \ldots)$$

then the successor to θ *in the* L-*expansion of* θ *is given by*

$$(7.9) \qquad \theta^L = E(j_1, j_2+1, j_3+1, j_4+1, \ldots) .$$

Equivalently, (7.2) *leads directly to the* EE-*expansion*

$$(7.10) \qquad \theta^L = EE(\ldots j_3+1, j_2+1, j_1+1, j_1, j_2+1, j_3+1, \ldots) .$$

To see (7.9), observe the drawing in Figure 4. The terms of the expansion ${a_1}^{j_1}$ represent j_1 steps along the X-axis with a rise of a_1 along the Y-axis each time, keeping just below the line $Y = \theta X$. After j_1 steps, a rise of (a_1+1) is required. The change from θ to θ^L means that the X-axis becomes the X^*-axis and the Y-axis becomes the Y^*-axis, where

$$(7.11) \qquad Y/X = a_1 + 1/(Y^*/X^*) ,$$

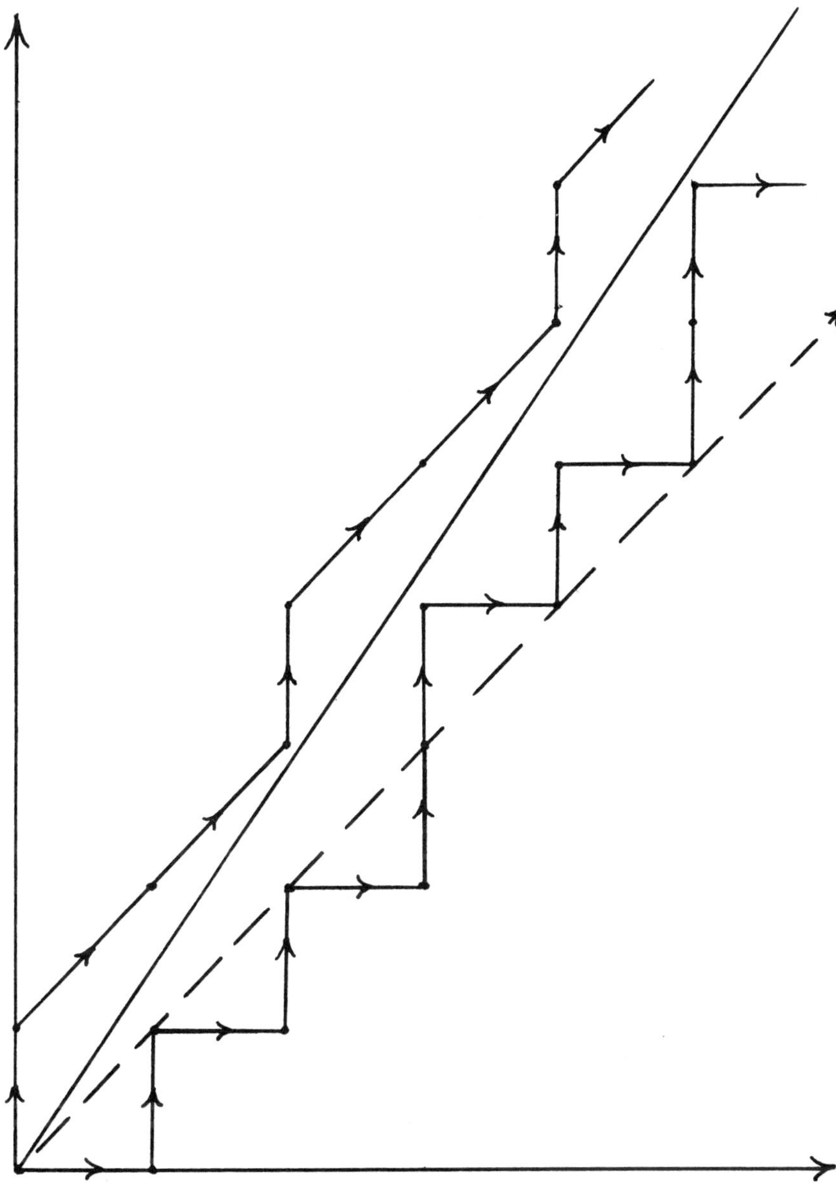

Fig. 4. Demonstration of how $\theta = E(a_1{}^2, (a_1+1), a_1, (a_1+1), \ldots)$ leads to $\theta^L = E(2, 2, \ldots)$. Note A′B′ corresponds to $a_1{}^2$ as AB corresponds to 2, while C′D′ corresponds to a_1 as CD corresponds to the second 2 in θ^L.

$$(7.12)\qquad \begin{cases} X = Y^* \\ Y = a_1 Y^* + X^* \end{cases} \qquad\qquad \begin{cases} X^* = Y - a_1 X \\ Y^* = X \end{cases}$$

Note the vital role of the asymmetry in the reversal of sign in (7.6).

Since $(\theta^L - 1)^L = \theta^{LL}$, the -1 is unimportant, so that Lemma 7.5 in essence equates $-G^M(\theta)$ with $G(\theta^L)$. If θ is real-quadratic, the iteration of the operator $\theta \mapsto \theta^L$ is ultimately periodic. Thus Lemma 7.5 defines an automorphism of F_2 leaving $G(\theta)$ invariant.

THEOREM 7.8. *For the real-quadratic case, the period of the* L*-expansion of* θ *defines the only automorphisms or anti-automorphisms of* F_2 *leaving* $G(\theta)$ *invariant. There will be an anti-automorphism if and only if the primitive period is odd; (then* $G(\theta)$ *will be reversed).*

The uniqueness of the period in defining automorphisms or anti-automorphisms will be seen later on.

§8. *Illustrations*

Several reduced quadratic surd values of θ are worked out in Table I. If we go back to the LL-expansion of $G(\theta)$ we find that the geodesic G has the limiting homotopy class

$$(8.1)\qquad \ldots V_1^{a_{-1}} V_2 V_1^{a_0} V_2 V_1^{a_1} V_2 \ldots \ .$$

Thus the symbols of the M-expansion, namely a_0 and a_1, correspond as follows:

$$(8.2)\qquad a_1 \mapsto V_1^{a_1} V_2, \qquad a_0 \mapsto V_1^{a_1+1} V_2 \ .$$

Hence the symbols a_0 and a_1 (numerically a_1+1 and a_1) are equally useful as generators of $F_2 = \{V_1, V_2\}$. Furthermore the anti-automorphism in the transformation from $G(\theta)$ to $G(\theta^L)$ is of the type

$\frac{1}{2}\left(1+5^{\frac{1}{2}}\right) = \theta$	$=$	$L((1,1))$
θ	$=$	$E(1,2;1,2,2;1,2;1,2,2;\ldots)$
$\theta = \theta^{L}$	$=$	$E(1,\ 2,\ 1,\ 2,\ 2,\ 1,\ldots)$
$\theta = \theta^{LL}$	$=$	$E(1,\ 2,\ 1,\ 2,\ldots)$
Aut $1 \mapsto 12,\ 2 \mapsto 122;\ T = \begin{pmatrix} 2 & 1 \\ 1 & 1 \end{pmatrix},\ \lambda = \frac{1}{2}\left(1+5^{\frac{1}{2}}\right).$		
$1+2^{\frac{1}{2}} = \theta$	$=$	$L((2,2))$
θ	$=$	$E(2,2,3,2,3;2,2,3,2,3;2,2,3,2,3,2,3;2,2,3\ldots)$
$\theta = \theta^{L}$	$=$	$E(\ 2,\ 2,\ 3,\ 2,\ 3,\ 2,\ 2,\ 3,\ldots)$
$\theta = \theta^{LL}$	$=$	$E(\ 2,\ 2,\ 3,\ldots)$
Aut $2 \mapsto 22323,\ 3 \mapsto 2232323;\ T = \begin{pmatrix} 3 & 1 \\ 2 & 1 \end{pmatrix},\ \lambda = \frac{1}{2}2^{\frac{1}{2}}$.		
$1+3^{\frac{1}{2}} = \theta$	$=$	$L((2,1))$
θ	$=$	$E(2,3,3;2,3,3,3;2,3,3,3;2,3,3;2,3,3,3;\ldots)$
$\frac{1}{2}\left(1+3^{\frac{1}{2}}\right) = \theta^{L}$	$=$	$E(1,\ 1,\ 2,\ 1,1,\ 2,\ 1,1,\ 2,\ 1,\ 2,\ 1,1,\ldots)$
$\theta = \theta^{LL}$	$=$	$E(\ 2,\ 3,\ 3,\ 2,\ 3,\ldots)$
Aut $2 \mapsto 233,\ 3 \mapsto 2333;\ T = \begin{pmatrix} 3 & 4 \\ 2 & 3 \end{pmatrix},\ \lambda = -1+3^{\frac{1}{2}}$.		

Table I. Expansions of some θ-geodesics for quadratic surds θ representing reduced periods. The semicolons are used for emphasis only, to better show the automorphisms.

$$j_1 \mapsto a_1^{\,j_1}(a_1+1),\ (j_1+1) \mapsto a_1^{\,j_1+1}(a_1+1)\ , \tag{8.3}$$

where $j_1 = [\theta^{L}]$ and the equivalent generators are now j_1 and (j_1+1), etc.

In composing elementary operations (8.3) it is easier to go from θ to θ^{LL} directly so as to deal with automorphisms. For example, in Table I, for $\theta = \frac{1}{2}\left(1+5^{\frac{1}{2}}\right)$, θ^{L} yields from θ

$$2 \mapsto 12,\ 1 \mapsto 2 \tag{8.4a}$$

which can be iterated to give the transformation in the tables only by *reversal* of multiplication, i.e., $2 \mapsto 122$, $1 \mapsto 12$. Likewise for $\theta = 1+2^{\frac{1}{2}}$, the anti-automorphism is

$$2 \mapsto 23, \quad 3 \mapsto 223 \; . \tag{8.4b}$$

Again, iteration with reversal yields $2 \mapsto 22323$, $3 \mapsto 2232323$.

We note that the abelianization of F_2 projects $\mathrm{Aut}\, F_2$ onto $GL_2(Z)$ and the anti-automorphisms and automorphisms onto matrices of mere tallies of symbols such as

$$T = \begin{pmatrix} t_{11} & t_{12} \\ t_{21} & t_{22} \end{pmatrix} \tag{8.5}$$

$$\begin{cases} \text{Card } (a_1+1) = t_{11} \text{ Card } (a_1+1) + t_{12} \text{ Card } a_1 \\ \text{Card } a_1 \quad\;\; = t_{21} \text{ Card } (a_1+1) + t_{22} \text{ Card } a_1 \; . \end{cases} \tag{8.6}$$

Thus if λ is the relative frequency $\text{Card } (a_1+1)/\text{Card } a_1$ in the E-expansion of θ, it satisfies the quadratic equation

$$T^T(\lambda) = (t_{11}\lambda + t_{21})/(t_{12}\lambda + t_{22}) = \lambda \; , \tag{8.7}$$

(where the *transpose* is used). The "average" of a_n as n varies is

$$\theta = a_1 + \lambda/(\lambda+1) \; . \tag{8.8}$$

As a by-product of these matrices, we note that in the projection $\mathrm{Aut}\, F_2 \to GL_2(Z)$ we find that the automorphisms and anti-automorphisms reduce to matrices and hence to continued fraction algorithms. This completes the proof of Theorem 7.8.

Finally, the automorphisms induced by the derivation process must be sufficient to generate all Markoff forms. Indeed this is the group Γ_N (described in [3]). It is isomorphic to the group Γ_F described in (5.10).

§9. *Generality of θ-geodesics*

It is clear from the general literature (see Koksma [6]) that not all

H^*-geodesics are θ-geodesics. Yet they represent a noncountable set of geodesics with all possible slopes. Indeed there is a trivial equivalence based on the dihedral group of the hexagon

(9.1) $$\theta \mapsto f(\theta);\ f(\theta) = \theta, 1/\theta, 1-\theta, \theta/(\theta-1), 1/(1-\theta), 1-1/\theta\ .$$

Otherwise there is a complicated pattern of equivalences in Aut F_2.

THEOREM 9.2. *If a* H^*-*geodesic* G *has the* M-*expansion*

(9.3) $$G = M(\ldots a_{-1}, a_0, a_1, a_2, \ldots)$$

where $0 \leq a_1 \in Z$, $a_0 = a_1 + 1$, $a_{-n} = a_{n+1}$ $(n > 0)$, *then there exists an irrational* θ *such that* $G = G(\theta)$, *i.e.*, $a_n = [n\theta] - [(n-1)\theta]$.

The proof consists in the fact that for a H^*-geodesic (by Theorem 4.5) repeated derivations can always be performed, as the special cases (2.9 ab) are inapplicable. But by (7.4) each derivation determines a term, a_1 (to start with), which is construed as the corresponding denominator of a continued fraction. By iteration, we obtain successive terms of an L-expansion; (j_1-1 is next in (7.4)). Thus θ is explicitly determined.

§10. *Fibonaccian constructions*

If we examine the E-expansion of (say) $\frac{1}{2}\left(1+5^{\frac{1}{2}}\right)$ we see it could be constructed in two different ways. What we did in Table I was to set up an automorphism for iteration, namely

(10.1a) $$1 \mapsto 12,\ 2 \mapsto 122\ .$$

Thus starting from 1, we iterate (10.1a) (using dots for emphasis):

(10.1b) $$1 \mapsto 12 \mapsto 12.122 \mapsto 12.122.12.122.122 \mapsto \ldots\ .$$

Another way is to write the infinite "Fibonaccian word"

(10.2a) $$w = w_1 w_2 w_3 w_4 w_5 w_6 \cdots;\ w_1 = 1,\ w_2 = 2,\ w_{n-1} w_n = w_{n+1}\ .$$

(10.2b) $$w = 1.2.12.212.12212.21212212.\ldots\ .$$

Actually, the automorphism construction is more easily obtained, but the Fibonaccian construction can be deduced from the anti-automorphism (8.4a). This matter can be explored further, but it is mentioned because, accidentally, the earliest case of a limiting geodesic was (10.2b), shown in [2].

MATHEMATICS DEPARTMENT
CITY COLLEGE OF NEW YORK
NEW YORK, NEW YORK 10031

REFERENCES

[1]. Cohn, H., *Approach to Markoff's minimal forms through modular functions*, Ann. of Math. 61 (1955), pp. 1-12.

[2]. Cohn, H., *Representation of Markoff's binary quadratic forms by geodesics on a perforated torus*, Acta Arith. 18 (1971), pp. 125-136.

[3]. Cohn, H., *Markoff forms and primitive words*, Math. Ann. 196 (1972), pp. 8-22.

[4]. Dickson, L. E., Studies in the Theory of Numbers, Chicago, 1930.

[5]. Fricke, R., *Über die Theorie der automorphen Modulgruppen*, Gött. Nach. (1896), pp. 91-101.

[6]. Koksma, J. F., Diophantische Approximationen, Berlin, 1936.

[7]. Magnus, W., Karrass, A., Solitar, D., Combinatorial Group Theory, New York, 1966.

[8]. Markoff, A. A., *Sur les formes binaires indefinies*, I. Math. Ann. 15 (1879), pp. 381-409; II. 17 (1880), pp. 379-400.

ON THE CARATHÉODORY METRIC IN TEICHMÜLLER SPACES

Clifford J. Earle*

§1. *Statement of results*

Let Δ be the open unit disk $\{z \in C; |z| < 1\}$ and Δ' the punctured disk $\Delta \setminus \{0\}$. Our main purpose is to prove the following extension theorem.

THEOREM 1. *Let* Γ *be any Fuchsian group. Given any holomorphic map* f *of* Δ' *into the Teichmüller space* $T(\Gamma)$, *there is a holomorphic map of* Δ *into* $T(\Gamma)$ *extending* f.

We prove Theorem 1 by studying the Carathéodory metric of $T(\Gamma)$. The crucial fact is

THEOREM 2. *The Carathéodory metric on* $T(\Gamma)$ *is complete.*

The author is grateful to Lipman Bers for bringing this extension problem to his attention.

§2. *The Carathéodory metric*

Let X be a bounded region in the complex Banach space E. We recall (see [4], p. 49) that the Carathéodory metric on X is defined by

$$d(p,q) = d_X(p,q) = \sup\{\rho(f(p), f(q)); f : X \to \Delta \text{ holomorphic}\}, \tag{2.1}$$

where ρ is the Poincaré metric

$$\rho(z,w) = \tanh^{-1} |(z-w)(1-\bar{w}z)^{-1}|$$

* This research was supported in part by N.S.F. Grant GP-28251.

on Δ. The classical Schwarz-Pick Lemma says that ρ is the Carathéodory metric of Δ. Further, since bounded holomorphic functions on Δ' extend to Δ, the Carathéodory metric of Δ' is ρ restricted to Δ'. We list some well-known properties of the Carathéodory metric in the following lemmas.

LEMMA 1 ([4], p. 50). *Let* X *and* Y *be bounded regions in complex Banach spaces. If* $f: X \to Y$ *is holomorphic, then*

$$d_Y(f(p), f(q)) \leqq d_X(p,q) \quad \textit{for all } p, q \in X .$$

LEMMA 2 (see Carathéodory [3]). *If* $\|x\| < r$ *for all* x *in* X, *then*

$$(2.2) \qquad d(x,y) \geqq \tanh^{-1}(\|x-y\|/2r) \quad \textit{for all } x, y \in X .$$

If X *is the ball* $\{x \in E; \|x-x_0\| < R\}$, *then*

$$(2.3) \qquad d(x_0,x) = \tanh^{-1}(\|x-x_0\|/R) \quad \textit{for all } x \in X .$$

Proof. Suppose $X = \{x \in E; \|x-x_0\| < R\}$. Let $x \in X$, $x \neq x_0$. Let ℓ be a linear functional of norm one with $\ell(x-x_0) = \|x-x_0\|$. Define $f: X \to \Delta$ by

$$f(p) = R^{-1}\ell(p-x_0) ,$$

and $g: \Delta \to X$ by $g(\lambda) = x_0 + R\lambda(x-x_0)/\|x-x_0\|$. Then $f(g(\lambda)) = \lambda$ for all λ in Δ, so

$$\rho(0,\lambda) = \rho(f(g(0)), f(g(\lambda))) \leqq d(g(0), g(\lambda)) \leqq \rho(0,\lambda) .$$

Therefore $d(x_0,g(\lambda)) = d(g(0),g(\lambda)) = \rho(0,\lambda)$ for all λ in Δ. Putting $\lambda = \|x-x_0\|/R$ we get (2.3).

Next, suppose $\|x\| < r$ for all x in X. Choose x in X and put $Y = \{y \in E; \|y-x\| < 2r\}$. Then $X \subset Y$, so Lemma 1 gives us

$$d(x,y) = d_X(x,y) \geqq d_Y(x,y) = \tanh^{-1}(\|y-x\|/2r)$$

for all y in X, by (2.3). That proves (2.2).

COROLLARY. *The Carathéodory metric induces the standard topology on X.*

In fact, if $x_n \to x$ in the Carathéodory metric, then (2.2) shows that $x_n \to x$ in norm. Conversely, suppose $x_n \to x$ in norm, where $x \in X$ and (x_n) is a sequence in X. Choose $r > 0$ so small that the ball

$$Y = \{y \in E;\ \|y-x\| < r\}$$

is contained in X. Then

$$d(x,x_n) = d_X(x,x_n) \leqq d_Y(x,x_n) = \tanh^{-1}(\|x_n-x\|/r)$$

for all sufficiently large n, so $x_n \to x$ in the Carathéodory metric.

§3. *Proof of Theorem* 1

It is well known ([4], p. 57) that Theorem 1 follows immediately from Theorem 2. The proof is easy. Lemma 1 says that $f : \Delta' \to T(\Gamma)$ is uniformly continuous in the Carathéodory metrics, so f extends to a map from the completion of Δ', which is Δ, to $T(\Gamma)$.

§4. *Proof of Theorem* 2

Let $T(\Gamma)$ be the Teichmüller space of the Fuchsian group Γ. Recall (see [1] or [2]) that $T(\Gamma)$ is a bounded region in the Banach space $B(\Gamma)$ of bounded quadratic differentials, containing the ball of radius 2 and contained in the ball $X = \{\phi \in B(\Gamma);\ \|\phi\| < 6\}$. Applying Lemma 1 to the inclusion map of $T(\Gamma)$ into X, we get

$$\tanh^{-1}(\|\phi\|/6) \leqq d(0,\phi) \tag{4.1}$$

for all ϕ in $T(\Gamma)$. Here d is the Carathéodory metric on $T(\Gamma)$, and we have used (2.3) to compute the Carathéodory metric on X. Since tanh is an increasing function, (4.1) implies

$$\|\phi\| \leqq 6 \tanh\ r \quad \text{if} \quad d(0,\phi) \leqq r\ . \tag{4.2}$$

We claim now that $S_0 = \{\phi \in T(\Gamma);\ d(0,\phi) \leqq r\}$ is complete in the metric d whenever

$$\tanh r < \frac{1}{3} \tag{4.3}$$

(that is, $r < \frac{1}{2}\log 2$). In fact, let (ϕ_n) be any d-Cauchy sequence in S_0. Then, by (2.2), (ϕ_n) is Cauchy with respect to the norm of $B(\Gamma)$, so ϕ_n converges (in norm) to some ϕ in $B(\Gamma)$. By (4.2) and (4.3), each ϕ_n satisfies

$$\|\phi_n\| \leqq 6 \tanh r < 2 .$$

Hence ϕ does also, and $\phi \in T(\Gamma)$. Therefore, by the Corollary to Lemma 2, $\phi_n \to \phi$ in the metric d. Since S_0 is closed, ϕ belongs to S_0 and S_0 is complete.

Our next claim is that every closed ball $S = \{\phi \in T(\Gamma); d(\phi_0, \phi) \leqq r\}$ is complete, provided that $\phi_0 \in T(\Gamma)$ and r satisfies (4.3). In fact there exist a Fuchsian group Γ' and a biholomorphic map $f: T(\Gamma') \to T(\Gamma)$ with $f(0) = \phi_0$. (f is an "allowable bijection" in the terminology of [2], Chapter II, Section 3.) The closed ball S in $T(\Gamma)$ is the isometric image under f of the closed ball of radius r and center 0 in $T(\Gamma')$. Hence S is complete.

It is now obvious that $T(\Gamma)$ is complete with respect to d. In fact any Cauchy sequence (ϕ_n) in $T(\Gamma)$ is eventually contained in a closed ball whose radius r satisfies (4.3).

§5. *Concluding remarks*

Since $T(\Gamma)$ is a bounded region in $B(\Gamma)$, any holomorphic map $f: \Delta' \to T(\Gamma)$ can obviously be extended to a map of Δ into $B(\Gamma)$. The point of Theorem 1 is that $f(0)$ belongs to $T(\Gamma)$, not merely to $B(\Gamma)$. Theorem 2 has a number of consequences in addition to Theorem 1. For instance, Theorem 2 implies the familiar result that finite dimensional Teichmüller spaces are domains of holomorphy (see [4], pp. 50 and 77). We are unable to answer the following two questions about finite dimensional Teichmüller spaces:

(i) Is the Carathéodory metric the same as the Teichmüller metric?

(ii) Is every closed Carathéodory-bounded set compact?

CORNELL UNIVERSITY

REFERENCES

[1]. Ahlfors, L. V., "Lectures on quasi-conformal mappings," Mathematical Studies No. 10, Van Nostrand, Princeton, 1966.

[2]. Bers, L., Uniformization, moduli, and Kleinian groups, *Bull. London Math. Soc.* 4 (1972), pp. 257-300.

[3]. Carathéodory, C., Über das Schwarzsche Lemma bei analytischen Funktionen von zwei komplexen Veränderlichen, *Math. Ann.* 97 (1926), pp. 76-98.

[4]. Kobayashi, S., "Hyperbolic manifolds and holomorphic mappings," Marcel Dekker, New York, 1970.

SOME REFINEMENTS OF THE POINCARÉ PERIOD RELATION

L. Ehrenpreis** and H. M. Farkas*

Introduction

This work is a continuation of the paper of the authors, H. Martens and H. E. Rauch, [EFMR], in which we showed how to derive Poincaré's asymptotic relation for the periods of the holomorphic differentials on a Riemann surface C of genus 4 from the existence of singular points on the theta divisor on the Jacobian, J(C). More explicitly, if π denotes a point in the Siegel upper half plane of degree 4 and z denotes a point in the torus corresponding to π, then in [EFMR] we obtained Poincaré's relation by studying the variation of the equations $\theta(z,\pi) = 0$ $\frac{\partial\theta}{\partial z_i}(z,\pi) = 0$ $i = 1,\dots,4$ near a point $z = 0$ $\pi = \pi_0$ diagonal. Here and for the remainder of this paper $\theta(z,\pi)$ refers to the first order theta function with characteristic $\begin{bmatrix}1111\\1111\end{bmatrix}$ i.e., $\theta\begin{bmatrix}1111\\1111\end{bmatrix}(z,\pi)$. This paper represents a refinement of the techniques employed in [EFMR].

In I we briefly describe the technique we use and give some rather simple examples which illustrate the ideas used. In II we apply these techniques to derive the Poincaré relation in a manner slightly different from that employed in [EFMR] and offer some evidence for a conjecture concerning the Schottky relation. In III we once again apply our techniques to construct another subvariety of the Siegel upper half plane of degree 4 of complex dimension 9 which we shall call the variety of quasi-Jacobians. This variety, Q, is characterized by the property that for generic $\pi \in Q$ near π_0, the theta divisor on the torus corresponding to π has precisely

* Research partially supported by NSF GP 19572 and A. P. Sloan Foundation Fellowship.

** Research partially supported by NSF GP 28499.

two singular points near $z = 0$. From II and III we get some idea of the complexity of the local variety defined by $\theta(z,\pi) = 0$ $\frac{\partial\theta}{\partial z_i}(z,\pi) = 0$ $i = 1, \ldots, 4$ near $\pi = \pi_0$ $z = 0$. Namely we see that this variety is singular and is highly reducible. It is for this reason that we needed the variational ideas in I. In IV we give a method of obtaining an explicit realization of the canonical curve for a generic surface of genus 4. In V we indicate how to apply our techniques to genus greater than 4.

I. *Examples*

Our analysis of the variety V of solutions z, π of $\theta(z,\pi) = 0$ $\frac{\partial\theta}{\partial z_i}(z,\pi) = 0$ $i = 1, \ldots, 4$ is based on a study of the family of analytic curves $(z(t), \pi(t))$ which are contained in V and pass through the point $(0, \pi_0)$. We expand $\theta(z,\pi)$ and $\frac{\partial\theta}{\partial z_i}(z,\pi)$ $i = 1, \ldots, 4$ in power series in t and study the equations that result when we equate each coefficient to zero.

In order to better understand how we use this technique we shall at this point state and illustrate three guiding principles which we use. In this paper we shall discuss only the formal aspects of power series. There is, however, no difficulty in modifying our methods to prove local convergence.

A. *Use of jets to distinguish components*

By a jet of order m we mean a combination of derivatives of order less than or equal to m. Suppose we wish to find the curves lying in the variety $x^2 - y^2 = 0$ which pass through the point (0,0). One way of doing this is to set $x(t) = \sum_{j=1}^{\infty} a_j t^j$, $y(t) = \sum_{j=1}^{\infty} b_j t^j$. Then in order for $(x(t), y(t))$ to be a curve in this variety it is necessary that $a_1^2 - b_1^2 = 0$ or that $a_1 = \pm b_1$. Hence it is the choice of sign which determines on which component of the variety $x^2 - y^2 = 0$ the curve $(x(t), y(t))$ lies. The lesson learned from this rather simple example is that we need to place an "initial condition" namely $a_1 = b_1$ or $a_1 = -b_1$ on the jet of order one in order to determine on which component our curve lies.

B. *Dimension of variety is determined by high order jets* (Low order phenomena may be misleading)

As an example let us study the analytic curves through $(0,0)$ on the variety $x^2 - y^3 = 0$. As in the previous example we let $x(t) = \sum_{j=1}^{\infty} a_j t^j$ and $y(t) = \sum_{j=1}^{\infty} b_j t^j$. The conditions now imposed on the power series coefficients are:

$$a_1^2 = 0$$

$$2a_1a_2 - b_1^3 = 0 \quad \text{or that} \quad b_1 = 0\ .$$

$$2a_1a_3 + a_2^2 - (3b_1^2b_2) = 0 \quad \text{or that} \quad a_2 = 0\ .$$

$$2a_1a_4 + 2a_2a_3 - (3b_1^2b_3 + 3b_1b_2^2) = 0 \quad \text{which gives us nothing new.}$$

$$2a_1a_5 + 2a_2a_4 + a_3^2 - (3b_1^2b_4 + 6b_1b_2b_3 + b_2^3) = 0$$
$$\text{or that} \quad a_3^2 - b_2^3 = 0\ .$$

$$2a_1a_6 + 2a_2a_5 + 2a_3a_4 - (3b_1^2b_5 + 6b_1b_2b_4 + 3b_1b_3^2 + 3b_2^2b_3) = 0$$
$$\text{or that} \quad 2a_3a_4 - 3b_2^2b_3 = 0\ .$$

These equations should be interpreted as follows: After the imposition of the conditions $a_1 = b_1 = a_2 = 0$ on the jets of order ≤ 2 we should think of the equation $a_3^2 = b_2^3$ as determining a_3 up to sign and then the equation $2a_3a_4 = 3b_2^2b_3$ as determining a_4 in terms of b_3 etc. Since the variety $x^2 - y^3$ is irreducible it turns out that the choice of sign for a_3 is irrelevant. Proceeding in this fashion we find that the dimension of jets of order one is zero; however, the dimension of the jets of any other order is one in conformity with the fact that the variety $x^2 - y^3 = 0$ is one dimensional. We finally remark that had we called derivatives of order m in y jets of order $m + 1$, then the above analysis would be more "homogeneous" i.e., the coefficient of t^{m+3} is used to find the relations in the jets of order m. Other examples of redefinition of the orders of certain jets appear in our work.

C. *The feasibility of explicitly solving infinitely many equations in infinitely many unknowns depends on the system being "almost triangular."*

We begin by explaining the concept of a system of equations being almost triangular. Consider the system of equations

$$\begin{aligned} &av_1 = 0 \\ &b_{21}v_1 + av_2 = 0 \\ &b_{31}v_1 + b_{32}v_2 + av_3 = 0 \\ &\quad\vdots \end{aligned}$$

We can write this system in matrix form as follows:

$$\begin{pmatrix} a & 0 & 0 & \cdots \\ b_{21} & a & 0 & 0 \cdots \\ b_{31} & b_{32} & a & \\ b_{41} & b_{42} & b_{43} & a \end{pmatrix} \begin{pmatrix} v_1 \\ v_2 \\ v_3 \\ v_4 \\ \vdots \end{pmatrix} = \begin{pmatrix} 0 \\ 0 \\ \vdots \end{pmatrix}$$

Here we understand the v_j to be vectors with a fixed number, say ℓ, of components and a, b_{ij} are $\ell \times \ell$ matrices. The simplest situation is of course when rank $(a) = \ell$. In this case we can solve the system of equations explicitly and find that $v_j = 0$ $j = 1, 2, \ldots$. We think of this system as being a triangular system in the sense that v_j is determined from the jth equation.

Let us now suppose that rank $a = \ell - 1$. In this case there exist vectors V_L and V_R, $V_L \neq 0$ and $V_R \neq 0$ such that $V_L a = aV_R = 0$. Here V_L is a row vector and V_R is a column vector.

The first equation is $av_1 = 0$ so that $v_1 = \lambda_1 V_R$ where λ_1 is any complex number. The second equation is $b_{21}v_1 + av_2 = 0$ or from the first equation we get $\lambda_1 b_{21} V_R + av_2 = 0$. We therefore find $\lambda_1 V_L b_{21} V_R = 0$ is a necessary condition for the second equation to have a solution. It therefore follows that if $V_L b_{21} V_R \neq 0$, $\lambda_1 = 0$ and $V_2 = \lambda_2 V_R$. In this situation we needed two steps in order to determine v_1. If $V_L b_{i+1,i} V_R \neq 0$ for all i then it is once again clear that $v_i = 0$ is the solution to the system. Such a system we call almost triangular since we need more than one step to determine v_1. The above procedure also applies to an inhomogeneous system.

In our applications the analogs of the operators b_{ij} are non-linear. Nevertheless the same general procedure here indicated for linear operators is applicable. We shall also meet in our applications analogs of the above where a is an $r \times \ell$ matrix with $r \leq \ell$. The most important case for us at the present time is $r = 1$. Here the condition rank $a = \ell$ is replaced by rank $a = r$. If this is the case the equation $av_1 = 0$ defines the relations among the components of v_1. The next equation $b_{21}v_1 + av_2 = 0$ then places no further restriction on v_1 and gives a relation among the components of v_1 and v_2. Proceeding in this way we successively obtain the relations among the components of v_1, the components of v_1 and v_2, the components of v_1, v_2 and v_3, etc. We refer to this as a triangular system of relations in $(v_1, v_2, \ldots)$. In a similar manner we can define almost triangular system of relations.

II. *Applications to the Schottky relation*

We are interested in the variety, V, defined by the equations $\theta(z,\pi) = 0$ $\frac{\partial\theta}{\partial z_i}(z,\pi) = 0$ $i = 1,\ldots,4$ in a neighborhood of $(0,\pi_o)$ where π_o is a diagonal matrix in the Siegel upper half plane of degree 4. In order to study this variety V we consider the collection of analytic curves $(z(t),\pi(t))$ through $(0,\pi_o)$ which lie in V. We impose Poincaré's initial condition on the curves which is that $\pi'_{ij}(0) = 0$ for $i \neq j$. This is an initial condition on the jets of order one. We begin by considering the equations

$$\frac{\partial\theta}{\partial z_i}(z(t),\pi(t)) = 0 \quad i = 1,\ldots,4 \; . \tag{1}$$

We expand (1) in power series using the fact that $z(t) = z'(0)t + z''(0)\frac{t^2}{2!} + \ldots$ and $\pi(t) = \pi_o + \pi'(0)t + \pi''(0)\frac{t^2}{2!} + \ldots$ and find that the vanishing of the coefficient of t^3 gives rise to the equation:

$$\Delta_2 z'(0) = (-4\pi i)\begin{pmatrix} z'_2(0)\, z'_3(0)\, z'_4(0) \\ z'_1(0)\, z'_3(0)\, z'_4(0) \\ z'_1(0)\, z'_2(0)\, z'_4(0) \\ z'_1(0)\, z'_2(0)\, z'_3(0) \end{pmatrix} \tag{2}$$

where $z'(0)$ is a column vector with components $z'_i(0)$ $i = 1, \ldots, 4$ and Δ_2 is a 4×4 symmetric matrix

$$\Delta_2 = \begin{pmatrix} 0 & \pi''_{34}(0) & \pi''_{24}(0) & \pi''_{23}(0) \\ & 0 & \pi''_{14}(0) & \pi''_{13}(0) \\ & & 0 & \pi''_{12}(0) \\ & & & 0 \end{pmatrix} \tag{3}$$

The important point about (2) is that we can solve for $z'(0)$ as a vector whose components are algebraic functions of the elements of Δ_2. Utilizing the same procedure as was used in [EFMR] we replace our system of equations for $z'_i(0)$ by the system

$$\begin{pmatrix} z'_1 & 0 & 0 & 0 \\ 0 & z'_2 & 0 & 0 \\ 0 & 0 & z'_3 & 0 \\ 0 & 0 & 0 & z'_4 \end{pmatrix} \Delta_2 z' = -4\pi i\, z'_1 z'_2 z'_3 z'_4 \begin{pmatrix} 1 \\ 1 \\ 1 \\ 1 \end{pmatrix} \tag{4}$$

where we now write z'_i for $z'_i(0)$ and z' for $z'(0)$.

It is now quite easy to see that a solution to (4) is given by

$$\pi''_{34} z'_1 z'_2 = \pi''_{12} z'_3 z'_4, \quad \pi''_{24} z'_1 z'_3 = \pi''_{13} z'_2 z'_4, \quad \pi''_{23} z'_1 z'_4 = \pi''_{14} z'_2 z'_3 \tag{5}$$

where we now write π''_{ij} for $\pi''_{ij}(0)$. It follows now quite easily that

$$(6)\quad z'_1 z'_2 = \sqrt{\frac{\pi''_{12}}{\pi''_{34}}}\,(T)^{\frac{1}{2}},\quad z'_1 z'_3 = \sqrt{\frac{\pi''_{13}}{\pi''_{24}}}\,(T)^{\frac{1}{2}},\quad z'_1 z'_4 = \sqrt{\frac{\pi''_{14}}{\pi''_{23}}}\,(T)^{\frac{1}{2}}$$

$$z'_2 z'_3 = \sqrt{\frac{\pi''_{23}}{\pi''_{14}}}\,(T)^{\frac{1}{2}},\quad z'_2 z'_4 = \sqrt{\frac{\pi''_{24}}{\pi''_{13}}}\,(T)^{\frac{1}{2}},\quad z'_3 z'_4 = \sqrt{\frac{\pi''_{34}}{\pi''_{12}}}\,(T)^{\frac{1}{2}}$$

where $T = z'_1 z'_2 z'_3 z'_4$. Furthermore, the first equation in (4) gives that

$$(7)\quad \sqrt{\pi''_{34}\pi''_{12}}\,(T)^{\frac{1}{2}} + \sqrt{\pi''_{24}\pi''_{13}}\,(T)^{\frac{1}{2}} + \sqrt{\pi''_{23}\pi''_{14}}\,(T)^{\frac{1}{2}} = -4\pi i\, T\ .$$

Hence it follows immediately that if $T \neq 0$

$$(8)\quad \sqrt{\pi''_{12}\pi''_{34}} + \sqrt{\pi''_{13}\pi''_{24}} + \sqrt{\pi''_{14}\pi''_{23}} = -4\pi i\,(T)^{\frac{1}{2}}\ .$$

It is now immediate from (6) and (8) that we can solve for z'_i as an algebraic function if the $\pi''_{k\ell}$, $k < \ell$. We illustrate this for z'_1.

It follows from the first two equations in (6) that $z'^2_1 z'_2 z'_3 = \sqrt{\frac{\pi''_{12}\pi''_{13}}{\pi''_{34}\pi''_{24}}}\,T$.

Hence $z'^2_1 = \sqrt{\frac{\pi''_{14}\pi''_{12}\pi''_{13}}{\pi''_{24}\pi''_{34}\pi''_{23}}}\,(T)^{\frac{1}{2}}$.

Using (8) we have

$$-4\pi i\, z'^2_1 = \pi''_{12}\sqrt{\frac{\pi''_{14}\pi''_{13}}{\pi''_{23}\pi''_{24}}} + \pi''_{13}\sqrt{\frac{\pi''_{12}\pi''_{14}}{\pi''_{34}\pi''_{23}}} + \pi''_{14}\sqrt{\frac{\pi''_{12}\pi''_{13}}{\pi''_{34}\pi''_{24}}}\ .$$

Up to this point we have only used the coefficient of t^3 in the power series for $\frac{\partial\theta}{\partial z_i}(z(t), \pi(t))$ $i = 1, \ldots, 4$. Equating to zero the coefficients of t^j in $\frac{\partial\theta}{\partial z_i}(z(t), \pi(t))$ gives us a system of equations for $z_i^{(j-2)}$ for $j \geq 4$. The matrix a of this system is symmetric and in fact

$$(9)\quad a = \begin{pmatrix} 0 & \pi''_{34} + 4\pi i\, z'_3 z'_4 & \pi''_{24} + 4\pi i\, z'_2 z'_4 & \pi''_{23} + 4\pi i\, z'_2 z'_3 \\ & 0 & \pi''_{14} + 4\pi i\, z'_1 z'_4 & \pi''_{13} + 4\pi i\, z'_1 z'_3 \\ & & 0 & \pi''_{12} + 4\pi i\, z'_1 z'_2 \\ & & & 0 \end{pmatrix}$$

By our previous remarks the elements of a are algebraic functions of the $\pi''_{k\ell}$ with $k < \ell$. This matrix a has a geometric significance. If we were to expand the Hessian matrix $H = (h_{ij})$, where $h_{ij} = \frac{\partial^2 \theta(z(t), \pi(t))}{\partial z_i \partial z_j}$, in power series a is the lowest order non-trivial term in the expansion. In so far as Jacobians are concerned it is known that generically Rank $H = 4$ at z where $\theta(z, \pi) = 0$ and $\frac{\partial \theta}{\partial z_i}(z, \pi) = 0$ $i = 1, \ldots, 4$. Hence it is reasonable to assume that rank $(a) = 4$ generically. This means that our system of equation for $z_i^{(j)}$, $j \geq 2$ falls into the category of a triangular system in the sense of Section I where however the $b_{k\ell}$ are now non-linear operators which depend on the terms z'_i which we have already agreed are algebraic functions of the $\pi''_{k\ell}$, $k < \ell$. More precisely, this means that we can solve for $z_i^{(j)}$ in terms of $\pi_{k\ell}^{(m)}$ for $2 \leq m \leq j+1$ for $k \neq \ell$ and $\pi_{kk}^{(m)}$ for $1 \leq m \leq j-2$. For the sake of homogeneity we shall call a term $\pi_{kk}^{(m)}$ a jet of order $m+3$. We can now say that we can solve for $z_i^{(j)}$ in terms of jets of order $j+1$ for $j \geq 2$. At any rate the system is triangular and we can find a solution which we shall denote by $z(\pi)$.

We now consider the function $\theta(z(\pi), \pi)$ where $z(\pi)$ is the solution to our system which satisfies

a) the Poincaré initial condition $\pi'_{ij} = 0$ for $i \neq j$.

b) $z'_1 z'_2 z'_3 z'_4 \neq 0$.

c) rank $a = 4$.

We make the following conjecture: $\theta(z(\pi), \pi) = k\, s(\pi)$ where k is a nowhere vanishing function of π and

$$
\begin{aligned}
s(\pi) = {} & \pm\left(\theta\begin{bmatrix}1100\\1100\end{bmatrix}\theta\begin{bmatrix}0111\\1110\end{bmatrix}\theta\begin{bmatrix}1011\\0011\end{bmatrix}\theta\begin{bmatrix}1111\\1001\end{bmatrix}\theta\begin{bmatrix}0000\\0001\end{bmatrix}\theta\begin{bmatrix}0100\\1011\end{bmatrix}\theta\begin{bmatrix}1000\\0110\end{bmatrix}\theta\begin{bmatrix}0011\\0100\end{bmatrix}\right)^{\frac{1}{2}} \\
& \pm\left(\theta\begin{bmatrix}1110\\0000\end{bmatrix}\theta\begin{bmatrix}0101\\0010\end{bmatrix}\theta\begin{bmatrix}1001\\1111\end{bmatrix}\theta\begin{bmatrix}1101\\0101\end{bmatrix}\theta\begin{bmatrix}0010\\1101\end{bmatrix}\theta\begin{bmatrix}0110\\0111\end{bmatrix}\theta\begin{bmatrix}1010\\1010\end{bmatrix}\theta\begin{bmatrix}0001\\1000\end{bmatrix}\right)^{\frac{1}{2}} \\
& \pm\left(\theta\begin{bmatrix}0000\\0010\end{bmatrix}\theta\begin{bmatrix}1011\\0000\end{bmatrix}\theta\begin{bmatrix}0111\\1101\end{bmatrix}\theta\begin{bmatrix}0011\\0111\end{bmatrix}\theta\begin{bmatrix}1100\\1111\end{bmatrix}\theta\begin{bmatrix}1000\\0101\end{bmatrix}\theta\begin{bmatrix}0100\\1000\end{bmatrix}\theta\begin{bmatrix}1111\\1010\end{bmatrix}\right)^{\frac{1}{2}}.
\end{aligned}
$$

Here $s(\pi)$ is one of the forms of the left hand side of Schottky's genus 4 relation $s(\pi) = 0$ [R].

If we expand $\theta(z(\pi),\pi)$ in power series in t and use assumptions a, b, c we find that the coefficients of t, t^2, t^3 vanish and as in [EFMR] the coefficient of t^4 is

$$(10)\qquad c[z'_1z'_2\pi''_{34}+\dots+z'_3z'_4\pi''_{12}+\frac{1}{4\pi i}(\pi''_{12}\pi''_{34}+\pi''_{13}\pi''_{24}+\pi''_{14}\pi''_{23})$$
$$+4\pi i\, z'_1z'_2z'_3z'_4]$$

where $c=\dfrac{d\theta\begin{bmatrix}1\\1\end{bmatrix}}{dz}(0,\pi_{11})\dfrac{d\theta\begin{bmatrix}1\\1\end{bmatrix}}{dz}(0,\pi_{22})\dfrac{d\theta\begin{bmatrix}1\\1\end{bmatrix}}{dz}(0,\pi_{33})\dfrac{d\theta\begin{bmatrix}1\\1\end{bmatrix}}{dz}(0,\pi_{44})$ and

π_{ii} are the diagonal entries of π_o.

It follows from (6), (7), (8) that (10) can be rewritten as

$$(11)\qquad -\frac{c}{2\pi i}\left(\sqrt{\pi''_{12}\pi''_{34}\pi''_{13}\pi''_{24}}+\sqrt{\pi''_{12}\pi''_{34}\pi''_{14}\pi''_{23}}+\sqrt{\pi''_{13}\pi''_{24}\pi''_{14}\pi''_{23}}\right)$$

and equating this to zero gives Poincaré's relation. On the other hand, Rauch [R] has shown that the lowest order term in $s(\pi)$ is also a multiple of Poincaré.

We can now go further. We examine the coefficient of t^j for $j>4$ in the expansion of $\theta(z(\pi),\pi)$. The coefficient involves jets of order $j-2$ in π and jets of order $j-4$ in z. Hence, it follows from the remarks above on the function $z(\pi)$ that this coefficient involves jets of order $j-2$ in π. The highest order term, i.e., the term involving $\pi^{(j-2)}_{k\ell}$ for $k\neq\ell$ is a multiple of

$$(12)\qquad \frac{\sqrt{\pi''_{23}\pi''_{14}\pi''_{13}\pi''_{24}}}{\pi''_{34}}\pi^{(j-2)}_{34}+\frac{\sqrt{\pi''_{12}\pi''_{34}\pi''_{14}\pi''_{23}}}{\pi''_{24}}\pi^{(j-2)}_{24}+\frac{\sqrt{\pi''_{12}\pi''_{34}\pi''_{13}\pi''_{24}}}{\pi''_{23}}\pi^{(j-2)}_{23}$$
$$+\frac{\sqrt{\pi''_{12}\pi''_{34}\pi''_{13}\pi''_{24}}}{\pi''_{14}}\pi^{(j-2)}_{14}+\frac{\sqrt{\pi''_{23}\pi''_{14}\pi''_{12}\pi''_{34}}}{\pi''_{13}}\pi^{(j-2)}_{13}+\frac{\sqrt{\pi''_{23}\pi''_{14}\pi''_{13}\pi''_{24}}}{\pi''_{12}}\pi^{(j-2)}_{12}$$

which is in fact a multiple of the coefficient of t^5 in $\theta(z(\pi),\pi)$. We have verified that the coefficient of t^5 in $\theta(z(\pi),\pi)$ and $s(\pi)$ agree except for a non-zero factor and that some terms in the coefficient of t^j agree in

$\theta(z(\pi),\pi)$ and $s(\pi)$ again up to non-zero multiples for all j. There is also further evidence for our conjecture which we shall not give here.

At any rate we can regard the system of equations obtained by equating to zero the coefficient of t^j in the expansion of $\theta(z(\pi),\pi)$ as a triangular system of equations for the relations in the jets of π. These equations may be regarded as higher order Poincaré relations, e.g.:

$$\sqrt{\pi''_{13}\pi''_{24}\pi''_{14}\pi''_{23}}\left(\frac{\pi'''_{34}}{\pi''_{34}}+\frac{\pi'''_{12}}{\pi''_{12}}\right)+\sqrt{\pi''_{12}\pi''_{34}\pi''_{14}\pi''_{23}}\left(\frac{\pi'''_{24}}{\pi''_{24}}+\frac{\pi'''_{13}}{\pi''_{13}}\right)$$
$$+\sqrt{\pi''_{12}\pi''_{34}\pi''_{13}\pi''_{24}}\left(\frac{\pi'''_{23}}{\pi''_{23}}+\frac{\pi'''_{14}}{\pi''_{14}}\right)=0$$

is the relation between the jets of orders 2 and 3.

III. *Applications to quasi-Jacobians*

We now abandon the Poincaré initial condition of $\pi'_{ij}=0$ for $i\neq j$ and again attempt to solve the system of equations $\frac{\partial\theta}{\partial z_i}(z,\pi)=0$ $i=1,\dots,4$ for z_i as a function of the $\pi_{k\ell}$. We proceed as before expanding z and π as power series in t, $z(t)=z'\,t+z''\frac{t^2}{2!}+\dots\ \pi(t)=\pi_0+\pi' t+\pi''\frac{t^2}{2!}+\dots$, where π_0 as before is a diagonal matrix. The vanishing of the coefficient of t^2 in $\frac{\partial\theta}{\partial z_i}(z(t),\pi(t))$ yields the equation

$$\Delta_1 z'=0 \tag{13}$$

the notation being the same as in Section II and Δ_1 is the symmetric matrix

$$\Delta_1=\begin{pmatrix}0 & \pi'_{34} & \pi'_{24} & \pi'_{23}\\ & 0 & \pi'_{14} & \pi'_{13}\\ & & 0 & \pi'_{12}\\ & & & 0\end{pmatrix}. \tag{14}$$

If rank $\Delta_1=4$ then $z'_i=0$ $i=1,\dots,4$. Moreover, the system of equations we obtain for $z^{(j)}$ is triangular and the only solution is $z^{(j)}=0$

for all j. In other words, $z(\pi) = 0$. Hence the equation $\theta(z(\pi), \pi) = 0$ becomes simply the equation $\theta(0, \pi) = 0$ which defines a variety, H, with the property $\pi \in H$ provided $\theta(0, \pi) = 0$. We call H the half period variety. It is well known that the variety H intersects the variety J of Jacobians in at most an 8 dimensional variety [F]. If, as usual, we expand π in a power series $\pi = \pi_o + \pi' t + \dots$ and then expand $\theta(0, \pi(t))$ in a power series in t. Equating to zero the coefficients of t^j defines a triangular system of relations for the $\pi^{(m)}_{k\ell}$. The coefficients of t^2, t^3 are respectively

(15) $$\frac{c}{(2\pi i)^2}[\pi'_{12}\pi'_{34} + \pi'_{13}\pi'_{24} + \pi'_{14}\pi'_{23}]$$

(16) $$\frac{c}{2(2\pi i)^2}[\pi'_{12}\pi''_{34} + \pi'_{13}\pi''_{24} + \dots + \pi'_{34}\pi''_{12}] + \frac{c}{(2\pi i)^3}\left[\frac{\theta'''_1\begin{bmatrix}1\\1\end{bmatrix}}{\theta'_3\begin{bmatrix}1\\1\end{bmatrix}}\pi'_{12}\pi'_{13}\pi'_{14}\right.$$

$$\left. + \frac{\theta'''_2\begin{bmatrix}1\\1\end{bmatrix}}{\theta'_2\begin{bmatrix}1\\1\end{bmatrix}}\pi'_{12}\pi'_{23}\pi'_{24} + \frac{\theta'''_3\begin{bmatrix}1\\1\end{bmatrix}}{\theta'_3\begin{bmatrix}1\\1\end{bmatrix}}\pi'_{13}\pi'_{23}\pi'_{34} + \frac{\theta'''_4\begin{bmatrix}1\\1\end{bmatrix}}{\theta'_4\begin{bmatrix}1\\1\end{bmatrix}}\pi'_{14}\pi'_{24}\pi'_{34}\right]$$

where $$\frac{\theta'''_i\begin{bmatrix}1\\1\end{bmatrix}}{\theta_i\begin{bmatrix}1\\1\end{bmatrix}} = \frac{\theta'''\begin{bmatrix}1\\1\end{bmatrix}(0, \pi_{ii})}{\theta'\begin{bmatrix}1\\1\end{bmatrix}(0, \pi_{ii})} .$$

The coefficient of t^4 is

(17) $$\left(\sum_{k\le\ell, i\le j}^{4} \frac{\partial^2\theta[0, \pi_o]}{\partial\pi_{ij}\,\partial\pi_{k\ell}}(4\pi'_{ij}\pi'''_{k\ell} + 3\pi''_{ij}\pi''_{k\ell}) + \right.$$

$$6\sum_{m\le n, k\le\ell\ i\le j=1}^{4} \frac{\partial^3\theta}{\partial\pi_{mn}\,\partial\pi_{k\ell}\,\partial\pi_{ij}}\pi''_{ij}\pi'_{k\ell}\pi'_{mn} +$$

$$\left.\sum_{p\le q\ m\le n\ k\le\ell\ i\le j=1}^{4} \frac{\partial^4\theta(0, \pi_o)}{\partial\pi_{pq}\,\partial\pi_{mn}\,\partial\pi_{k\ell}\,\partial\pi_{ij}}\pi'_{ij}\pi'_{k\ell}\pi'_{mn}\pi'_{pq}\right)\frac{1}{4!} .$$

The reader should observe that we have not bothered to simplify the coefficient of t^4 as we have done for the coefficient of t^2 and t^3. The only point we need mention here is that it is in this coefficient that the terms π'_{ii} appear for the first time.

We now impose the rank 3 initial condition, namely that the rank of the matrix Δ_1 is precisely 3. Now the system of equations for $z_i^{(j)}$ becomes an almost triangular system in the sense of Section I. To see how things work here let $z' = \lambda_1 z_0$ where z_0 is a fixed non-trivial solution of $\Delta_1 z' = 0$. For example

$$Z_0 = \begin{pmatrix} \pi'_{12}\pi'_{13}\pi'_{14} \\ \pi'_{12}\pi'_{34} \\ \pi'_{13}\pi'_{24} \\ \pi'_{14}\pi'_{23} \end{pmatrix}. \tag{18}$$

We can show that no component of z_0 is zero and hence, $\pi'_{ij} \neq 0$ for $i \neq j$. We now equate the coefficient of t^3 in $\frac{\partial\theta}{\partial z_i}(z(t), \pi(t))$ to zero and obtain the system of equations

$$\Delta_1 z'' + B_{21} z' = 0 \tag{19}$$

where B_{21} is a non-linear operator and z' is understood to be $\lambda_1 z_0$. Since Δ_1 is symmetric and $\Delta_1 z_0 = 0$, ${}^t z_0 \Delta_1 = 0$ also. Therefore, in order that there be any solution z'' the equation

$${}^t z_0 B_{21} z' = 0 \,. \tag{20}$$

It turns out that (20) becomes a cubic equation for λ_1 where the coefficient of $\lambda_1 \neq 0$. Hence this equation has at most 3 solutions and generically has 3 solutions. It turns out that $\lambda_1 = 0$ is one of the solutions and we denote the other two solutions by $\lambda_1 = \pm\alpha_1$. The important point here is that system is continuable. The solution $\lambda = 0$ gives rise to the solution $z' = z'' = \ldots = 0$ and we obtain nothing new. The solution $\lambda_1 = \alpha_1$ gives rise to a unique $z^{(j)}$ for $j \geq 2$ and $\lambda_1 = -\alpha_1$ gives rise to a unique $-z^{(j)}$. Hence in this case we obtain two functions $\pm z(\pi)$.

We now consider the equation $\theta(z(\pi),\pi)=0$ which is the same as $\theta(-z(\pi),\pi)=0$ since θ is an even function of z. This gives a triangular system of relations for $\pi_{k\ell}^{(m)}$. The initial condition of rank 3 gives the equation

$$\pi_{12}'^{2}\pi_{34}'^{2}+\pi_{13}'^{2}\pi_{24}'^{2}+\pi_{14}'^{2}\pi_{23}'^{2}-2(\pi_{12}'\pi_{34}'\pi_{13}'\pi_{24}'+\pi_{12}'\pi_{34}'\pi_{14}'\pi_{23}'+\pi_{13}'\pi_{24}'\pi_{14}'\pi_{23}')=0 \tag{21}$$

the coefficients of t^2 and t^3 in $\theta(z(\pi),\pi)$ are precisely the same as the coefficients of t^2 and t^3 in $\theta(0,\pi)$. The coefficients of t^4 is the same as the coefficient of t^4 in $\theta(0,\pi)$ except for one term which is a multiple of the product of the components of z_0 which is not zero. We therefore note the following: the equation $\theta(z(\pi),\pi)=0$ defines a variety, Q, in the Siegel upper half plane of degree 4 with the property that for each π near π_0, $\theta(z,\pi)$ has generically two singular points near zero. We call this the variety of quasi-Jacobians. We further note that $\dim Q = 9$ as one sees from the jets of order ≥ 2. (Here a term $\pi_{pp}^{(m)}$ is considered as a jet of order $m+2$.) The jets of order 1 form an 8-dimensional family. Finally, the jets of order 1 are a subvariety of the jets of order 1 for the variety H. The jets of order 2 for Q and H are precisely the same. It is only at order 3 that the varieties Q and H really separate. This is probably the reason the variety Q was not discovered before now.

IV. *The canonical curve*

It is well-known that the canonical curve for a non-hyperelliptic Riemann surface of genus 4 can be represented as the intersection of a quadric and cubic in complex projective 3 space. We can derive explicit expressions for the quadric and cubic using the theta function and the singular points. The quadric relation is classical but the expression for the cubic seems to be new.

We think of the Riemann surface as a curve C embedded in its Jacobian and then may interpret the canonical curve as the curve defined by the tangents to C. It is a classical result that

(22) $$\theta(u-v \pm z(\pi), \pi) = 0 \quad \text{for all} \quad u, v \in C\ .$$

Here $\pm z(\pi)$ are the singular points of θ and we are thinking of $z(\pi)$ being the solution to the system of equations (1) where π' satisfies the Poincaré initial conditions. Let ζ be a local parameter on C at v and let $u = u(\zeta)$. We differentiate (22) with respect to ζ and the result is not particularly interesting. We take a second derivative and find

(23) $$\sum_{i,j=1}^{4} \frac{\partial^2\theta(z(\pi),\pi)}{\partial z_i \partial z_j} \frac{\partial u_i}{\partial \zeta} \frac{\partial u_j}{\partial \zeta} = 0\ .$$

In (23) $\frac{\partial u_i}{\partial \zeta}$ is evaluated at $\zeta = 0$ or at $u = v$. Since v is an arbitrary point on C (23) is the relation among the products of holomorphic differentials on C and is the quadric relation we spoke of. We now take a third derivative of (22) with respect to ζ and find

(24) $$\sum_{i,j,k=1}^{4} \frac{\partial^3\theta(z(\pi),\pi)}{\partial z_i \partial z_j \partial z_k} \frac{\partial u_i}{\partial \zeta} \frac{\partial u_j}{\partial \zeta} \frac{\partial u_k}{\partial \zeta} + \sum_{i,j=1}^{4} \frac{\partial^2\theta(z(\pi),\pi)}{\partial z_i \partial z_j} \frac{\partial^2 u_i}{\partial \zeta_i} \frac{\partial u_j}{\partial \zeta} = 0\ .$$

Assuming now as we may that $z(\pi)$ is not zero then $-z(\pi)$ is also a singular point. Using the fact that θ is even and the relations we obtain using $z(\pi)$ and $-z(\pi)$ we obtain easily that

(25) $$\sum_{i,j,k=1}^{4} \frac{\partial^3\theta(z(\pi),\pi)}{\partial z_i \partial z_j \partial z_k} \frac{\partial u_i}{\partial \zeta} \frac{\partial u_j}{\partial \zeta} \frac{\partial u_k}{\partial \zeta} = 0$$

which is the cubic relation we spoke of.

The quadric and the cubic in complex projective 3-space are therefore simply $\sum_{i,j=1}^{4} \frac{\partial^2\theta}{\partial z_i \partial z_j}(z(\pi),\pi)\, u_i u_j = 0$ and $\sum_{i,j,k=1}^{4} \frac{\partial^3\theta(z(\pi),\pi)}{\partial z_i \partial z_j \partial z_k} u_i u_j u_k = 0$ where (u_1, u_2, u_3, u_4) are the coordinates in CP_3.

The only further item that needs to be verified is that the quadric and cubic are independent. The Riemann-Roch Theorem tells us that the dimension of the space of holomorphic cubic differentials on C is 15. There are five independent linear relations among the cubics. Four of these are trivial in the sense that they are products of the quadric with a linear form. We therefore need show that the cubic we have written is not such a product.

We have not actually written down a proof of this fact but using our trick of expanding the coefficients in power series in t and checking the lowest order terms indicate that the cubic we have is generically not such a product. It does not seem hard to prove the general theorem.

V. *Genus greater than or equal to five*

A serious difficulty arises in genus 5 because for a generic Jacobian in genus 5 the theta function has a one-dimensional set of singular points on the theta divisor. In terms of Section III this means that we want to study those π in the Siegel upper half plane of degree 5 for which the set of z satisfying $\theta(z,\pi) = 0$ and $\frac{\partial\theta}{\partial z_i}(z,\pi) = 0$ $i = 1, \ldots, 5$ is one-dimensional.

It is now important to analyze our previous discussion in Section III where we treated $g = 4$. There we studied the variety $\theta(z,\pi) = 0$ $\frac{\partial\theta}{\partial z_i}(z,\pi) = 0$ $i = 1, \ldots, 4$ in a neighborhood of $(0, \pi_0)$. In order to know that our work is actually related to Jacobians we have to know that as $\pi \to \pi_0$ through Jacobians, the singular points on the Jacobian approaches zero. Poincaré's initial condition even describes the rate of approach to zero. Unfortunately, we have no proof of the fact although we have evidence for it.

In genus 5 since we have a whole one dimensional set of singular points it might well be the case that as $\pi \to \pi_0$ in addition to the singular point at zero for θ we also obtain other singular points z which are limits of singular points on the Jacobians. In particular, suppose other half periods appear as limits of singular points and suppose further that Poincaré's initial condition is satisfied at each of these points. If we assume that for each j such a half period is $\frac{\pi_{jj}}{2}$ where as before π_{jj} are the diagonal entries of π_0 we can apply the technique of Section III to obtain the five relations:

(26) $\sqrt{\pi''_{12}\pi''_{34}\pi''_{13}\pi''_{24}} + \sqrt{\pi''_{12}\pi''_{34}\pi''_{14}\pi''_{23}} + \sqrt{\pi''_{13}\pi''_{24}\pi''_{14}\pi''_{23}} = 0$

(27) $\sqrt{\pi''_{25}\pi''_{34}\pi''_{35}\pi''_{24}} + \sqrt{\pi''_{25}\pi''_{34}\pi''_{45}\pi''_{23}} + \sqrt{\pi''_{35}\pi''_{24}\pi''_{45}\pi''_{23}} = 0$

(28) $\sqrt{\pi''_{15}\pi''_{34}\pi''_{13}\pi''_{45}} + \sqrt{\pi''_{15}\pi''_{34}\pi''_{14}\pi''_{35}} + \sqrt{\pi''_{13}\pi''_{45}\pi''_{14}\pi''_{35}} = 0$

(29) $\sqrt{\pi''_{12}\pi''_{45}\pi''_{15}\pi''_{24}} + \sqrt{\pi''_{12}\pi''_{45}\pi''_{14}\pi''_{25}} + \sqrt{\pi''_{15}\pi''_{24}\pi''_{14}\pi''_{25}} = 0$

(30) $\sqrt{\pi''_{12}\pi''_{35}\pi''_{13}\pi''_{25}} + \sqrt{\pi''_{12}\pi''_{35}\pi''_{15}\pi''_{23}} + \sqrt{\pi''_{13}\pi''_{25}\pi''_{15}\pi''_{23}} = 0\ .$

It is easy to see that these equations define a variety of codimension 3 in $\pi''_{k\ell}$ space. These are the relations initially discovered by Poincaré and rederived by Garabedian [G] when $g = 5$. A similar procedure works for any $g > 5$.

Perhaps a more satisfying way to attack the problem is to study directly the condition that there be a $(g-4)$-dimensional set of z satisfying $\theta(z,\pi) = 0$ and $\frac{\partial\theta}{\partial z_i}(z,\pi) = 0$ $i = 1,\ldots,g$. Even in the case $g = 5$ the algebra becomes very complicated and we have not yet really penetrated it.

BELFER GRADUATE SCHOOL OF SCIENCE
YESHIVA UNIVERSITY
STATE UNIVERSITY OF NEW YORK
STONY BROOK, NEW YORK

REFERENCES

[EFMR]. Ehrenpreis, L., Farkas, H. M., Martens, H., Rauch, H. E., "On the Poincaré Relation" (to appear).

[F] Farkas, H. M., "Special Divisors and Analytic Subloci of Teichmüller Space," Amer. J. of Math. (1966), pp. 881-901.

[G] Garabedian, P., "Asymptotic identities among Periods of Integrals of the first Kind," Amer. J. of Math. (1951), pp. 107-121.

[R] Rauch, H. E., "Schottky Implies Poincaré," "Advances in the theory of Riemann Surfaces," Ann. of Math. Studies (66), Princeton University Press, (1971), pp. 355-364.

REMARKS ON AUTOMORPHISMS OF COMPACT RIEMANN SURFACES

Hershel M. Farkas*

Introduction

In the first part of this paper we prove a technical theorem about automorphisms of a compact Riemann surface from which well known theorems about automorphisms of a hyperelliptic surface emerge as simple corollaries. Along the way we show that the maximal number of fixed points for an automorphism of a non-hyperelliptic surface is $2g-1$ and that this result is best possible. This result improves Lewittes' generalization [L1] of Hurwitz's theorem [H]. We then give further applications of our theorem along the lines of Accola's results in [A1, A2].

In the second part of the paper we make the observation that when a compact Riemann surface has automorphisms this property is reflected in the divisor of zeros of the theta function associated with the surface. Our principal results in this direction are Theorems 4 and 5. In Theorem 4 we prove that if $(S,(\Gamma,\Delta))$ is a compact Riemann surface of genus $g \geq 2$ together with a canonical homology basis and T is an automorphism of prime order N with $g \equiv 1 \bmod N$, then if p is any fixed point of $T, K(p)$, the vector of Riemann constants is a zero of the theta function of order $\geq \frac{N-1}{2N} N(T)$. We also indicate when the order is exactly $\frac{N-1}{2N} N(T)$. In Theorem 5 we prove that under the same hypothesis made in Theorem 4 along with the hypothesis that the quotient surface S factored by the group generated by T have genus ≥ 1, then the theta function vanishes at a half period to order $\geq \frac{N-1}{2N} N(T)$.

* Research partially supported by NSF GP 19572 and A. P. Sloan Foundation Fellowship.

In the last part of the paper we consider the case of surfaces which have automorphisms of order 2 with fixed points and describe their period matrices in a special case. We briefly discuss the theta functions associated with these surfaces and leave the reader with some unanswered questions.

The author is particularly indebted to Professor R. D. M. Accola for many informative conversations and access to as yet unpublished work. The results of this paper has a large intersection with Accola's work but the methods used here are different. Some of the remarks in part three of this paper were obtained independently by Accola, Fay and Schiller.

I. We introduce the following notation: S will always denote a compact Riemann surface of genus $g \geq 2$. $A(S)$ will denote the group of automorphisms (conformal homeomorphisms) of S and for $T \in A(S)$, $N(T)$ and $O(T)$ will denote the number of fixed points of T and the order of T respectively. $<T>$ will denote the group generated by T and as usual $S/<T>$ will denote the surface obtained by identifying points on S which are equivalent under the action of $<T>$ on S.

Our first lemma is a slight variant of an idea of Hurwitz and is motivated by Theorem 1 of [A2].

LEMMA 1. *Let* f *be a meromorphic function on* S *and let* $O(f)$ *denote the number of poles of* f. *Let* $T \in A(S)$ *and assume* $N(T) > 2\,O(f)$. *Then* $f = f \circ T$ *and* $O(f)$ *is a multiple of* $O(T)$. *If in addition* $O(f)$ *is prime, then* $O(T) = O(f)$ *and* $S/<T>$ *is the sphere.*

Proof. Consider $g = f - f \circ T$. g is a meromorphic function on S and $O(g) \leq 2\,O(f) - r$ where $0 \leq r \leq O(f)$ and r poles of f are fixed points of T. Clearly each fixed point of T which is not a pole of f is a zero of g. Hence our hypothesis on $N(T)$ yields that g has more than $2\,O(f) - r$ zeros. Hence $g \equiv 0$ so that $f = f \circ T$.

Since $f = f \circ T$ there is a well defined function $\tilde{f}$ on $\tilde{S} = S/\langle T \rangle$ and f is simply the lift of $\tilde{f}$ to S. Hence $O(f) = O(\tilde{f}) \cdot O(T)$ which is our second assertion. If in addition $O(f)$ is prime since $O(T) > 1$ $O(\tilde{f}) = 1$ and $O(f) = O(T)$. Hence $O(\tilde{f}) = 1$ and therefore $\tilde{S}$ is the sphere.

Our next lemma is an immediate consequence of the Riemann-Hurwitz relation. Recall that if S_1 and S_2 are compact Riemann surfaces of genus g_1 and g_2 and if f is an analytic map from S_1 onto S_2 then $2g_1-2 = n(2g_2-2) + V$ where V is the total number of branch points and n is the number of points in $f^{-1}(p)$ for $p \in S_2$ where a branch point is counted as many times as its multiplicity.

LEMMA 2. $N(T) \leq 2 + \dfrac{2g}{O(T)-1}$.

Proof. Applying the Riemann-Hurwitz relation to S and $S/\langle T \rangle = \tilde{S}$ we find $2g-2 = O(T)(2\tilde{g}-2) + \sum_{i=1}^{O(T)-1} N(T^i)$. Since $N(T) \leq N(T^i)$ we have $2g-2 \geq O(T)(2\tilde{g}-2) + (O(T)-1)N(T)$. Hence $N(T) \leq 2 + \dfrac{2(g-O(T)\tilde{g})}{O(T)-1} \leq 2 + \dfrac{2g}{O(T)-1}$.

It is now an immediate consequence of Lemma 2 that the maximum number of fixed points of an automorphism is $2g+2$ and that this occurs only in the case $\tilde{g} = 0$ and $O(T) = 2$. Such a surface is a hyperelliptic surface, i.e., a two-sheeted cover of the sphere with $2g+2$ branch points. This theorem was proved by Hurwitz [H]. Lewittes, on the other hand, has shown that on a non-hyperelliptic surface $N(T) \leq 2g$ [L]. As a consequence of Lemma 2 we prove

COROLLARY 1. *On a non-hyperelliptic surface* $N(T) \leq 2g-1$ *and this result is best possible.*

Proof. The proof of Lemma 2 shows that an automorphism of order 2 has precisely $2g+2-4\tilde{g}$ fixed points. Hence there are no automorphisms of order 2 with $2g$ fixed points. If an automorphism is to have $2g$ fixed points it therefore must have order greater than or equal to 3. In this case, Lemma 2 yields $N(T) \leq g+2$. When $g > 2$, $2g > g+2$ hence the only possibility is $g = 2$ which is a hyperelliptic surface. We therefore conclude that on a non-hyperelliptic surface $N(T) \leq 2g-1$. Finally, there is a non-hyperelliptic surface of genus 3 with $2g-1 = 5$ fixed points. The surface with the property is the Riemann surface of $w^3 = (z-z_1)^2 (z-z_2)\dots(z-z_5)$. This surface is obviously not hyperelliptic and has an automorphism of order 3 which has five fixed points.

THEOREM 1. *Let* S *be a compact Riemann surface of genus* $g \geq 2$ *and* $\tilde{T} \in A(S)$ *such that* $O(\tilde{T}) = N$ *with* N *prime. Let* $\tilde{g}$ *be the genus of* $\tilde{S} = S/<\tilde{T}>$ *so that* $N(\tilde{T}) = 2 + \dfrac{2(g-N\tilde{g})}{N-1}$. *Assume* $g > N^2\tilde{g} + (N-1)^2$. *Then if there is a* $T \in A(S)$ *with* $N(T) > 2N(\tilde{g}+1)$

i) *Each fixed point of* $\tilde{T}$ *is a fixed point of* T *and* $O(T) \leq O(\tilde{T})$.

ii) *If* $O(T) = O(\tilde{T})$ *then* $T \in <\tilde{T}>$.

iii) $<\tilde{T}>$ *is normal in* $A(S)$.

iv) *If* $N = 2$, $\tilde{T}$ *commutes with each element of* $A(S)$.

Proof. Our first observation is that our hypothesis on g implies that $N(\tilde{T}) > 0$ so let $p_1, \dots, p_{N(\tilde{T})}$ denote the fixed points of T on S. With each fixed point p_i we can associate a meromorphic function f_i on S with the property that f_i has a pole of order at most $N(\tilde{g}+1)$ at p_i and is otherwise finite. This follows by considering the surface $\tilde{S} = S/<\tilde{T}>$. Let $\tilde{p}_i$ be the image of p_i under the projection map $\pi: S \to \tilde{S}$. There is certainly a function $\tilde{f}_i$ on $\tilde{S}$ with a pole of order at most $\tilde{g}+1$ at $\tilde{p}_i$ and otherwise finite. Let f_i be the lift of $\tilde{f}_i$, and f_i has the desired property. If $T \in A(S)$ and $N(T) > 2N(\tilde{g}+1)$ we have by Lemma 1 that $f_i \circ T = f_i$. It therefore follows that T fixes each fixed point of $\tilde{T}$ and that $N(T) \geq N(\tilde{T})$.

We have thus far not used the hypothesis that $g > N^2\tilde{g} + (N-1)^2$ except to be sure that $N(\tilde{T}) > 0$. This would be true as long as $g > N\tilde{g} - (N-1)$, which is a much weaker assertion. We now, however, use our hypothesis on g. $g > N^2\tilde{g} + (N-1)^2$ gives $Ng > N^2\tilde{g} + (N-1)^2 + (N-1)g$ which gives $N(g-N\tilde{g}) > (N-1)(N-1+g)$ or that $\frac{g-N\tilde{g}}{N-1} > \frac{g+N-1}{N}$. The condition $N(T) \geq N(\tilde{T})$ now gives that

$$N(T) \geq 2 + \frac{2(g-N\tilde{g})}{N-1} > 2 + 2\left(\frac{g+N-1}{N}\right) = 2 + \frac{2g}{N} + 2\left(\frac{N-1}{N}\right) .$$

Hence $N(T) > 2 + \frac{2g}{N}$. It follows from Lemma 2 now that $O(T) \leq N = O(\tilde{T})$. This concludes the proof of statement i).

Suppose now that $O(T) = O(\tilde{T})$. By what we have already shown if p_i is any fixed point of $\tilde{T}$, p_i is also a fixed point of T. Let S_{p_i} denote the subgroup of A(S) which fixes p_i. It is well known that S_{p_i} is a cyclic group. Hence if $O(T) = O(\tilde{T})$ and T and $\tilde{T}$ are both elements of S_{p_i} it follows that $T \in \langle \tilde{T} \rangle$. This gives us statement ii).

To prove iii) let $B \in A(S)$ and consider $C = B^{-1}\tilde{T}B$. $O(C) = O(\tilde{T})$ and $N(C) = N(\tilde{T})$. The condition $g > N^2\tilde{g} + (N-1)^2$ yields that $N(\tilde{T}) > 2N(\tilde{g}+1)$ and therefore $N(C) > 2N(\tilde{g}+1)$. It follows now from ii) that $C \in \langle \tilde{T} \rangle$ and therefore $\langle \tilde{T} \rangle$ is normal in A(S).

Statement iv) now follows immediately from iii) since when $N = 2$ the statement that $\langle \tilde{T} \rangle$ is normal in A(S) is equivalent to the statement that $\tilde{T}$ commutes with each element of A(S).

We now give some applications of our theorem. Our first application will be to hyperelliptic surfaces. If S is a hyperelliptic surface of genus $g \geq 2$ then there is an automorphism $J \in A(S)$ such that $O(J) = 2$, $N(J) = 2g+2$, and $S/\langle J \rangle = \tilde{S}$ is the sphere. It therefore follows from our theorem that if there is a $T \in A(S)$ with $N(T) > 4$, $T = J$, and J commutes with every element of A(S). Another application, not of our theorem, but rather of Lemma 1, is the following: Suppose f is a meromorphic function on S with $O(f) < g+1$. Then O(f) is even. This follows from Lemma 1 since $N(J) > 2\,O(f)$ implies O(f) is a multiple of $O(J) = 2$. These results are, of course, well known. We now go on to some applications which are not so well known. We will begin with the following definition.

DEFINITION 1. S will be called $\tilde{g}$-hyperelliptic provided S can be represented as a two sheeted covering of a compact Riemann surface of genus $\tilde{g}$.

We now observe that if S is a $\tilde{g}$-hyperelliptic surface, then there is an automorphism $J_{\tilde{g}} \epsilon A(S)$, $O(J_{\tilde{g}}) = 2$ and $N(J_{\tilde{g}}) = 2g+2-4\tilde{g}$. If $\tilde{g} = 0$, then S is hyperelliptic. At any rate, we have the following theorem:

THEOREM 2. *Let* S *be a* $\tilde{g}$*-hyperelliptic surface of genus* $g > 4\tilde{g}+1$.

i) *If* $T \epsilon A(S)$ *and* $N(T) > 4(\tilde{g}+1)$ *then* $T = J_{\tilde{g}}$.

ii) $J_{\tilde{g}}$ *commutes with every element of* $A(S)$.

iii) *If* f *is a meromorphic function on* S *with* $O(f) < g+1-2\tilde{g}$ *then* $O(f)$ *is even.*

Proof. The hypothesis of i) is the same as the hypothesis of Theorem 1, hence $O(T) \leq O(\tilde{J})$. Since $O(J_{\tilde{g}}) = 2$, we have $O(T) = 2$ and therefore by ii) of Theorem 1, $T = J_{\tilde{g}}$. This is statement i). ii) is just iv) of Theorem 1. To prove iii) we use Lemma 1. If $O(f) < g+1-2\tilde{g}$ then by Lemma 1, $f = f \circ J_{\tilde{g}}$ and $O(f)$ is a multiple of $O(J_{\tilde{g}})$.

COROLLARY 2. *If* $g > 4\tilde{g}+1+2r$ *a surface cannot be both* $\tilde{g}$*-hyperelliptic and* $\tilde{g}+r$*-hyperelliptic. In particular, if* $g > 3$ *a surface cannot be both hyperelliptic and* 1*-hyperelliptic.*

Proof. Suppose S is $\tilde{g}$-hyperelliptic, and $g > 4\tilde{g}+1+2r$. Then by Theorem 2, if there is a $T \epsilon A(S)$ with $N(T) > 4(\tilde{g}+1)$, $T = J_{\tilde{g}}$. If S were $\tilde{g}+r$-hyperelliptic, then there would be an automorphism $J_{\tilde{g}+r} \epsilon A(S)$ with $N(J_{\tilde{g}+r}) = 2g+2-4(\tilde{g}+r)$. Since $g > 4\tilde{g}+1+2r$, $N(J_{\tilde{g}+r}) > 4(\tilde{g}+1)$. Hence $J_{\tilde{g}+r} = J_{\tilde{g}}$ which shows that there cannot be any $J_{\tilde{g}+r}$.

The particular case of Corollary 2 mentioned, i.e., $\tilde{g} = 0$ and $r = 1$ has been observed by Accola. Furthermore, if $g = 3$ there are surfaces which are both hyperelliptic and 1-hyperelliptic. Any smooth two sheeted cover of a surface of genus 2 is an example of such a surface.

II. Recall that a divisor on S is a formal symbol $p_1^{\alpha_1} \cdots p_k^{\alpha_k}$ where $p_i \in S$ and α_i are integers. A divisor is called integral or positive if $\alpha_i > 0$ for each i. The degree of the divisor is $\sum_{i=1}^{k} \alpha_i$. A divisor ζ is said to be a multiple of a divisor ω provided there is an integral divisor δ such that $\zeta = \omega\delta$. $\Omega(\zeta)$ is the vector space of holomorphic differentials on S whose divisors are multiples of ζ and $i(\zeta) = \dim \Omega(\zeta)$. $L(\zeta)$ is the vector space of meromorphic functions on S whose divisors are multiples of ζ and $r(\zeta) = \dim L(\zeta)$. The Riemann-Roch theorem says that $r\left[\frac{1}{\zeta}\right] = \deg \zeta - g + 1 + i(\zeta)$, where if $\zeta = p_1^{\alpha_1} \cdots p_k^{\alpha_k}$, $1/\zeta = p_1^{-\alpha_1} \cdots p_k^{-\alpha_k}$. In particular, if $\deg \zeta = g-1$, $r\left[\frac{1}{\zeta}\right] = i(\zeta)$.

THEOREM 3. *Let* S *be a compact Riemann surface of genus* $g \geq 2$. *Let* $T \in A(S)$, $N(T) > 0$, $O(T) = N$ *with* N *prime, and assume that* $g \equiv 1 \bmod N$. *Then there is an integral divisor,* ζ, *of degree* $g-1$ *on* S *with* $i(\zeta) \geq \frac{N-1}{2N} N(T)$. *In fact* $\zeta = p^{g-1}$ *where* p *is any fixed point of* T.

Proof. Let p be a fixed point of T and let $\tilde{S} = S/\langle T \rangle$. Let $\tilde{p}$ be the projection of p on $\tilde{S}$. It is easy to see that $g = N(\tilde{g}-1) + \frac{N-1}{2} N(T) + 1$ and that if $\tilde{f} \in r\left[\frac{1}{\tilde{p}^{\frac{g-1}{N}}}\right]$, f, the lift of $\tilde{f}$ is in $r\left[\frac{1}{p^{g-1}}\right]$. Now $\frac{g-1}{N} = \tilde{g}-1 + \frac{N-1}{2N} N(T)$ so that $r\left[\frac{1}{\tilde{p}^{\frac{g-1}{N}}}\right] = \frac{N-1}{2N} N(T) + i\left[p^{\frac{g-1}{N}}\right] \geq \frac{N-1}{2N} N(T)$. Hence $r\left[\frac{1}{p^{g-1}}\right] = i[p^{g-1}] \geq \frac{N-1}{2N} N(T)$.

In certain cases we can do better than Theorem 3 in the sense that we can say that $i(\zeta) = \frac{N-1}{2N} N(T)$. This occurs when $S/\langle T \rangle$ has genus 0 and $N \leq 3$ and when S is $\tilde{g}$-hyperelliptic and $\pi(p) \in S/\langle J_{\tilde{g}} \rangle$ is not a Weierstrass point on $S/\langle J_{\tilde{g}} \rangle$ and $g \geq 2\tilde{g}+1$. In order to prove the above statement we recall the Weierstrass gap theorem. The gap theorem states that if p is any point on S there are exactly g numbers $\gamma_1, \ldots, \gamma_g$

between one and $2g$ such that there is no meromorphic function on S where only singularity is a pole of order γ_i, $i = 1, \ldots, g$ at p.

We are now assuming that the hypothesis of Theorem 3 is fulfilled and in addition, assume that the genus of $S/\langle T\rangle$ is zero. Let p be a fixed point of T. Then clearly there exists on S meromorphic functions whose only singularity is at p with the following orders. $N, 2N, \ldots, N\left(\frac{N-1}{2N} N(T)-1\right) = g-1$, $N\left(\frac{N-1}{2N} N(T)\right), \ldots, N\left(\frac{N-1}{N} N(T)-2\right) = 2g-2, \ldots$. These functions exist since they are the lifts of functions on the sphere with poles of order $1, 2, \ldots$ at $\pi(p)$, where $\pi(p)$ is the projection of p to $S/\langle T\rangle$. We see already at this point that $r\left[\frac{1}{p^{g-1}}\right] = i(p^{g-1}) \geq \frac{N-1}{2N} N(T)$. The question is, are there any more functions with poles of order $\leq g-1$ at p and otherwise finite which we have not listed. We show that the answer is no if $N \leq 3$. Suppose there were one other function with a pole of order $\leq g-1$ which we have not listed. We would then find that there are at least $\frac{3(N-1)}{2N} N(T)-2$ functions in our list with orders $\leq 2g-2$. Now in the case we are considering, we have $N(T) = 2 + \frac{2g}{N-1}$. Hence we find $\frac{3(N-1)}{2N} N(T)-2 \geq g$ as long as $N \leq 3$. This contradicts the gap theorem so we get our result.

Now let us once again assume the hypothesis of Theorem 3 with the genus of $S/\langle T\rangle = \tilde{g} > 0$ and that the projection of the fixed point of T is not a Weierstrass point on $S/\langle T\rangle$. Recall that a point is not a Weierstrass point if the gaps are the integers $1, 2, \ldots, g$. In this case we find that there exist meromorphic functions on S whose only singularity is at p with the orders $N(\tilde{g}+1), N(\tilde{g}+2), \ldots, N(\tilde{g}+\frac{N-1}{2N} N(T)-1) = g-1, N(\tilde{g}+\frac{N-1}{2N} N(T)), \ldots, N(2\tilde{g}-2+\frac{N-1}{N} N(T)) = 2g-2$. These functions exist since they are the lifts of functions on $S/\langle T\rangle$ with poles of order $\tilde{g}+1, \tilde{g}+2, \ldots$ at $\pi(p)$. As in the previous case we see that $r\left[\frac{1}{p^{g-1}}\right] = i(p^{g-1}) \geq \frac{N-1}{2N} N(T)$, and the question as before is whether there are any more functions with poles of order $\leq g-1$ and otherwise finite which we have not listed. The argument is the same as before. If there were one more function we would find at least $\frac{3(N-1)}{2N} N(T)+2g-2$ functions in our list with orders $\leq 2\tilde{g}-2$.

In the case we are considering we have $N(T) = 2 + \frac{2(g-N\tilde{g})}{N-1}$ and we find $\frac{3(N-1)}{2N} N(T) + 2\tilde{g}-2 \geq g$ as long as $N = 2$. This once again contradicts the gap theorem and we get our result. We therefore get

COROLLARY 3. *Let* S *be a* $\tilde{g}$*-hyperelliptic surface of odd genus and assume that* $J_{\tilde{g}}$ *has at least one fixed point* p *whose projection on* $S/\langle J_{\tilde{g}}\rangle$ *is not a Weierstrass point. Then* $i(p^{g-1}) = \frac{g+1-2\tilde{g}}{2}$. *In particular if* $\tilde{g} = 0$ *or* $\tilde{g} = 1$, $i(p^{g-1}) = \frac{g+1}{2}$ *or* $\frac{g-1}{2}$.

COROLLARY 4. *Let* S *be a compact Riemann surface of genus* $g \equiv 1$ mod 3. *Suppose there is a* $T \in A(S)$, $O(T) = 3$ *and* $N(T) = 2+g$. *Let* p *be a fixed point of* T. *Then* $i(p^{g-1}) = \frac{g+2}{3}$.

DEFINITION 2. The first order g-dimensional theta function $\theta(z,\tau)$ is defined by the following series

$$\theta(z,\tau) = \sum_{n_1,\ldots,n_g=-\infty}^{\infty} \exp 2\pi i\left(\frac{1}{2}\sum_{i,j=1}^{g} n_i\tau_{ij}n_j + \sum_{i=1}^{g} n_i z_i\right)$$

where $\tau = (\tau_{ij})$ is a $g\times g$ symmetric matrix with positive definite imaginary part and $z = z_1,\ldots,z_g$ is an element in complex g-spaces.

We shall denote the collection of all $g\times g$ symmetric matrices with positive definite imaginary part by $\mathfrak{S}_g$, and call $\mathfrak{S}_g$ the Siegel upper half space of degree g. Complex g-space will as usual be denoted by C^g. Since the series defining $\theta(z,\tau)$ converges absolutely and uniformly on compact subsets of $C^g\times\mathfrak{S}_g$ it follows that $\theta(z,\tau)$ is a holomorphic function on $C^g\times\mathfrak{S}_g$ and in particular for fixed $\tau \in \mathfrak{S}_g$, $\theta(z,\tau)$ is an entire function in C^g.

Suppose now that S is a compact Riemann surface of genus g and (Γ,Δ), $\Gamma = \gamma_1,\ldots,\gamma_g$, $\Delta = \delta_1,\ldots,\delta_g$ is a canonical homology basis on S. The space of holomorphic differentials on S, $A_1(S)$, is a g-dimensional vector space over the complex numbers and if $\omega_1,\ldots,\omega_g$ is any basis for $A_1(S)$, we can form the matrix (A, B) where $A = (a_{ij})$, $i, j = 1, \ldots, g$, $B = (b_{ij})$, $i, j = 1, \ldots, g$, and $a_{ij} = \int_{\gamma_j} \omega_i$ and $b_{ij} = \int_{\delta_j} \omega_i$. Since A can be shown to be non-singular, we can always find a basis for $A_1(S)$ such that $A = I$ where I is the $g \times g$ identity matrix. It then can be shown that $B = \Pi$ where $\Pi \in \mathfrak{S}_g$. Hence we see that given a pair $(S, (\Gamma, \Delta))$ we get $\Pi \in \mathfrak{S}_g$.

DEFINITION 3. The Riemann theta function associated with $(S, (\Gamma, \Delta))$ is the g-dimensional theta function of Definition 1 with τ replaced by π where π is uniquely determined by $(S, (\Gamma, \Delta))$.

Given $(S, (\Gamma, \Delta))$ with $p_o \in S$, we define a holomorphic multivalued map ϕ, of $S \to C^g$ with $\phi(p) = \int_{p_o}^{p} \Phi$ where $\Phi = (\phi_1 \ldots \phi_g)^t$ where ϕ_i is the ith element of the basis for $A_1(S)$ and $\int_{\gamma_j} \phi_i = \delta_{ij}$ where δ_{ij} is the Kronecker δ. This map is multivalued for the image depends on the path of integration one takes from p_o to p. It is clear, however, that any two images of p under this map can differ only by an integral linear combination of the columns of the $g \times 2g$ matrix (I, π). If we identify, therefore, all points in C^g which differ by an integral linear combination of the columns of (I, π) our map becomes single valued. The map ϕ can be extended in a natural way to divisors on S. If $\zeta = p_1^{\alpha_1} \ldots p_k^{\alpha_k}$ we defined $\phi(\zeta) = \sum_{i=1}^{k} \alpha_i \phi(p_i)$. C^g identified under the group of translations generated by the columns of the matrix (I, π) will be denoted by $J(S, (\Gamma, \Delta))$. The theta function $\theta(z, \pi)$ is an entire function in C^g, but is not well defined on $J(S, (\Gamma, \Delta))$; however, the zeros of $\theta(z, \pi)$ are well defined on $J(S, (\Gamma, \Delta))$ since if z_1 and z_2 differ by an integral linear combination of the columns of (I, π) it is easy to see that $\theta(z_1, \pi)$ differs from $\theta(z_2, \pi)$ by an exponential multiplier. It is known [12] that there is a vector $K(p_o(\Gamma, \Delta))$ in $J(S, (\Gamma, \Delta))$ such that the set of zeros of $\theta(z, \pi)$ is exactly the image of the set of integral divisors of degree $g-1$ on $(S, (\Gamma, \delta))$ translated by $K(p_o, (\Gamma, \Delta))$. Furthermore, if ζ is an integral divisor of degree $g-1$ on $(S(\Gamma,\Delta))$, then, $\phi(\zeta) + K(p_o, (\Gamma,\Delta)) = z_1$ is a zero of order $i(\zeta)$. If we choose $\zeta = p_o^{g-1}$ we have $K(p_o, (\Gamma,\Delta))$ is always a zero of $\theta(z,\pi)$ and has order $i(p_o^{g-1})$.

THEOREM 4. *Let* S *be a compact Riemann surface of genus* $g \geq 2$. *Let* $T \in A(S), N(T) > 0$, $O(T) = N$ *with* N *prime and assume* $g \equiv 1 \bmod N$. *Then if* p *is a fixed point of* T, $K(p)$ *is a zero of* $\theta(z,\pi)$ *of order* $\geq \frac{N-1}{2N} N(T)$. *In particular, if* S *is hyperelliptic* (1-*hyperelliptic*) *and* g *is odd and* p *is a fixed point of* $J(J_1)$ *then* $K(p)$ *is a zero of* $\theta(z,\pi)$ *of order exactly* $\frac{g+1}{2}\left(\frac{g-1}{2}\right)$. *If* $\tilde{g} > 1$ *and* p *is a fixed point of* $J_{\tilde{g}}$ *whose image under the projection map* $\pi: S \to S/\langle J_{\tilde{g}}\rangle$ *is not a Weierstrass point then* $K(p)$ *is a zero of* $\theta(z,\pi)$ *of order exactly* $\frac{g+1-2\tilde{g}}{2}$. *Lastly, if* $N = 3$ *and* $S/\langle T\rangle$ *has genus* 0, *then if* p *is a fixed point of* T, $K(p)$ *is a zero of* $\theta(z,\pi)$ *of order exactly* $\frac{g+2}{3}$.

Proof. The proof is simply the proofs of Theorem 3, Corollary 3 and 4 and the remarks preceding the statement of Theorem 4.

We have already defined $J(S,(\Gamma,\Delta))$ to be C^g identified under the group of translations generated by the $2g$ columns of the matrix (I,π).

DEFINITION 4. A period in C^g is the zero element in $J(S,(\Gamma,\Delta))$ and a half period in C^g is a point $e \in C^g$ such that $2e$ is the zero element of $J(S,(\Gamma,\Delta))$.

Theorem 4 has shown us that the set of zeros of $\theta(z,\pi)$ includes points such that the order of vanishing of $\theta(z,\pi)$ at those points is $\geq \frac{N-1}{2N} N(T)$. These points are the points $K(p)$ where p is a fixed point of an automorphism T of S with $O(T) = N$ prime and $N(T) > 0$ where the genus of S satisfies $g \equiv 1 \bmod (N)$. We shall now show that this order of vanishing is always attained at a half period in $J(S,(\Gamma,\Delta))$ when the genus $\tilde{g}$ of $S/\langle T\rangle$ is greater than zero.

THEOREM 5. *Let* $(S,(\Gamma,\Delta))$ *be a compact Riemann surface of genus* $g \geq 2$ *together with a canonical homology basis on* S. *Let* (I,π) *be the* $g\times 2g$ *period matrix. If there is a* $T \in A(S)$ *with* $N(T) > 0$, $O(T) = N$ *with* N *prime and the genus of* $S/\langle T\rangle > 0$. *Then if* $g \equiv 1 \mod N$, $\theta(z,\pi)$ *vanishes to order at least* $\frac{N-1}{2N} N(T)$ *at a half period of* $J(S,(\Gamma,\Delta))$.

Proof. The proof depends on the following facts: Δ is the divisor of any abelian differential on S if and only if $\phi(\Delta) \equiv -2K$ and degree of $\Delta = 2g-2$. Hence if we could find a holomorphic differential on S with double zeros say with divisor $p_1^2 \cdots p_{g-1}^2$ we would have $\phi(p_1 \cdots p_{g-1}) \equiv -K+e$ where e is a half period. By our previous remarks $\theta(z,\pi)$ vanishes at e to order $i(p_1 \cdots p_{g-1})$. To prove our theorem it therefore suffices to show that $i(p_1 \cdots p_{g-1}) \geq \frac{N-1}{2N} N(T)$.

Our hypothesis on S implies that there is a holomorphic differential, ω, on S which is the lift of a holomorphic differential on $S/\langle T\rangle$. This differential has divisor $p_1^{N-1} \cdots p_{N(T)}^{N-1} \Delta\Delta' \dots \Delta^{(N-1)}$ where Δ is the divisor of a holomorphic differential on $S/\langle T\rangle$ which we may assume contains none of the points $\pi(p_1) \dots \pi(p_{N(T)})$ where π is the projection map $\pi: S \to S/\langle T\rangle$ and $\Delta' \dots \Delta^{(n-1)}$ are the points conjugate to the points of Δ under $\langle T\rangle$. Now the differential ω is invariant under $\langle T\rangle$. Furthermore, there is a differential ω' on S which satisfies the condition $\omega' \circ T = e^{\frac{2\pi i k}{N}} \omega'$ for some k, $1 \leq k \leq N-1$. Now the quotient $f = \omega'/\omega$ is a function on S which satisfies $f \circ T = e^{\frac{2\pi i k}{N}} f$ and therefore projects to an N-valued function on $S/\langle T\rangle$. The important point here is that if ω' vanishes at any of the points p_i, its order there is congruent to $t-1 \mod N$ where $1 \leq t \leq N-1$ [L1]. In particular, therefore, it does not have order $N-1$ at any of the points p_i. Therefore, the divisor of ω' is

$$p_1^{r_1 N + t_1 - 1} \cdots p_{N(T)}^{r_{N(T)} N + t_{N(T)} - 1} XX' \dots X^{(N-1)}$$

and the divisor of f is of the form

$$\frac{p_1^{r_1N+t_1-1}\cdots p_{N(T)}^{r^N_{N(T)}N+t_{N(T)}-1}\,XX'\ldots X^{(n-1)}}{p_1^{N-1}\cdots p_{N(T)}^{N-1}\,\Delta\Delta'\ldots\Delta^{(N-1)}}$$

where we, of course, have to perform the necessary cancellation of zeros and poles. This function, when projected to $S/\langle T\rangle$, has divisor

$$\frac{p_1^{r_1+\frac{t_1-1}{N}}\cdots p_{N(T)}^{r_{N(T)}+\frac{t_{N(T)}-1}{N}}\,X}{p_1^{\frac{N-1}{N}}\cdots p_{N(T)}^{\frac{N-1}{N}}\,\Delta}$$

again after performing the cancellation of zeros and poles. Here fractional order has the obvious meaning. In any event, it is clear that on $S/\langle T\rangle$ the divisor, ζ, of the projected function satisfies $\phi(\zeta)$ is an Nth period in $J(S/\langle T\rangle(\Gamma,\Delta))$. Conversely, given any function on $S/\langle T\rangle$ with fractional order $rN+K/N$ at the points $p_1\ldots p_{N(T)}$ with r an integer, $1\leq K<N-1$, with ζ equal the divisor of f and $\phi(\zeta)$ equal to an Nth period, then f is a well defined meromorphic function on an N-sheeted cover branched over $p_1\ldots p_{N(T)}$.

Let S satisfy the hypotheses of Theorem 5. Then there is a surface $\tilde{S}=S/\langle T\rangle$ of genus $\tilde{g}>0$ such that S is an N-sheeted cover of $\tilde{S}$ branched over the points $\tilde{p}_1\ldots\tilde{p}_{N(T)}$ on $\tilde{S}$ and an Nth period such that if $\tilde{f}$ is a function on $\tilde{S}$ with fractional order $rN+K/N$ at each point $\tilde{p}_i$ and $\phi(\zeta)$ equal to the given N^{th} period where ζ is the divisor of $\tilde{f}$ then f, the lift of $\tilde{f}$, is a well defined meromorphic function on S.

We construct a function g on $\tilde{S}$ in $L\left(\dfrac{1}{\tilde{p}_1^{N-1}\cdots\tilde{p}_{N(T)}^{N-1}\Delta^N}\right)$ where Δ is the divisor of a holomorphic differential on $\tilde{S}$. The degree of $\tilde{p}_1^{N-1}\cdots\tilde{p}_{N(T)}^{N-1}\Delta^N$ is $N(2\tilde{g}-2)+(N-1)N(T)$. We wish to find a divisor X on $\tilde{S}$ with degree of $X=\tilde{g}-1+\dfrac{(N-1)N(T)}{2N}$ and such that $\phi(X^{2N})\equiv\phi\left(\tilde{p}_1^{N-1}\cdots\tilde{p}_{N(T)}^{N-1}\Delta^N\right)$. We wish to do this subject to the additional condition that $\phi(X^2)-\phi\left(\tilde{p}_1^{\frac{N-1}{N}}\cdots\tilde{p}_{N(T)}^{\frac{N-1}{N}}\Delta\right)$ is congruent to our given N^{th}

period. If we construct such a function g then $g^{1/N} = \tilde{f}$ is a function which "lives" on S and the divisor of f, the lift of $\tilde{f}$ on S will be

$$\frac{X^2(X^2)'\ldots(X^2)^{(N-1)}}{p_1^{N-1}\cdots p_{N(T)}^{N-1}\Delta\Delta'\ldots\Delta^{(N-1)}}.$$

Now the "polar" divisor of f is the divisor of a holomorphic differential on S and, in fact, the divisor of a differential which is invariant under $\langle T \rangle$. Abel's theorem therefore gives that the "zero" divisor of f is also the divisor of a holomorphic differential. The words polar and zero being in quotation marks since there may be cancellation due to common zeros and poles but this is irrelevant to our argument. If we denote now the divisor $X(X)'\ldots(X)^{(N-1)}$ by Y we have $2\phi(Y) \equiv -2K$ on S and therefore $\phi(Y) + K \equiv e$ where e is a half period in $J(S,(\Gamma,\Delta))$. We need therefore show that $i(Y) \geq \frac{N-1}{2N}N(T)$.

In order to see that $i(Y) \geq \frac{N-1}{2N}N(T)$, we show that the divisor X has at least $\frac{N-1}{2N}N(T) - 1 = q$ free points in the sense that given any points $p_1\cdots p_q$ we can find a divisor Z with degree Z = degree X, Z contains the points $p_1\cdots p_q$ and $\phi(Z) \equiv \phi(X)$. This will imply that the divisor Y on S will have at least q free points and then elementary estimates give $i(Y) \geq \frac{N-1}{2N}N(T)$.

In order to show that X has at least q free points, recall that degree $X = \tilde{g} + \frac{N-1}{2N}N(T) - 1$. The solution of the Jacobi inversion problem says that we can solve the equation $\phi(X) \equiv a$ with $a \in C^g$ and X a divisor of degree g on a surface of genus g. Therefore, the equation $\phi(X^{2N}) \equiv 2N\phi(X) \equiv \phi(\tilde{p}_1^{N-1}\cdots\tilde{p}_{N(T)}^{N-1}\Delta^N)$ can always be solved with $\frac{N-1}{2N}N(T) - 1$ free points. Furthermore, there are $(2N)^{2\tilde{g}}$ different solutions corresponding to all the $(2N)^{th}$-periods. Not all of these are good, however, since our side condition dictates that we can only take those solutions which correspond to $(2N)^{th}$-periods, e_i, such that $2e_i$ is congruent to our given N^{th} period. There are, of course, $2^{2\tilde{g}}$ such. At any rate we see that Y has at least $\frac{N-1}{2N}N(T) - 1$ free points and hence $i(Y) \geq \frac{N-1}{2N}N(T)$ which gives us our theorem.

We should remark immediately that our theorem has little content unless $\frac{N-1}{2N} N(T) \geq 2$. This is so since there are always half periods where the theta function has a simple zero. Our proof used the fact that the genus of $S/\langle T\rangle$ is greater than zero. One can ask whether the theorem could hold with $S/\langle T\rangle$ of genus zero. To show that the answer is no, we point out that the Riemann surfaces S_1 and S_2 of the algebraic functions $\omega^3 = (z-z_1)^2(z-z_2)^2 \ldots (z-z_6)^2$ and $\omega^3 = (z-z_1)^2(z-z_2)^2 (z-z_3)^2(z-z_4)(z-z_5)(z-z_6)$ are both surfaces of genus 4 with automorphisms of period 3 such that the quotient is the sphere. The former has the property that $\theta(z,\pi)$ vanishes at a half period to order $2 = \frac{N-1}{2N} N(T)$ while the latter does not. They both have the property that $\theta(z,\pi)$ vanishes to order exactly 2 at the points $K(p_i)$.

III. In this section we shall construct the period matrix of a surface S of genus $g = 2\tilde{g} + t-1$ which permits an automorphism T of order 2 with $2t$ fixed points such that the quotient surface $S/\langle T\rangle$ has genus $\tilde{g} \geq 1$. We shall assume that $t \geq 1$ since the case $t = 0$ has already been discussed in [F2]. Furthermore, if g is odd, then we know from the previous section that the theta function associated with this surface vanishes at points of order two (half periods) to order at least $t/2$. The condition that g be odd is, of course, that t be even. We recall that when $t = 0$, $\tilde{g} \geq 2$, it was already proved in [F1] that the associated theta vanishes to order greater than or equal to two at $2^{\tilde{g}-1}(2^{\tilde{g}-1}-1)$ half periods and the main results of [F2] exploited this fact.

Here we shall be constructing a period matrix of a surface of genus $g = 2\tilde{g} + t-1$ $\tilde{g} \geq 1$ $t \geq 1$ which is a two sheeted branched cover of a fixed surface $\tilde{S}$, of genus $\tilde{g}$, branched over the points $\tilde{p}_1, \ldots, \tilde{p}_{2t}$ on $\tilde{S}$. We have remarked in II that there are $2^{2\tilde{g}}$ such surfaces and here we are considering the one which corresponds to the half period zero.

We begin with a surface $\tilde{S}$, of genus $\tilde{g}$, with $2t$ distinguished points $\tilde{p}_1, \ldots, \tilde{p}_{2t}$ and consider two copies of this surface as indicated in the following, Figure 1.

In Figure 1, 1b is obtained from 1a by rotating 1a about the x_3-axis. By the usual method of cutting and pasting, we slit both copies of $\tilde{S}$ along the indicated arcs joining the pairs of points $\tilde{p}_i$ and paste the surfaces together obtaining a surface of genus $2\tilde{g}+t-1$ as indicated in Figure 2.

It is clear that the surface in Figure 2 has genus $2\tilde{g}+t-1$ and that this surface permits an automorphism T of order 2 (rotation about x_3-axis) with fixed points p_i $i=1,\ldots,2t$. Furthermore, the automorphism T acts on the homology and satisfies $T(\gamma_i)=-\gamma_i$ $i=1,\ldots,t-1$, $T(\delta_i)$ is homologous to $-\delta_i$ $i=1,\ldots,t-1$,

$$T(\gamma_{t+i})=\gamma_{t+\tilde{g}+i} \quad i=0,\ldots,\tilde{g}-1 \text{ and } T(\delta_{t+i})=\delta_{t+\tilde{g}+i}, \quad i=0,\ldots,g-1.$$

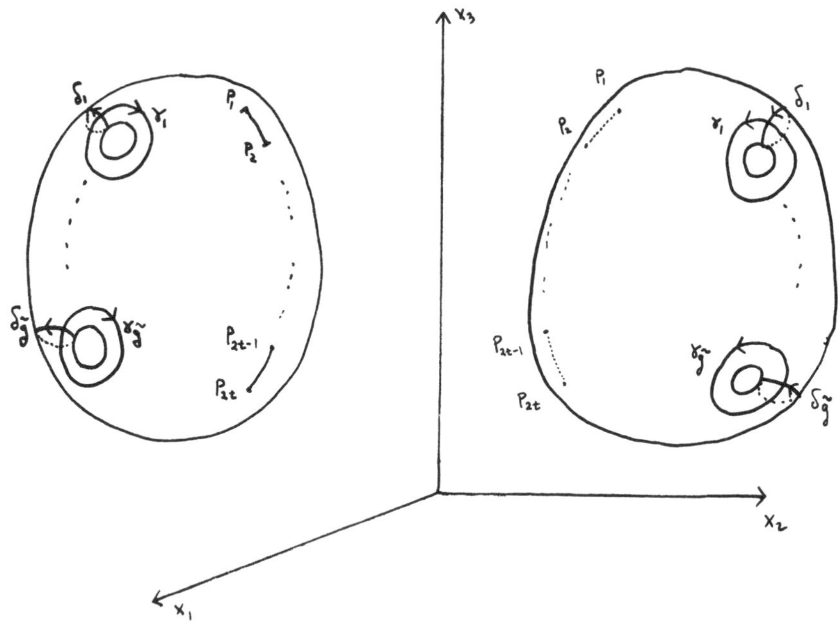

Fig. 1.

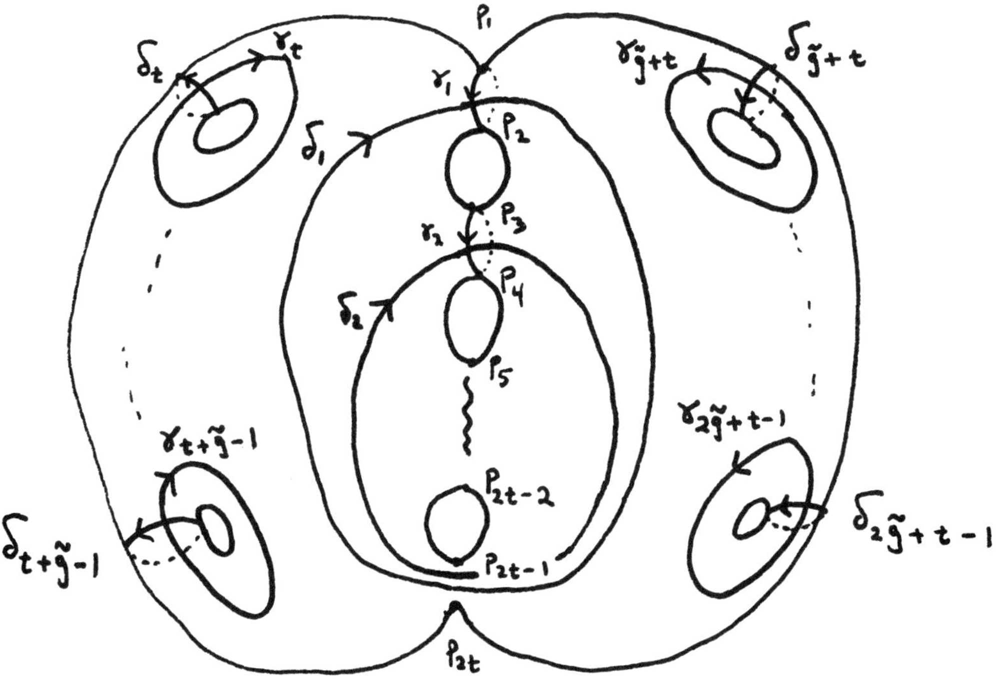

Fig. 2.

We now use the fact [L1] that there is a direct sum decomposition of the vector space of holomorphic differentials on S into subspaces A_1^+ and A_1^- of dimensions $\tilde{g}$ and $\tilde{g}+t-1$. The former is canonically isomorphic to the vector space of holomorphic differentials on $\tilde{S} = S/\langle T\rangle$ and satisfies the condition that for each $\omega \in A_1^+, \omega \circ T = \omega$ while the latter satisfies the condition that for each $\omega \in A_1^-$, $\omega \circ T = -\omega$. We shall call the elements of A_1^+ T-invariant differentials and the elements of A_1^- anti-T-invariant differentials.

Let $\theta_1, \ldots, \theta_{\tilde{g}+t-1}$ be a basis for the anti-T-invariant differentials and let $\theta_{\tilde{g}+t}, \ldots, \theta_{2\tilde{g}+t-1}$ be a basis for the T-invariant differentials such that the period matrix of θ_i $i = 1, \ldots, 2\tilde{g}+t-1$ with respect to the given canonical homology basis has the following form:

	$\gamma_1,\ldots,\gamma_{t-1}$	$\gamma_t,\ldots,\gamma_{\tilde{g}+t-1}$	$\gamma_{\tilde{g}+t},\ldots,\gamma_{2\tilde{g}+t-1}$	$\delta_1,\ldots,\delta_{t-1}$	$\delta_t,\ldots,\delta_{\tilde{g}+t-1}$	$\delta_{\tilde{g}+t},\ldots,\delta_{2\tilde{g}+t-1}$
θ_1 . . . θ_{t-1}	$I_{t-1\times t-1}$	0	0	2L	$M_{t-1\times\tilde{g}}$	$-M_{t-1\times\tilde{g}}$
θ_t . . . $\theta_{\tilde{g}+t-1}$	0	$I_{\tilde{g}\times\tilde{s}}$	$-I_{\tilde{s}\times\tilde{g}}$	2R	$S_{\tilde{s}\times\tilde{g}}$	$-S_{\tilde{s}\times\tilde{g}}$
$\theta_{\tilde{g}+t}$ . . . θ_{2g+t-1}	0	$I_{\tilde{g}\times\tilde{s}}$	$I_{\tilde{s}\times\tilde{g}}$	0	$\tilde{\pi}_{\tilde{s}\times\tilde{g}}$	$\tilde{\pi}_{\tilde{g}\times\tilde{s}}$

It is clear that we can choose such a basis. We now choose a new basis ϕ_i $i = 1, \dots, 2\tilde{g}+t-1$ in the following way: We let $\phi_i = \theta_i$ $i = 1, \dots, t-1$, $\phi_i = \frac{\theta_i + \theta_{\tilde{g}+i}}{2}$ $i = t, \dots, \tilde{g}+t-1$ and $\phi_{\tilde{g}+i} = \frac{\theta_{\tilde{g}+i} - \theta_i}{2}$ $i = t, \dots, \tilde{g}+t-1$. A simple calculation then gives the period matrix:

$$\begin{pmatrix} & 2L & M & -M \\ I, & R & \frac{\tilde{\pi}+s}{2} & \frac{\tilde{\pi}-s}{2} \\ & -R & \frac{\tilde{\pi}-s}{2} & \frac{\tilde{\pi}+s}{2} \end{pmatrix}$$

and since we know that the right hand block is symmetric we have $R = {}^tM$. It is also clear that the matrix $\tilde{\pi}$ which appears is the period matrix of $S/\langle T\rangle$ with respect to the homology basis $S/\langle T\rangle$ inherits from S in the natural way. As a consequence of this we have

THEOREM 6. *A necessary condition for* S *of genus* g, *to be a two sheeted cover of* $\tilde{S}$, *a surface of genus* $\tilde{g}$, *branched over* $2t$ *points on* S, *is that* S *have a period matrix of the form*

$$\begin{pmatrix} & 2L & M & -M \\ I, & {}^tM & \frac{\tilde{\pi}+s}{2} & \frac{\tilde{\pi}-s}{2} \\ & -{}^tM & \frac{\tilde{\pi}-s}{2} & \frac{\tilde{\pi}+s}{2} \end{pmatrix}$$ *where* I *is the* $g \times g$ *identity matrix,* L *is a* $t-1 \times t-1$ *matrix,* M *is a* $t-1 \times \tilde{g}$ *matrix, and* $\tilde{\pi}$ *and* s *are* $\tilde{g} \times \tilde{g}$ *matrices. Furthermore,* $\tilde{\pi}$ *is the period matrix for* $\tilde{S}$ *with respect to the homology basis on* $\tilde{S}$ *determined by the natural map from* $S \to \tilde{S}$.

Recall that there is a natural action on $\mathfrak{S}_g$ by $2g \times 2g$ real matrices $\begin{pmatrix} A & B \\ C & D \end{pmatrix}$ where A, B, C, D are real $g \times g$ matrices and $\mathfrak{S}_g \ni \pi \mapsto (A\pi+B)(C\pi+D)^{-1}$. In particular the Siegel modular group of $2g \times 2g$ matrices

$\begin{pmatrix} A & B \\ C & D \end{pmatrix}$ where A, B, C, D are matrices of integers, A^tB and C^tD are symmetric and $A^tD - B^tC = I$, acts on $\mathfrak{S}_g$ in the above fashion. It is well known that if $\pi \in \mathfrak{S}_g$ is a period matrix of a surface then no conjugate of π under the Siegel modular group is equivalent to a diagonal matrix or even a matrix made up of blocks on the diagonal. There are, however, $2g \times 2g$ integer matrices which act on $\mathfrak{S}_g$ as above which are similar to elements of the Siegel modular group in that we require that A^tB and C^tD be symmetric; however, we relax the requirement that $A^tD - B^tC = I$ to be $A^tD - B^tC = nI$ for some integer $n \geq 1$. The elements of the Siegel modular group are sometimes called linear transformations while the more general type of transformation is called an nth order transformation.

THEOREM 7. *There is a second order transformation of* $\mathfrak{S}_{2\tilde{g}+t-1}$ *which maps the period matrix of* S *to the matrix* $\begin{pmatrix} \tau & 0 \\ 0 & \pi \end{pmatrix}$ *where* $\tau = \begin{pmatrix} L & M \\ {}^tM & s \end{pmatrix}$.

Proof. The proof is by computation. We take $\begin{pmatrix} A & B \\ C & D \end{pmatrix}$ where

$$A = \begin{pmatrix} I_{t-1} & 0 & 0 \\ 0 & I_{\tilde{g}} & -I_{\tilde{g}} \\ 0 & I_{\tilde{g}} & I_{\tilde{g}} \end{pmatrix}$$

$B = C = 0$ and

$$D = \begin{pmatrix} 2I_{t-1} & 0 & 0 \\ 0 & I_{\tilde{g}} & -I_{\tilde{g}} \\ 0 & I_{\tilde{g}} & I_{\tilde{g}} \end{pmatrix}.$$

A simple computation then gives that $A^tD = 2I$ and

$$A \begin{pmatrix} 2L & M & -M \\ {}^tM & \frac{\tilde{\pi}+s}{2} & \frac{\tilde{\pi}-s}{2} \\ -{}^tM & \frac{\tilde{\pi}-s}{2} & \frac{\tilde{\pi}+s}{2} \end{pmatrix} D^{-1} = \begin{pmatrix} L & M & 0 \\ {}^tM & s & 0 \\ 0 & 0 & \tilde{\pi} \end{pmatrix}.$$

COROLLARY 5. *The matrix* $\tau = \begin{pmatrix} L & M \\ {}^tM & S \end{pmatrix}$ *is an element of* $\mathfrak{S}_{\tilde{g}+t-1}$.

We remark at once that the importance of Theorem 7 and Corollary 5 rests in the fact that now one can define a theta function using τ as in Definition 2, and the subsequent theory parallels the theory developed in [F2] where the case of a Riemann surface which has an automorphism of order two with no fixed points was discussed. Before pursuing this discussion we pause for the definition of theta function with characteristic.

DEFINITION 5. A g-characteristic is a $2\times g$ matrix of integers $\begin{bmatrix} \varepsilon \\ \varepsilon' \end{bmatrix} = \begin{bmatrix} \varepsilon_1,\dots,\varepsilon_g \\ \varepsilon_1',\dots,\varepsilon_g' \end{bmatrix}$.

DEFINITION 6. The first order g-dimensional theta function with g-characteristic $\begin{bmatrix} \varepsilon \\ \varepsilon' \end{bmatrix}$, $\theta\begin{bmatrix} \varepsilon \\ \varepsilon' \end{bmatrix}(z,\tau)$ is defined by the following series

$$\theta\begin{bmatrix} \varepsilon \\ \varepsilon' \end{bmatrix}(z,\tau) = \sum_{\eta_1,\dots,\eta_g=-\infty}^{\infty} \exp 2\pi i \left[\frac{1}{2}\sum_{i,j=1}^{g} (\eta_i+\varepsilon_i/2)(\eta_j+\varepsilon_j/2)\tau_{ij} + \sum_{i=1}^{g} (\eta_i+\varepsilon_i/2)(z_i+\varepsilon_i'/2)\right].$$

We now consider the Riemann theta function with characteristic associated with S and the matrix we have constructed. Our next theorem states that the theta functions with matrices

$$\pi = \begin{pmatrix} 2L & M & -M \\ {}^tM & \frac{\tilde{\pi}+S}{2} & \frac{\tilde{\pi}-S}{2} \\ -{}^tM & \frac{\tilde{\pi}-S}{2} & \frac{\tilde{\pi}+S}{2} \end{pmatrix}, \quad \tilde{\pi} \text{ and } \tau = \begin{pmatrix} L & M \\ {}^tM & S \end{pmatrix} \text{ are}$$

related by a simple formula.

THEOREM 8. *Let* $\begin{bmatrix} g & \varepsilon & \delta \\ g' & \varepsilon' & \delta' \end{bmatrix}$ *be a* $2\tilde{g}+t-1$ *characteristic where*

$$\begin{bmatrix} g \\ g' \end{bmatrix} = \begin{bmatrix} g_1,\ldots,g_{t-1} \\ g_1',\ldots,g_{t-1}' \end{bmatrix}, \quad \begin{bmatrix} \varepsilon \\ \varepsilon' \end{bmatrix} = \begin{bmatrix} \varepsilon_1,\ldots,\varepsilon_{\tilde{g}} \\ \varepsilon_1',\ldots,\varepsilon_{\tilde{g}}' \end{bmatrix}, \text{ and } \begin{bmatrix} \delta \\ \delta' \end{bmatrix} = \begin{bmatrix} \delta_1,\ldots,\delta_{\tilde{g}} \\ \delta_1',\ldots,\delta_{\tilde{g}}' \end{bmatrix}.$$

Then

$$\theta\begin{bmatrix} g & \varepsilon & \delta \\ g' & \varepsilon' & \delta' \end{bmatrix}(z_1,\ldots,z_{2\tilde{g}+t-1},\pi) = \sum_{J\epsilon(\mathbf{Z}/2)^{\tilde{g}}} \theta\begin{bmatrix} g & \frac{\varepsilon-\delta}{2}+J \\ g' & \varepsilon'-\delta' \end{bmatrix}(w_1,\ldots,w_{\tilde{g}+t-1},2\tau)$$

$$\theta\begin{bmatrix} \frac{\varepsilon+\delta}{2}+J \\ \varepsilon'+\delta' \end{bmatrix}(w_{\tilde{g}+t},\ldots,w_{2\tilde{g}+t-1},2\tilde{\pi})$$

where

$$z_1 = w_1,\ldots,z_{t-1} = w_{t-1} \quad z_t = \frac{w_t+w_{\tilde{g}+t}}{2},\ldots,z_{\tilde{g}+t-1} = \frac{w_{\tilde{g}+t-1}+w_{2\tilde{g}+t-1}}{2}$$

$$z_{\tilde{g}+t} = \frac{w_{\tilde{g}+t}-w_t}{2},\ldots,z_{2\tilde{g}+t-1} = \frac{w_{2\tilde{g}+t-1}-w_{\tilde{g}+t-1}}{2}.$$

Furthermore, if $\begin{bmatrix} \varepsilon \\ \varepsilon' \end{bmatrix} = \begin{bmatrix} \delta \\ \delta' \end{bmatrix}$ *the above reduces to*

$$\theta\begin{bmatrix} g & \varepsilon & \varepsilon \\ g' & \varepsilon' & \varepsilon' \end{bmatrix}(z,\pi) = \sum_{J\epsilon(\mathbf{Z}/2)^{\tilde{g}}} (-1)^{(\varepsilon+J)\cdot\varepsilon'}\,\theta\begin{bmatrix} g & J \\ g' & 0 \end{bmatrix}(w_1,\ldots,w_{\tilde{g}+t-1},2\tau)$$

$$\theta\begin{bmatrix} \varepsilon+J \\ 0 \end{bmatrix}(w_{\tilde{g}+t},\ldots,w_{2\tilde{g}+t-1},2\tilde{\pi}).$$

Proof. The proof of the above is a calculation essentially the same as the calculations made in [F2] and is therefore omitted.

It is an easy exercise to show that for theta functions with characteristic $\theta\begin{bmatrix}\varepsilon\\ \varepsilon'\end{bmatrix}(-z,\pi) = (-1)^{\sum \varepsilon_i \varepsilon_i'}\theta\begin{bmatrix}\varepsilon\\ \varepsilon'\end{bmatrix}(z,\pi)$, and that the theta function $\theta\begin{bmatrix}\varepsilon\\ \varepsilon'\end{bmatrix}(z,\pi)$ is simply the theta function with $\begin{bmatrix}\varepsilon\\ \varepsilon'\end{bmatrix} = \begin{bmatrix}0\\ 0\end{bmatrix}$ where z is translated by a half period in $C^g/\langle I,\pi\rangle$ multiplied by an exponential function. Hence, an immediate corollary of Theorem 8 is

COROLLARY 6.

$$\theta\begin{bmatrix}g & \varepsilon & \varepsilon\\ g' & \varepsilon' & \varepsilon'\end{bmatrix}\left(0,\ldots,0,\frac{w_{\tilde g+t}}{2},\ldots,\frac{w_{2\tilde g+t-1}}{2},\frac{w_{\tilde g+t}}{2},\ldots,\frac{w_{2\tilde g+t-1}}{2},\pi\right) = 0$$

whenever $\begin{bmatrix}g\\ g'\end{bmatrix}$ *is such that* $\sum g_i \cdot g_i' \equiv 1 \bmod 2$.

Proof. According to Theorem 8 we have

$$\theta\begin{bmatrix}g & \varepsilon & \varepsilon\\ g' & \varepsilon' & \varepsilon'\end{bmatrix}\left(w_1,\ldots,w_{t-1},\frac{w_t+w_{\tilde g+t}}{2},\ldots,\frac{w_{\tilde g+t-1}+w_{2\tilde g+t-1}}{2},\frac{w_{\tilde g+t}-w_t}{2},\ldots,\right.$$

$$\left.\frac{w_{2\tilde g+t-1}-w_{\tilde g+t-1}}{2},\pi\right) = \sum_{J\in(\mathbf{Z}/2)^{\tilde g}} (-1)^{(\varepsilon+J)\cdot\varepsilon'}\,\theta\begin{bmatrix}g & J\\ g' & 0\end{bmatrix}(w_1,\ldots,w_{\tilde g+t-1},2\tau)$$

$$\theta\begin{bmatrix}\varepsilon+J\\ 0\end{bmatrix}(w_{\tilde g+t},\ldots,w_{2\tilde g+t-1},2\tilde\pi)\ .$$

Hence, if we set $w_1 = \ldots = w_{\tilde g+t-1} = 0$ and let $g\cdot g' \equiv 1 \bmod 2$, we get the result.

We conclude with the following remarks: The case $\tilde g = 0$ is the usual case of a hyperelliptic surface and in this case the period matrix is simply the matrix $2L$. In this case one can compute from the matrix $2L$ over which points on the sphere the branching takes place. One would therefore expect that when $\tilde g > 0$ the information over which points of the surface $\tilde S$ does the branching take place should be contained somehow in the matrix $\begin{bmatrix}2L & 2M\\ {}^t2M & 2S\end{bmatrix}$.

We have seen in part II of this paper that the theta function with matrix π vanishes at half periods to order at least $\frac{t}{2}$. If we take $t=4k$ and take into account the remarks preceding Corollary 6, we have at least $2^{2\tilde{g}}$ even characteristics $\begin{bmatrix} e \\ e' \end{bmatrix}$ for which $\theta\begin{bmatrix} e \\ e' \end{bmatrix}(0,\pi) = 0$. This together with Theorem 8 would give equations connecting the theta functions with matrices $2\tilde{\pi}$ and 2τ. In the case of a two-sheeted smooth covering these equations were the key to obtaining the Schottky relations. It would be interesting to see whether any information or proportionalities would follow from the equations here.

REFERENCES

[A1]. Accola, R. D. M., Riemann surfaces with automorphism groups admitting partitions, Proc. of the A.M.S., (1969), pp. 477-482.

[A2]. ———, Strongly branched coverings of closed Riemann surfaces, Proc. of A.M.S. (1970), pp. 315-322.

[F1]. Farkas, Hershel M., Automorphisms of compact Riemann surfaces and the vanishing of theta constants, Bull. of A.M.S. (1967), pp. 231-232.

[F2]. ———, On the Schottky relation and its generalization to arbitrary genus, Annals of Math. (1970), pp. 57-86.

[H]. Hurwitz, A., Uber algebraische gebilde mit eindentigen transformationen in sich, Mathematische Werke Bd. 1, Birkhauser, Basil (1932), pp. 241-259.

[L1]. Lewittes, J., Automorphisms of Riemann surfaces, Amer. Journal of Math. (1963), pp. 734-752.

[L2]. ———, Riemann surfaces and the theta function, Acta Mathematica (1964).

THE STRUCTURE OF $PSL_2(R)$; R, THE RING OF INTEGERS IN A EUCLIDEAN QUADRATIC IMAGINARY NUMBER FIELD

Benjamin Fine

Introduction

1. If R is any commutative ring, $SL_2(R)$ is the special linear 2×2 group over R; the group of 2×2 matrices with entries from R and determinant $+1$. $PSL_2(R)$ denotes the projective special linear group $SL_2(R)/Z(SL_2)$ where $Z(SL_2)$ is the center of SL_2. If R is the ring of integers in a quadratic imaginary number field $Q(\sqrt{-d})$, with d a positive square-free integer, it has been conjectured that the group theoretical structure of $PSL_2(R)$ can be described in terms of free products with amalgamated subgroups [7]. We will denote the ring of integers in $Q(\sqrt{-d})$ as $I(-d)$. We will examine the structure of $PSL_2(I(-d))$ and show that such a description is possible in the five cases where $I(-d)$ is a Euclidean domain; $d = 1, 2, 3, 7, 11$.

2. P. M. Cohn has shown how to obtain presentations of $SL_2(R)$ for certain types of rings R [2]. These results can be applied to the rings $I(-d)$ when $I(-d)$ is a Euclidean domain, $d = 1, 2, 3, 7, 11$ [2]. This is done in Section 1. There, finite presentations for the five groups in question are derived. The finiteness of the presentation is deduced from the finite generation over the integers and the finiteness of units in these rings. Each of these groups can be presented on 3 generators, however for simplicity in $SL_2I(-1)$ and $SL_2I(-3)$ we have included a fourth generator. For $d = 2, 7, 11$ the 3 generators are

$$a = \begin{pmatrix} 0 & -1 \\ 1 & 0 \end{pmatrix}, \quad t = \begin{pmatrix} 1 & 1 \\ 0 & 1 \end{pmatrix}, \quad u = \begin{pmatrix} 1 & \omega \\ 0 & 1 \end{pmatrix}$$

where $\omega = i\sqrt{2}$ for $d = 2$, $\omega = \frac{1+i\sqrt{7}}{2}$ for $d = 7$, and $\omega = \frac{1+i\sqrt{11}}{2}$ for $d = 11$.

For $d = 1$, the 4 generators for $SL_2I(-1)$ are

$$a = \begin{pmatrix} 0 & -1 \\ 1 & 0 \end{pmatrix}, \quad t = \begin{pmatrix} 1 & 1 \\ 0 & 1 \end{pmatrix}, \quad u = \begin{pmatrix} 1 & i \\ 0 & 1 \end{pmatrix}, \quad \ell = \begin{pmatrix} i & 0 \\ 0 & -i \end{pmatrix},$$

while for $d = 3$, the 4 generators for $SL_2I(-3)$ are

$$a = \begin{pmatrix} 0 & -1 \\ 1 & 0 \end{pmatrix}, \quad t = \begin{pmatrix} 1 & 1 \\ 0 & 1 \end{pmatrix}, \quad u = \begin{pmatrix} 1 & \omega \\ 0 & 1 \end{pmatrix}, \quad \ell = \begin{pmatrix} \omega^2 & 0 \\ 0 & \omega \end{pmatrix} \text{ with } \omega = \frac{-1+i\sqrt{3}}{2}.$$

The projective groups $PSL_2I(-d)$ correspond to the groups formed by letting each matrix equal its negative. These can also be considered as groups of linear fractional transformations $Z' = \frac{az+b}{cz+d}$ of the extended complex plane. Presentations for the projective groups are also derived in Section 1.

Now $E_2(R)$ denotes the elementary 2×2 matrix group over ring R. That is the subgroup of $GL_2(R)$ generated by all matrices of the form $\begin{pmatrix} x & 1 \\ -1 & 0 \end{pmatrix}$, $x \in R$. When R is commutative and has a Euclidean algorithm then $E_2(R) \equiv SL_2(R)$ [3]. In a related result in Section 1 we will show that $\forall\, d \neq 1, 2, 3, 7, 11$, the elementary matrix groups $E_2(I(-d))$ are isomorphic. Further if $d \neq 1, 2, 3, 7, 11$ it is found that

$$E_2(I(-d)) = \{j, a, t, u;\ j^2 = 1,\ j \text{ central},\ tu = ut\ a^2 = (at)^3 = j\}$$

with $a = \begin{pmatrix} 0 & -1 \\ 1 & 0 \end{pmatrix}$, $t = \begin{pmatrix} 1 & 1 \\ 0 & 1 \end{pmatrix}$, $u = \begin{pmatrix} 1 & \omega \\ 0 & 1 \end{pmatrix}$ where $1, \omega$ constitute an integral base for $I(-d)$.

3. It is well known that the group of unimodular linear fractional transformations $PSL_2(Z)$ is a free product of a 2-cycle and a 3-cycle [5]. H. Waldinger [9] has extended this to the Picard group $PSL_2(I(-1))$. He has shown that $PSL_2(I(-1))$ contains a subgroup H of finite index which is a free product with amalgamated subgroup. In Section 2 this extension is continued to the five groups considered in Section 1. We show

that if $d = 1, 2, 3, 7, 11$ then $PSL_2(I(-d))$ is a semidirect product of a group G_d by a cyclic group. If $d = 1, 3$ this cyclic group is finite; of order 2 if $d = 1$, and trivial for $d = 3$. If $d = 2, 7, 11$, it is infinite. Further for each of the five d, the structure of G_d is describable in terms of free products with amalgamated subgroup. For $d = 2, 7, 11$, G_d is itself a free product of 2 groups with an amalgamated subgroup. However in each of these cases, G_d can be described more simply as a tree product of groups U_{d_n}, $n = 0, \pm 1, \pm 2, \ldots$ with amalgamations. These groups U_{d_n} are in turn free products of finite groups with amalgamated subgroups. The concept of a tree product and several of its properties are also discussed in Section 2. For $d = 1, 3$, G_d contains a subgroup H_d of finite index, which is a free product with amalgamated subgroup. These results are then used to discuss the abstract word problem for these five groups.

It is also proven in Section 2 that for $d \neq 1, 2, 3, 7, 11$ $E_2(I(-d))$ is a free product of 2 groups M_1 and M_2 with a subgroup M amalgamated. $M_1 \cong$ modular group of matrices $SL_2(Z)$ while $M_2 \cong$ direct product of a free abelian group of rank 2 and a cyclic group of order 2, $(Z \times Z \times Z_2)$. The amalgamated subgroup M is isomorphic to the direct product of an infinite cyclic group, and a cyclic group of order 2 $(Z \times Z_2)$.

4. We let $\mathfrak{F}_2$ denote the fundamental group of a Riemann surface of genus 2. This has the presentation $a, b, c, d;\ abcda^{-1} b^{-1} c^{-1} d^{-1} = 1$. The fundamental groups of Riemann surfaces of finite genus > 2 appear as subgroups of finite index in $\mathfrak{F}_2$ [6]. J. Mennicke has given a representation of $\mathfrak{F}_2$ in the Picard group [6]. His results are geometric and the groups obtained are non-loxodromic. Using our results we can give a method by which we obtain algebraically 30 distinct faithful representations of $\mathfrak{F}_2$ in the Picard group. Further the subgroups of $PSL_2(I(-1))$ obtained as isomorphic images of $\mathfrak{F}_2$ contain loxodromic substitutions and thus are non-Fuchsian. The results in this subsection will be published later.

Section 1

1. If R is any ring, let U(R) denote the group of units of R. For $x \in R$

$$E(x) = \begin{pmatrix} x & 0 \\ -1 & 0 \end{pmatrix} \text{ and if } \alpha, \beta, \mu \in U(R)$$

$$[\alpha, \beta] = \begin{pmatrix} \alpha & 0 \\ 0 & \beta \end{pmatrix}, \quad D(\mu) = \begin{pmatrix} \mu & 0 \\ 0 & \mu^{-1} \end{pmatrix}.$$

Let $E_2(R)$ be the subgroup of $GL_2(R)$ generated by all $E(x)$, $x \in R$ and $D_2(R)$ the subgroup of $GL_2(R)$ generated by all $[\alpha, \beta]$, $\alpha, \beta \in U(R)$. $GE_2(R)$ denotes the subgroup generated by $E_2(R)$ and $D_2(R)$. If R has a Euclidean algorithm it follows that $GE_2 \cong GL_2$ [1]. Further if R is commutative as well as Euclidean $E_2(R) \cong SL_2(R)$ [1].

The result of Cohn's that is of the most importance here is the following [2].

PROPOSITION. *Let* R *be a subring of the complex numbers* C, *with the usual absolute value* ∋ *range of values is well ordered and if* $a \in R$, *with* $|a| < 2$ *then either* $a = 0$ *or* $|a| = 1$ *with* $a \in U(R)$ *or* $|a| = \sqrt{\rho}$ *with* $\rho = 2$ *or* 3 *and then* $\bar{a} \in R$. *Then a complete set of relations for* $E_2(R)$ *in terms of the generators* $E(x)$, $x \in R$ *is*

(1) $E(x)E(0)E(y) = -E(x+y)$, $x, y \in R$.

(2) $E(x)D(\mu) = D(\mu^{-1})E(\mu x \mu)$, $x \in R, \mu \in U(R)$.

(3) $(E(\bar{a})E(a))^p = -I \ \forall \ a \in R$ *with* $|a| = \sqrt{\rho}$, $\rho = 2$ *or* 3.

(4) $E(\mu)E(\mu^{-1})E(\mu) = -D(\mu)$, $\mu \in U(R)$.

Now if $d = 1, 2, 3, 7, 11$ the ring of integers $I(-d)$ in $Q(\sqrt{-d})$ is Euclidean [2], so we have $E_2(I(-d)) \cong SL_2(I(-d))$. Applying the proposition to these rings we get

PROPOSITION. *If* $d = 1, 2, 3, 7, 11$, $SL_2 I(-d)$ *has the presentation; generated by all* $E(x)$, $x \in I(-d)$ *with relations*

(1) $E(x)E(0)E(y) = -E(x+y)$,

(2) $E(x)D(\mu) = D(\mu^{-1})E(\mu\times\mu)$

(3) $(E(\overline{\alpha})E(\alpha)) = -I$ *for each* $\alpha \in I(-d)$ *with* $|\alpha| = \sqrt{\rho}$, $\rho = 2$ or 3

(4) $E(\mu)E(\mu^{-1})E(\mu) = -D(\mu)$ *for each* $\mu \in U(R)$.

Now we can proceed to find finite presentations for these five groups.

PROPOSITION A.

$$SL_2I(-1) \cong \{a,\ell,t,u,j;\ a^2 = \ell^2 = (at)^3 = (a\ell)^2 = (t\ell)^2 = (u\ell)^2 = (ua\ell)^3 = j$$
$$j^2 = 1,\ j \text{ central } tu = ut\}$$

with $j = -I$, $a = \begin{pmatrix} 0 & -1 \\ 1 & 0 \end{pmatrix}$, $t = \begin{pmatrix} 1 & 1 \\ 0 & 1 \end{pmatrix}$, $u = \begin{pmatrix} 1 & i \\ 0 & 1 \end{pmatrix}$, $\ell = \begin{pmatrix} i & 0 \\ 0 & -i \end{pmatrix}$.

PROPOSITION B.

$$SL_2(I(-3)) \cong \{j,t,u,\ell,a;\ j^2 = 1,\ j \text{ central } tu = ut,$$
$$\ell^3 = 1,\ a^2 = (a\ell)^2 = (at)^3 = (ua\ell)^3 = j,\ \ell^{-1}u\ell = t,\ \ell^{-1}t\ell = t^{-1}u^{-1}$$

with

$$j = -I,\ \ell = \begin{pmatrix} \omega^2 & 0 \\ 0 & \omega \end{pmatrix},\ t = \begin{pmatrix} 1 & 1 \\ 0 & 1 \end{pmatrix},\ u = \begin{pmatrix} 1 & \omega \\ 0 & 1 \end{pmatrix},\ a = \begin{pmatrix} 0 & -1 \\ 1 & 0 \end{pmatrix}$$

with $\omega = \dfrac{-1+i\sqrt{3}}{2}$.

Equivalently

$$\cong \{j,a,d,e;\ j^2 = 1,\ j \text{ central},\ a^2 = (da)^2 = (ade)^3 = j,\ d^3 = e^3 = (dae^{-1})^3 = 1,$$
$$(e^{-1}ded)a(ed^{-1}e^{-1}d^{-1})a^{-1} = 1$$

where

$$j = -I,\quad a = \begin{pmatrix} 0 & 1 \\ -1 & 0 \end{pmatrix},\quad d = \begin{pmatrix} \omega & 0 \\ 0 & \omega^2 \end{pmatrix},\quad e = \begin{pmatrix} -1 & 1 \\ -1 & 0 \end{pmatrix}.$$

(the second presentation will be used later).

PROPOSITION C. $SL_2I(-2)$ *has the presentation*

$$\{j,t,u,a;\ j^2=1,\ j \text{ central},\ tu=ut,\ a^2=(at)^3=(u^{-1}a^{-1}ua)^2=j\}$$

with $j=-I,\ t=\begin{pmatrix}1&1\\0&1\end{pmatrix},\ a=\begin{pmatrix}0&-1\\1&0\end{pmatrix},\ u=\begin{pmatrix}1&\omega\\0&1\end{pmatrix},\ \omega=i\sqrt{2}.$

PROPOSITION D. $SL_2I(-7)$ *has the presentation*

$$\{j,t,u,a;\ j^2=1,\ j \text{ central}\ \ tu=ut,\ a^2=(at)^3-(u^{-1}a^{-1}uat)^2=j\}$$

$$j=-I,\ t=\begin{pmatrix}1&1\\0&1\end{pmatrix},\ u=\begin{pmatrix}1&\omega\\0&1\end{pmatrix},\ a=\begin{pmatrix}0&-1\\0&0\end{pmatrix},\ \omega=\frac{1+i\sqrt{7}}{2}.$$

PROPOSITION E. $SL_2I(-11)$ *has the presentation*

$$\{j,t,u,a;\ j^2=1,\ j \text{ central},\ tu=ut,\ a^2=(at)^3=(u^{-1}auat)^3=j\}$$

$$j=-I,\ a=\begin{pmatrix}0&-1\\1&0\end{pmatrix},\ t=\begin{pmatrix}1&1\\0&1\end{pmatrix},\ u=\begin{pmatrix}1&\omega\\0&1\end{pmatrix},\ \omega=\frac{1+i\sqrt{11}}{2}.$$

Proof. Let $1,\omega$ be an integral base for $I(-d)$. So $\omega=i,\ i\sqrt{2},\ \frac{-1+i\sqrt{3}}{2},\ \frac{1+i\sqrt{7}}{2},\ \frac{1+i\sqrt{11}}{2}$ respectively in the five cases.

Then the units are $\pm1,\pm i$ in $I(-1),\pm1,\ \pm\omega,\pm\omega^2$ in $I(-3)$ and ±1 in $I(-2)$, $I(-7)$, $I(-11)$, while the elements of norm between 1 and 2 are $1\pm i,-1\pm i$ in $I(-1)$, $\pm(1+\omega)\omega^n, n=0,1,2$ in $I(-3)$, $\pm\omega,1\pm\omega,-1\pm\omega$ in $I(-2)$, $\pm\omega,1-\omega,-1+\omega$ in $I(-7)$ and $\pm\omega,1-\omega,-1+\omega$ in $I(-11)$. Using this data we can set up relation systems for these groups from the proposition. From the relations $E(x)E(0)E(y)=-E(x+y)$ we can reduce the generators in each case to $E(0)$, $E(1)$, $E(\omega)$, and we can reduce the relations through a long calculation to the below finite set involving just $E(0)$, $E(1)$, $E(\omega)$.

(1) $(-I)^2=I,\ -I$ central, $E(0)^2=E(1)^3=-I$

(2) $E(1)E(0)E(\omega)=E(\omega)E(0)E(1)$

(3) $(E(\overline{\alpha})E(\alpha))^p=-I$ for each $\alpha\in I(-d)$ with $|\alpha|=\sqrt{p}$. $p=2$ or 3 where $E(\alpha)$ and $E(\overline{\alpha})$ are written in "standard forms" involving $E(1)$, $E(0)$, $E(\omega)$.

(4)* $D(i)^2 \equiv (E(0)D(i))^2 = -I,\ (D(i)E(i))^3 = I$ for $I(-1),\ \omega = i$.

(4)** $D(\omega)^3 = (E(0)D(\omega))^2 = -I$ for $I(-3),\ \omega = \frac{-1+i\sqrt{3}}{2}$.

The details are carried out in [4].

We then rewrite the presentations to include as generators the generators of the modular group $a = \begin{pmatrix} 0 & -1 \\ 1 & 0 \end{pmatrix} = E(0)^{-1}$, $t = \begin{pmatrix} 1 & 1 \\ 0 & 1 \end{pmatrix} = E(0)E(1)^{-1}$ and the translation group $u = \begin{pmatrix} 1 & \omega \\ 0 & 1 \end{pmatrix} = E(0)E(\omega)^{-1}$ to obtain our results.

2. Now $PSL_2(R)$ the projective special linear group is $SL_2(R)$ modulo its center $Z(SL_2(R))$. It is easy to see that the centers of each of these groups is precisely $\{I, -I\} = \{1, j\}$. Then we get the following.

COROLLARY A. $PSL_2I(-1)$ (Picard Group) $= \{a,\ell,t,u;\ a^2 = \ell^2 = (a\ell)^2 = (t\ell)^2 = (u\ell)^2 = (at)^3 = (ua\ell)^3 = 1,\ tu = ut\}$.

COROLLARY B. $PSL_2I(-3) = \{a,d,e;\ a^2 = (da)^2 = (ade)^3 = d^3 = e^3 = (dae^{-1})^3 = e^{-1}dedaed^{-1}e^{-1}d^{-1}a = 1\}$.

COROLLARY C. $PSL_2I(-2) = \{a,t,u;\ tu = ut,\ a^2 = (at)^3 = (u^{-1}aua)^2 = 1\}$.

COROLLARY D. $PSL_2I(-7) = \{a,t,u;\ tu = ut,\ a^2 = (at)^3 = (u^{-1}auat)^2 = 1\}$.

COROLLARY E. $PSL_2I(-11) = \{a,t,u;\ tu = ut,\ a^2 = (at)^3 = (u^{-1}auat)^3 = 1\}$.

These all follow from letting $j = 1$ in the respective presentations for $SL_2I(-d)$.

3. We have only considered $d = 1, 2, 3, 7, 11$ since these are the only cases $d > 0$, where $I(-d)$ is Euclidean; cases where it follows $E_2 \cong SL_2$. However, for $d > 0$, d squarefree, $d \neq 1, 2, 3, 7, 11$ we get

THEOREM. *If* d *is a squarefree positive integer* $d \neq 1, 2, 3, 7, 11$ *then* $E_2I(-d) \cong \{j,a,t,u;\ j^2 = 1,\ j$ central, $tu = ut,\ a^2 = (at)^3 = j\}$ *with* $a = \begin{pmatrix} 0 & -1 \\ 1 & 0 \end{pmatrix}$,

$t = \begin{pmatrix} 1 & 1 \\ 0 & 1 \end{pmatrix}$, $u = \begin{pmatrix} 1 & \omega \\ 0 & 1 \end{pmatrix}$, $j = -I$ *where* $\omega = i\sqrt{d}$ *if* $d \not\equiv 3 \bmod 4$ *or* $= \frac{1+i\sqrt{d}}{2}$ *if* $d \equiv 3 \bmod 4$.

Proof. If $d \neq 1, 2, 3, 7, 11$, $I(-d)$ has no units other than ± 1 and no elements of norm between 1 and 2 [4]. Further an integral base for $I(-d)$ is 1, where

$$\omega = i\sqrt{d} \text{ if } d \not\equiv 3 \bmod 4 \text{ or}$$

$$\omega = \frac{1+i\sqrt{d}}{2} \text{ if } d \equiv 3 \bmod 4 \ [4].$$

Applying the proposition we get the general presentation in terms of generators $E(x)$

(1) $$E(x)E(0)E(y) = -E(x+y)$$

(2) $$-I \text{ central}, \ (-I)^2 = I$$

(3) $$E(1)^3 = -I, \ E(-1)^3 = I$$

which we can reduce as before [4] to

$$\{E(0), E(1), E(\omega);\ E(0)^2 = E(1)^3 = -I,\ -I \text{ central},\ (-I)^2 = I$$
$$E(1)E(0)E(\omega) = E(\omega)E(0)E(1)\}.$$

Letting $a = E(0)^{-1}$, $t = E(0)E(1)^{-1}$, $u = E(0)E(\omega)^{-1}$ we get our result.

To extend this to $SL_2I(-d)$ for general positive d, the relationship between E_2 and SL_2 in the non-Euclidean case must be determined.

4. We will note that Swan [8], using a completely different approach has given a method for computing $SL_2(I(-d))$ in general. His work is geometric. To calculate he employs a work of Bianchi, who found fundamental domains for these groups. The difficulty of the calculation increases with the discriminant. His results agree of course in the Euclidean cases. In addition he gives presentations for $SL_2I(-5)$, $SL_2I(-6)$ and $SL_2I(-10)$.

Section 2

1. It is well known that the group of unimodular matrices $SL_2(Z)$ is a free product of a cyclic group of order 6 and a cyclic group of order 4 with a 2-cycle amalgamated [5]. If we then look at the projective group $PSL_2(Z)$, the unimodular group of linear fractional transformations $z' = \frac{az+b}{cz+d}$, $ad-bc = 1$ with integral entries, it becomes a free product of a 2-cycle and a 3-cycle [5]. H. Waldinger [9] has extended this to the Picard group $PSL_2I(-1)$. He has shown that $PSL_2I(-1)$ contains a subgroup H of finite index which is a free product with amalgamated subgroup. J. P. Serre has conjectured that $PSL_2I(-d)$ $\forall$ quadratic imaginary fields $Q(\sqrt{-d})$ can be described in terms [7] of free products with amalgamations. In this chapter we will extend the results to each of the 5 groups discussed in Section 1. Basically we will show that $PSL_2(I(-d))$, $d = 1,2,3,7,11$ is a split extension of a group G_d by a cyclic group; finite for $d = 1,3$, infinite otherwise. Further in each case, the structure of G_d is given in terms of free products with amalgamations. We will also show a similar statement for $E_2(I(-d))$, $d \neq 1,2,3,7,11$. As a by-product of this, we will be able to say something about the abstract word problem in each of the 5 groups.

2. In this section we will introduce the concept of a tree product with amalgamations and then give some of its outstanding properties.

Now say for each n, $n = 0, \pm 1, \pm 2, \ldots$ we have a group G_n with the presentation

$$G_n \cong \left\{ g_{1_n}, g_{2_n}, \ldots; R_{1_n}, R_{2_n}, \ldots \right\} .$$

Let $J_{1_n}, J_{2_n}, \ldots, J_{k(n)_n}$, $K_{1_n}, K_{2_n}, \ldots, K_{\ell(n)_n}$ be 2 sets of words in the generators $\{g_{i_n}\}$ of G_n. Say J_n is the subgroup of G_n generated by the J_{i_n}'s and K_n the subgroup generated by the K_{i_n}.

Then if $k(n) = \ell(n-1)$ $\forall n$ and $\forall n$ the correspondences $J_{i_n} \leftrightarrow K_{i_{(n-1)}}$, $i = 0, \ldots, \ell(n)$ induce isomorphisms between J_n and K_{n-1} the group G with the presentation

$$G \cong \left\{ g_{1_n}, g_{2_n}, \dots \; n = 0, \pm 1, \pm 2, \dots ; \; R_{1_n}, R_{2_n}, \dots \; n = 0, \pm 1, \dots \right.$$
$$\left. J_{i_n} = K_{i_{(n-1)}}, \quad i = 0, \dots, \ell(n), \; n = 0, \pm 1, \dots \right\}$$

is called the *tree product* of the groups G_n with G_n amalgamated to G_{n+1} via the subgroup K_n.[1] A tree product with amalgamated subgroups reduces to a free product of 2 groups with an amalgamated subgroup.

Our results on tree products depend on the following:

THEOREM. *If* G *is the tree product of groups* G_n *with* G_n *amalgamated to* G_{n+1} *via* K_n *(that is* G *has the presentation above and* $G_n \cong \{g_{1_n}, g_{2_n}, \dots ; R_{1_n}, R_{2_n}, \dots\}$*); then the subgroup of* G *generated by* $g_{1_n}, g_{2_n} \dots$ *is* G_n *(has the presentation* $\{g_{1_n}, g_{2_n}, \dots ; R_{1_n}, R_{2_n}, \dots\}$.

Proof. If all but a finite number of the groups G_n were trivial, then G would be a finite string of free products with amalgamations.

Then the result would follow through induction directly from the corresponding property holding in the free product of 2 groups with an amalgamated subgroup.

Also in a finite string it follows as the above that if G is the tree product of groups G_k, $k = 1, \dots, m$, with G_k amalgamated to G_{k+1} via H_k, $k = 1, \dots, m$, the subgroup of G generated by $\{g_{i_1}, g_{i_2}, \dots\}$ $i = 1, \dots, \ell$ (where $G_k = \{g_{1_k}, g_{2_k}, \dots ; R_{1_k}, R_{2_k}, \dots\}$), $\ell \leq n$ has a presentation as a tree product of groups G_i, $i = 1, \dots, \ell$, with G_i amalgamated to G_{i+1} via H_i, with the same amalgamations as in G.

The carryover to the infinite case is accomplished through the following lemma.

---

1 This is not the most general type of tree product but will suffice for our purposes.

LEMMA. *Let* $S_0 \subset S_1 \subset \ldots \subset S_n \subset \ldots$ *be an increasing sequence of symbols,* $T_0 \subset T_1 \subset \ldots \subset T_n$ *words in* S_i. *Say* $G_n \{S_n; T_n\}$ *and say for each* n *the subgroup generated by* S_{n-1} *in* G_n *has the presentation* $\{S_{n-1}; T_{n-1}\}$. *Then let* $S = \bigcup_{i=0}^{\infty} S_i$, $T = \bigcup_{i=0}^{\infty} T_i$. *If* $G = \{S; T\}$ *the subgroup of* G *generated by the symbols of* S_n *is* G_n.

Proof. This is given as a problem in [5] and the proof as well as the remaining details of this theorem can be found in [4].

COROLLARY. *If* G *is a tree product of groups* G_n *with* G_n *amalgamated to* G_{n+1} *via* K_n, *then* G *is the free product of 2 groups with an amalgamated subgroup.*

Proof. The two groups are formed by splitting the infinite string at any finite n.

COROLLARY. *Say* G *is a tree product of groups* G_n *with* G_n *amalgamated to* G_{n+1} *via* K_n. *Say* $\forall n$ *the word problems for* G_n *and* H_n *are solvable and the extended word problem for* $G_n \bmod H_n$ *and* $G_n \bmod H_{n-1}$ *is solvable and* $\exists$ *effective procedure for computing the images of the isomorphisms given by the amalgamations, then the word problem for* G *is solvable.*

Proof. This follows immediately in finite strings from the similar property in the free product of 2 groups with amalgamated subgroup. But any word W in G lies in a finite string, and from the corollary above a presentation for the finite string is the finite tree product.

3. Now we will show that the structure of $PSL_2(I(-d))$, $d = 1,2,3,7,11$ is describable in terms of free products with amalgamations. We will also show that in a way this structure is determined by the way the unimodular group $PSL_2(Z)$ is embedded in $PSL_2(I(-d))$.

THEOREM A. *If* $\Gamma_1 = PSL_2(I(-1))$ *then* Γ_1 *is a semidirect product of a group* G_1 *and a cyclic group of order* 2. *Further* G_1 *contains a subgroup* H *of index* 24 *which is a free product of 2 isomorphic groups* H_1, H_2 *with an amalgamated subgroup. For* $i = 1,2$, H_i *is the free product of 2 free abelian groups of rank* 2 *while the amalgamated subgroup is free of rank* 3. *Further* G_1 *is the normal closure in* Γ_1 *of the modular group,* $PSL_2(Z)$.

Proof. From Corollary A in Section I we have

$$PSL_2(I(-1)) = \{a,t,\ell,u;\ a^2 = (at)^3 = \ell^2 = (a\ell)^2 = (u\ell)^2 = (t\ell)^2 = (ua\ell)^3 = 1,\ tu = ut\},$$

where a is the substitution $z' = -1/z$, t is $z' = z+1$, ℓ is $z' = -z$ and u is $z' = z+i$.

i) Now $a = (z' = -1/z)$ and $t = (z' = z+1)$ generate the modular group [5] $(PSL_2(Z) \cong \sigma(2) * \sigma(3)$.

j) Consider $N(a,t) = G_1$, the normal closure in Γ_1 of the modular group. Then

$$\Gamma_1/G_1 \cong \{a,t,\ell,u;\ a=t=1,\ \ell^2 = u^2 = (u\ell)^2 = (u\ell)^3 = 1\} \cong \{\ell;\ \ell^2 = 1\} \cong Z_2$$

so $[\Gamma_1 : G_1] = 2$.

ji) Further $a,t \in G_1$ and $u\ell \in G_1$ so a,t,ℓ,u are in $G_1 \cdot \{\ell\}$ so $\{\ell\}$ and G_1 generate Γ_1.

jii) Also the extension splits giving a semidirect product for $G_1 \cap \{\ell\} = \{1\}$ since ℓ is of order 2 and $\ell \notin G_1$. So Γ_1 is a semidirect product.

ii) The subgroup H is the subgroup found in [9]. We find it by obtaining successively in Γ_1 the commutator subgroup Γ', which is contained in and of index 2 in G_1; (since Γ_1/G_1 is abelian and $[\Gamma_1 : \Gamma_1'] = 4$) the second commutator subgroup Γ_1'', and then F the subgroup of Γ_1'' consisting of all words of even exponent sum in the generators of Γ_1''. Finally H is the subgroup of F, consisting of all words of even exponent sum in the generators of F. Q.E.D.

THEOREM B. *If* $\Gamma_3 = PSL_2(I(-3))$ *then* Γ_3 *has a subgroup* H *of index* 144. H *is the free product of 2 isomorphic groups* G_1, G_2 *with an amalgamated subgroup* K. *For* $i = 1, 2,$

$$G_i \cong \{x,y,\mu,\sigma,\lambda,\varepsilon;\ x^3 = y^3 = (xy)^3 = \mu^3 = \sigma^3 = (\mu\sigma)^3 = 1,\ \lambda\varepsilon = \varepsilon\lambda\}$$

while K *is the subgroup generated by* $\lambda, \varepsilon, y^{-1}x, \sigma^{-1}\mu, \mu\epsilon y^{-1}, \sigma\lambda x^{-1}$. G_i *is then a free product of 2 isomorphic groups of Euclidean motions* $\{u,v;\ u^3 = v^3 = (uv)^3 = 1\}$ *and a free abelian group of rank* 2. *The amalgamated subgroup* K *is a free product of a free group on 4 generators and a free abelian group of rank* 2. *Further the normal closure in* Γ_3 *of the modular group* $PSL_2(Z)$ *is* Γ_3.

Proof. From Corollary B in Section I,

$$PSL_2(I(-3)) \cong \{a,d,e;\ a^2 = d^3 = e^3 = (da)^2 = (ade)^3 = (dae^{-1})^3 = 1$$
$$e^{-1}dedaed^{-1}e^{-1}d^{-1}a = 1\},$$

where $a = (z' = -1/z)$, $d = (z' = \omega^2 z)$, $e = \left(z' = \frac{-z+1}{-z}\right)$.

i) Now a, e generate the modular group.

Setting $a = e = 1$ in presentation above we find that

$$\Gamma_3/N(a,t) = \{1\} \Longrightarrow \Gamma_3 \cong N(a,t)$$

so the normal closure in Γ_3 of the modular group is Γ_3.

ii) To find the subgroup H we find successively the following subgroups.

(1) H_1 = commutator subgroup of Γ_3

(2) H_2 = commutator subgroup of H_1

(3) H_3 = a normal subgroup of H_2 of index 3 determined as the kernel of a homomorphism into Z_3.

(4) H = the subgroup of H_3 generated by the squares, that is, $H = H_3(X^2)$.

LEMMA 1. $[\Gamma_3 : H_1] = 3$ *and a presentation for*

$$H_1 = \{j,m,t,u;\ j^2 = m^2 = t^2 = u^2 = (tj)^3 = (mtu)^3 = (umj)^3 = (mt)^2 = 1\}$$

with $t = da$. $m = e^{-1}dae$, $u = eae^{-1}$, $j = mam$ *where* a, d, e *are the generators of* Γ_3.

Proof. Abelianizing Γ_3 gives us $\Gamma_3/H_1 \cong \{e : e^3\}$. Choosing coset representatives $1, e, e^{-1}$ and using the method for presentations of subgroups we get our result.

LEMMA 2. $[H_1 : H_2] = 4$ *and* $H_2 = \{\ell,b,d;\ \ell^3 = b^3 = d^3 = (\ell b)^3 = (\ell d)^3 = (\ell bd)^3 = 1\}$ *with* $\ell = tmu$, $b = umj$, $d = jum$.

Proof. Abelianizing H_1 we get that

$$H_1/H_2 \cong \{j,m;\ j^2 = m^2 = (jm)^2 = 1\} \cong Z_2 \times Z_2\ .$$

Choosing coset representatives 1, j, m, jm we get 8 generators $a = tj$, $\beta = umj$, $\gamma = jt$, $\delta = jum$, $\varepsilon = (mj)^2$, $\phi = muj$, $\xi = (jm)^2$, $\mu = jmu$ along with 18 relations.

We eliminate 4 of the generators and are left with $a, \beta, \delta, \varepsilon$ and relations

$$a^3 = \beta^3 = \delta^3 = (\delta\varepsilon)^3 = (a\beta^{-1})^3 = (a\varepsilon^{-1})^3 = (a\delta)^3 = (\beta\varepsilon^{-1})^3 = \delta\epsilon\beta^{-1} = 1$$

eliminating ε and letting $\beta = b$, $\delta = d$ and $\ell = a\beta^{-1}$ we get our result

$$H_2 = \{\ell,b,d;\ \ell^3 = b^3 = d^3 = (\ell b)^3 = (\ell d)^3 = (\ell bd)^3 = 1\}\ .$$

Now in H_2 we map $\ell \to \sigma$, $d \to \sigma$, $b \to \sigma$, where $(\sigma; \sigma^3)$ is the cyclic group of order 3.

Letting H_3 be the kernel, we choose coset representatives $1, \ell, \ell^{-1}$ with $\overline{b} = \overline{d} = \ell$. We get the following.

LEMMA 3. $[H_2 : H_3] = 3$ *and*

$$H_3 = \{u,v,x,z;\ xz = zx,\ uv = vu,\ (zv)^3 = (xu)^3 = (uvxz)^3 = 1\}\ .$$

Now we are ready to get H, which we find as the square subgroup of H_3.

i) Now H_3/H $\{u,v;\ u^2 = v^2 = (uv)^2 = 1\}$ $Z_2 \times Z_2$ so $[H_3 : H] = 4$. We take as coset representatives $1, u, v, uv$ with $\overline{z} = v$, $\overline{x} = u$.

ii) Then we get 10 generators $\alpha = zv^{-1}$, $\beta = xu^{-1}$, $\gamma = uzv^{-1}u^{-1}$, $\delta = ux$, $\varepsilon = u^2$, $\lambda = v^2$, $\phi = vz$, $\xi = vxu^{-1}v^{-1}$, $\mu = uvzu^{-1}$, $\sigma = uvxv^{-1}$ and 19 relations which we will write as

(1) $(\alpha\lambda)^3$ (2) $(\beta\varepsilon)^3$ (3) $(\alpha\lambda\beta\varepsilon)^3$ (4) σ^3 (5) μ^3 (6) $(\sigma\mu)^3$
(7) $\lambda\varepsilon = \varepsilon\lambda$

(1)′ $(\gamma\lambda_1)^3$ (2)′ σ^3 (3)′ $(\gamma\lambda\delta)^3$ (4)′ ϕ^3 (5)′ $(\xi\varepsilon_1)^3$ (6)′ $(\phi\xi\varepsilon_1)^3$
(7)′ $\lambda_1\varepsilon_1 = \varepsilon_1\lambda_1$

(a) $\gamma\xi^{-1} = \beta^{-1}\alpha$ (b) $\mu\beta^{-1} = \xi^{-1}\phi$ (c) $\alpha\sigma^{-1} = \delta^{-1}\gamma$
(d) $\phi\delta^{-1} = \sigma^{-1}\mu$ (e) $\lambda = \lambda_1$ (f) $\varepsilon = \varepsilon_1$.

Let

$$G_1 = \{\alpha,\beta,\delta,\mu,\lambda,\varepsilon;\ (\alpha\lambda)^3 = (\beta\varepsilon)^3 = (\alpha\lambda\beta\varepsilon)^3 = \sigma^3 = \mu^3 = (\mu\sigma)^3 = 1\ \ \lambda\varepsilon = \varepsilon\lambda\}.$$

This is a free product of 2 isomorphic groups of Euclidean motions $(a,b;\ a^3, b^3, (ab)^3)$[2] and a free abelian group of rank 2. Letting $x = \alpha\lambda$, $y = \beta\varepsilon$ then

$$G_1 = \{x,y,\lambda,\varepsilon,\mu,\sigma;\ x^3 = y^3 = (xy)^3 = 1,\ \mu^3 = \sigma^3 = (\mu\sigma)^3 = 1,\ \lambda\varepsilon = \varepsilon\lambda\}\ .$$

---

2 The group $P = \{a,b;\ a^3 = b^3 = (ab)^3 = 1\}$ is the group generated by 2 translations of equal length and making an angle with each other of $2\pi/3$, and a rotation about the origin of $2\pi/3$. That P plays a large role in the structure of Γ_3 is to be expected, for as generators of P we can choose $z' = z + \omega$, $z' = z + \omega^2$ and $z' = \omega z$ where $\omega = [-1+i\sqrt{3}]/2$ is part of integral base of $I(-3)$. But $z' = z + \omega$ is u, and $z' = \omega z$ is d which were generators of $PSL_2 I(-3)$ [3].

Then the subgroup generated by $\beta^{-1}\alpha$, $\sigma^{-1}\mu$, $\mu\beta^{-1}$, $\alpha\sigma'$, $\lambda_1\varepsilon$ is the subgroup generated by $y^{-1}x$, $\sigma^{-1}u$, $\mu\epsilon y^{-1}$, $\sigma\lambda x^{-1}$, λ, ε which is a free product of a free group on 4 generators $(y^{-1}x, \sigma^{-1}\mu, \mu\epsilon y^{-1}, \sigma\lambda x^{-1})$ and a free abelian group of rank 2. (That those 4 generators generate a free group can be seen by showing that in any relator none of them can appear nontrivially.) Similarly

$$G_2 = \{\gamma,\delta,\phi,\xi,\varepsilon_1,\lambda_1;\ \gamma^3 = (\gamma\lambda_1)^3 = (\delta\gamma\lambda_1)^3 = \lambda_1\varepsilon_1 = \varepsilon_1\lambda_1,\ \phi^3 = (\xi\varepsilon_1)^3 = (\phi\xi\varepsilon_1)^3 = 1\}\ .$$

Letting $v = \gamma\lambda_1$, $w = \xi\varepsilon_1$ we get

$$G_2 = \{v,w,\delta,\phi,\lambda,\varepsilon_1;\ \delta^3 = v^3 = (v\delta)^3 = \phi^3 = w^3 = (\phi w)^3 = 1,\ \lambda_1\varepsilon_1 = \varepsilon_1\lambda_1\}\ .$$

So $G_2 \cong G_1$ and the subgroup generated by $\gamma\xi^{-1}$, $\xi^{-1}\phi$, $\delta^{-1}\gamma$, $\phi\delta^{-1}$, λ_1, ε_1 is the subgroup generated by λ_1, ε_1, $\delta^{-1}v$, $w^{-1}\phi$, $\phi\delta^{-1}$, $v\lambda^{-1}$, εw^{-1} which is also a free product of a free group of rank 4 and a free abelian group of rank 2.

Therefore the amalgamations given by (a), (b), (c), (d), (e) and (f) induce an isomorphism between the subgroups and thus H is a free product of G_1, and G_2 with the subgroup amalgamated.

Further $[\Gamma_3 : H] = [\Gamma_3 : H_1] \cdot [H_1 : H_2][H_2 : H_3][H_3 : H] = 3 \times 4 \times 3 \times 4 = 144$.

THEOREM C. *If* $\Gamma_2 = PSL_2(I(-2))$ *then* Γ_2 *is the semidirect product of a group* G_2 *and a free group* F *on 1 generator.* G_2 *is the tree product of groups* U_n *with* U_n *amalgamated to* U_{n+1} *via isomorphic subgroup* V_n. *For each* n, U_n *is the free product of* $Z_2 \times Z_2$ *and* A_4 *(the alternating group) with a* 2*-cycle amalgamated. For each* n, V_n *is the subgroup of* U_n *generated by a* 2*-cycle from* Z_2 *and a* 3*-cycle from* A_4 *and is isomorphic to the modular group. Further* G_2 *is the normal closure in* Γ_2 *of the modular group.*

Proof. From Corollary C in Section 1,

$$\Gamma_2 = PSL_2(I(-2)) = \{a,t,u;\ tu = ut,\ a^2 = (at)^3 = (u^{-1}aua)^2 = (uau^{-1}ata)^3 = (uau^{-1}at^{-1}a)^3 = 1\}\ .$$

i) Now a,t generate the modular group so consider $G_2 = N(a,t)$ the normal closure.

j) Then $\Gamma_2/G_2 \cong \{u\}$ infinite cyclic a,t,u generate Γ_2 so u and G_1 generate Γ_2. This extension splits since for no $n \in Z$ does $u^n \to 1$ in the homomorphism defined by $\Gamma_2 \to \Gamma_2/G_2$. Therefore $G_1 \cap \{u\} = \{1\}$ so Γ_2 is a semidirect product.

ii) Using coset representatives u^n, $n = 0, \pm 1, \pm 2, \ldots,$ we obtain a presentation for G_2.

j) It has generators, $\alpha_n = u^n a u^{-n}$, $n = 0, \pm 1, \pm 2, \ldots$ and $t_n = u^n t u^{-n}$. However $t_n = t_0\ \forall n$ since $(u,t) = 1 \Longrightarrow G_2$ has generators α_n, t, $n = 0, \pm 1, \pm 2, \ldots$.

jj) The relations in terms of α_n, t are $\alpha_n^2 = (\alpha_n t)^3 = (\alpha_n \alpha_{n+1})^2 = 1$, since $a^2 \Longrightarrow \alpha_n^2$, $(at)^3 \Longrightarrow (\alpha_n t)^3$, $(u^{-1}aua)^2 \Longrightarrow (\alpha_n \alpha_{n+1})^2$ while $(uau^{-1}ata)^3 \Longrightarrow (\alpha_{n+1}\alpha_n\alpha_{n+1} t \alpha_n)^3$ derivable, and similarly for $(uau^{-1}at^{-1}a)^3$.

So G_2 has the presentation

$$\left\{\alpha_n, t;\ \alpha_n^2 = (\alpha_n t)^3 = (\alpha_n \alpha_{n+1})^2 = 1,\ \ n = 0, \pm 1, \pm 2, \ldots\right\}.$$

iii) Let $\beta_n = \alpha_n t\ \forall n$ and add the derivable relations $(\beta_{n+1}\beta_n^{-1})^2$ since

$$\beta_{n+1}\beta_n^{-1} = \alpha_{n+1}\alpha_n\ .$$

Then

$$G_2 = \left\{\alpha_n, \beta_n;\ \alpha_n^2 = \beta_n^3 = (\alpha_n\alpha_{n+1})^2 = (\beta_{n+1}\beta_n^{-1})^2 = 1\,, \beta_{n+1}\beta_n^{-1} = \alpha_{n+1}\alpha_n\right\}.$$

j) Let $U_n = \left\{\alpha_n, \alpha_{n+1}, \beta_n, \beta_{n+1};\ \alpha_n^2 = \alpha_{n+1}^2 = (\alpha_n\alpha_{n+1})^2 = 1,\ \beta_n^3 = \beta_{n+1}^3 = (\beta_{n+1}\beta_n^{-1})^2 = 1,\ \beta_{n+1}\beta_n^{-1} = \alpha_{n+1}\alpha_n\right\}.$

This is a free product of

$$\left\{\alpha_n,\alpha_{n+1};\ \alpha_{n+1}^2 = (\alpha_n\alpha_{n+1})^2\right\} \cong Z_2 \times Z_2$$

and

$$\left\{\beta_n,\beta_{n+1};\ \beta_n^3 = \beta_{n+1}^3 = (\beta_{n+1}\beta_n^{-1})^2 = 1\right\} \cong A_4 \cong PSL(2,3)$$

with the amalgamation

$$\beta_{n+1}\beta_n^{-1} = \alpha_{n+1}\alpha_n$$

which is a 2-cycle in both cases.

ji) Further U_n is amalgamated to U_{n+1} via the identifications $\alpha_{n+1} = \alpha_{n+1}$, $\beta_{n+1} = \beta_{n+1}$, so V_n the amalgamated subgroup is the group generated by $\alpha_{n+1} = u^{n+1}au^{-n+1}$, $\beta_{n+1} = u^{n+1}atu^{n+1}$ which is just a transform of the modular group $\Longrightarrow V_n = (\alpha_{n+1}, \beta_{n+1};\ \alpha_{n+1}^2 = \beta_{n+1}^3 =) \cong$ modular group. Q.E.D.

THEOREM D. *$\Gamma_7 = PSL_2(I(-7))$ is the semidirect product of a group G_7 by a free group F on 1 generator. G_7 is the tree product of groups U_n with U_n amalgamated to U_{n+1} via subgroup V_n. For each n, U_n is the free product of S_3 (the symmetric group) with S_3 with a 2-cycle amalgamated. V_n is the subgroup of U_n generated by a 2-cycle from one of the factors S_3 and a 3-cycle from the other factors. For each n, V_n is then isomorphic to the modular group. Further G_7 is the normal closure in Γ_n of the modular group.*

Proof. From Corollary D in Section 1 we have

$$\Gamma_7 = PSL_2I(-7) = \{a,t,u;\ a^2 = (at)^3 = (u^{-1}auat)^2 = (uau^{-1}ta)^2 = 1,\ tu = ut\}\ .$$

i) Again a, t generate the modular group so let $G_7 = N(a,t)$ and we get that

$$\Gamma_7/G_7 \cong \{u\}\ ,$$

and as before this is a split extension $\Longrightarrow \Gamma_7 \cong G_7 \dot{\times} \{u\}$.

ii) Choosing coset representatives u^n, $n = 0, \pm 1, \dots$ for G_7 we get as a presentation:

$$G_7 = \left\{\alpha_n, t;\ \alpha_n^2 = (\alpha_n t)^3 = (\alpha_n \alpha_{n+1} t)^2 = 1,\ n = 0, \pm 1, \pm 2, \dots\right\}$$

with $\alpha_n = u^n a u^{-n}$, $t = t$.

k) Let

$$U_n = \left\{\alpha_n, \alpha_{n+1}, t;\ \alpha_n^2 = \alpha_{n+1}^2 = (\alpha_n t)^3 = (\alpha_{n+1} t)^3 = (\alpha_n \alpha_{n+1} t)^2 = 1\right\}.$$

Then G is the tree product of the U_n with U_n amalgamated to U_{n+1} via amalgamations $\alpha_{n+1} = \alpha_{n+1}$; $t = t$ which obviously give an isomorphism.

ki) Further

$$U_n = \left\{\alpha_n, \alpha_{n+1}, t;\ \alpha_n^2 = \alpha_{n+1}^2 = (\alpha_n t)^3 = (\alpha_{n+1} t)^3 = (\alpha_n \alpha_{n+1} t)^2 = 1\right\}$$

$$\left\{\alpha_n, \beta_n, \alpha_{n+1}, \beta_{n+1};\ \alpha_n^2 = \beta_{n+1}^3 = (\alpha_n \beta_{n+1})^2 = 1,\ \alpha_{n+1}^2 = \beta_n^3\right.$$

$$\left.(\alpha_{n+1}\beta_n)^2 = 1),\ \alpha_n \beta_{n+1} = \alpha_{n+1}\beta_n\right\}$$

by letting $\beta_n = \alpha_n t$, $\beta_{n+1} = t\,\alpha_{n+1}$.

But this is a free product of 2 copies of S_3, $\{u, v;\ u^2 = v^3 = (uv)^2 = 1\}$, with a 2-cycle amalgamated.

kii) Also V_n is the subgroup of G_n generated by $\alpha_{n+1} = u^{n+1} a u^{-n}$ and $t = u^{n+1} t\, u^{n+1}$ which is just a transform of the modular group and so is isomorphic to the modular group. Q.E.D.

The structure of G_7 becomes even simpler if we look at its commutator subgroup G_7'.

COROLLARY. G_7', *the commutator subgroup of* G_7 *has index* 2 *and is a tree product of groups* U_n *with* U_n *amalgamated to* U_{n+1} *via* V_n. $V_n\ U_n$ *is a free product of two 3-cycles and* V_n *is also the free product of two 3-cycles.*

Proof. Now $G_7 = \left\{\alpha_n, t;\ \alpha_n^2 = (\alpha_n t)^3 = (\alpha_n \alpha_{n+1} t)^2 = 1,\ n = 0, \pm 1, \dots\right\}$. Abelianizing G_7 we get $G_7/G_7' \cong \{t; t^2\}$; so $[G_7 : G_7'] = 2$. Using $1, t$ as coset representatives we get

$$G_7' = \left\{\gamma_n, \beta_n, s;\ \gamma_n^3 = (\beta_n s)^3 = \beta_n \gamma_{n+1} \gamma_n \beta_{n+1} s = \gamma_n \beta_n = 1,\ n = 0, \pm 1, \dots\right\}$$

with $\gamma_n = t\alpha_n$, $\beta_n = \alpha_n t^{-1}$, $s = t^2$.

Eliminating the β_n's we have

$$G_7' = \left\{\gamma_n;\ \gamma_n^3 = 1,\ \gamma_n^{-1}\gamma_{n+1}\gamma_n\gamma_{n+1}^{-1} = \gamma_{n+1}^{-1}\gamma_{n+2}\gamma_{n+1}\gamma_{n+2}^{-1}\right\}$$

so we let $U_n = \left\{\gamma_n, \gamma_{n+1};\ \gamma_n^3 = \gamma_{n+1}^3 = 1\right\}$; and the U_n is amalgamated to U_{n+1} via

$$\gamma_{n+1} = \gamma_{n+1} \quad \text{and} \quad \gamma_n^{-1}\gamma_{n+1}\gamma_n\gamma_{n+1}^{-1} = \gamma_{n+1}^{-1}\gamma_{n+2}\gamma_{n+1}\gamma_{n+2}^{-1},$$

and the subgroup generated by γ_{n+1} and $\gamma_n^{-1}\gamma_{n+1}\gamma_n\gamma_{n+1}^{-1}$ is the subgroup generated by γ_{n+1} and $\gamma_n^{-1}\gamma_{n+1}\gamma_n$ and is also the free product of two 3-cycles. Q.E.D.

THEOREM E. *If $\Gamma_{11} = PSL_2I(-11)$ then Γ_{11} is the semidirect product of a group G_1 and a free group of rank 1. G_{11} is the tree product of groups, U_n with U_n amalgamated to U_{n+1} via V_n. For each n, U_n is the free product of 2 copies of A_4 with a 3-cycle amalgamated. For each n, V_n is the subgroup of U_n generated by a 2-cycle from 1 factor A_4 and a 3-cycle from the second factor of A_4 and is isomorphic to the modular group. Further G_{11} is the normal closure in Γ_{11} of the modular group.*

Proof. From Corollary E, Section 1:

$$PSL_2I(-11) = \Gamma_{11} = \{a, t, u;\ a^2 = (at)^3 = (u^{-1}a\ uat)^3 = (uau^{-1}ta)^3 = 1$$
$$tu = ut\}.$$

As before let $G_{11} = N(a, t)$ then $\Gamma_{11}/G_1 \cong \{u\}$ and the extension splits.

i) We find as a presentation for G_{11}:

$$G_{11} = \left\{\alpha_n, t;\ \alpha_n^2 = (\alpha_n t)^3 = (\alpha_n \alpha_{n+1} t)^3 = 1,,\ n = 0, \pm 1, \pm 2, \ldots\right\}$$

with $\alpha_n = u^n a u^{-n}$, $t = t$.

ii) Let

$$U_n = \left\{\alpha_n, \alpha_{n+1}, t;\ \alpha_n^2 = \alpha_{n+1}^2 = (\alpha_n t)^3 = (\alpha_{n+1} t)^3 = (\alpha_n \alpha_{n+1} t)^3 = (\alpha_{n+1} t \alpha_n)^3 = 1\right\} .$$

Then G is the tree product of the U_n with U_n amalgamated to U_{n+1} by $\alpha_{n+1} = \alpha_{n+1}$, $t = t$.

iii) Now

$$U_n \cong \left\{\alpha_n, \beta_{n+1}, \alpha_{n+1}, \beta_n;\ \alpha_n^2 = (\beta_{n+1})^3 = (\alpha_n \beta_{n+1})^3 = 1 \right.$$
$$\left. \alpha_{n+1}^2 = \beta_n^3 = (\alpha_{n+1}\beta_n)^3 = 1,\ \alpha_{n+1}\beta_n = \beta_{n+1}\alpha_n\right\}$$

by letting $\beta_n = t\alpha_n$, $\beta_{n+1} = \alpha_{n+1} t$; but this is a free product of 2 copies of $\{u, v;\ u^2 = v^3 = (uv)^3\} \cong A_4$ [3] with a 3-cycle generated by uv amalgamated.

iv) V_n is the subgroup of U_n, generated by α_{n+1}, t which is again a transform of the modular group and so is isomorphic to the modular group.

4. Now we will consider the groups $E_2(I(-d))$ where $d \neq 1, 2, 3, 7, 11$. From the theorem in Section 1,

$$E_2(I(-d)) = \{a, t, u, j;\ a^2 = (at)^3 = j,\ \ j^2 = 1,\ \ j \text{ central},\ \ tu = ut\} ;$$

so we get

THEOREM. *Say* d *is a positive squarefree integer* $d \neq 1, 2, 3, 7, 11$; *then*

(1) $E_2(I(-d)$ *is the free product of the modular group of matrices* $SL_2(Z)$ *and a direct product of a free abelian group of rank* 2 *and a cyclic group of order* 2 $(Z \times Z \times Z_2)$ *with a subgroup isomorphic to* $Z \times Z_2$ *amalgamated.*

(2) $PE_2(I(-d)) = E_2/Z(E_2)$ *is the free product of the modular group* $PSL_2(Z)$ *and a free abelian group of rank* 2 *with a free group of rank* 1 *amalgamated.*

(3) $E_2(I(-d))$ *is the semidirect product of a group* G_d *and a free group* F *of rank* 1. G_d *is the tree product of groups* U_n *with* U_n *amalgamated to* U_{n+1} *via* H_n. *For each* n, $U_n \cong SL_2(Z)$ *and* V_n *is a subgroup isomorphic to* $Z \times Z_2$.

Proof. (1) $E_2 = \{a,t,u;\ a^2 = (at)^3 = j,\ j^2 = 1,\ tu = ut,\ j \text{ central}\}$

i) Let

$$(E_2)_1 = \{a,t,j;\ a^2 = (at)^3 = j,\ j^2 = 1,\ j \text{ central}\} \cong SL_2(Z)$$

and

$$(E_2)_2 = \{t,u,j;\ j^2 = 1,\ j \text{ central},\ tu = ut\} \cong Z \times Z \times Z_2$$

then E_2 is the free product of $(E_2)_1$ and $(E_2)_2$ with amalgamations $t = t$, $j = j$, which generate a subgroup isomorphic to $Z \times Z_2$.

(2) The center of E_2 is $\{1, j\}$ so $PE_2 = \{a,t,u;\ a^2 = (at)^3 = 1,\ tu = ut\}$ = free product of $\{a,t;\ a^2 = (at)^3 = 1\} \cong$ modular group and $\{t,u;\ tu = ut\} \cong$ free abelian of rank 2 with the amalgamation $t = t$, which generates an infinite cyclic subgroup.

(3) Let $G_d = N(a,t)$ the normal closure in E_2 of the modular group.

i) Then $E_2/G_d \cong \{u\}$ and as in the other cases this extension splits.

ii) Using u^n, $n = 0, \pm 1, \pm 2, \ldots$ as coset representatives we get

$$G_d = \{\alpha_n, t;\ \alpha_n^2 = (\alpha_n t)^3 = 1,\ n = 0, \pm 1, \pm 2, \ldots\}$$

which is the tree product of

$$U_n = \left\{\alpha_n, t;\ \alpha_n^2 = (\alpha_n t)^3 = 1\right\}$$

with U_n amalgamated to U_{n+1} via $t = t \Longrightarrow$ each $U_n \cong$ modular group and V_n is infinite cyclic. Q.E.D.

5. In this section we will examine the abstract word problem for these groups in the light of the above results.

For $d = 2, 7, 11$, $PSL_2I(-d)$ are tree products of groups with amalgamations. The component groups are free products of finite groups with amalgamated subgroups. Since the word problem in a tree product is dependent on the word problem for each level n, we would expect the word problem for these groups to be solvable. We will prove

PROPOSITION. *For* $d = 2, 7, 11$, *the abstract word problem for the presentation given for* $PSL_2I(-d)$ *is solvable.*

Proof. Now for $d = 2, 7, 11$, $PSL_2I(-d)$ is a split extension of a group G_d by a free group of rank 1. First we will show that the word problem is solvable in G_d and then show that the result can be extended to the whole group.

LEMMA. *For* $d = 2, 7, 11$ *the abstract word problem for* G_d *is solvable.*

Proof. Now G_d is a tree product of *isomorphic* groups U_n with U_n amalgamated to U_{n+1} via V_n where V_n is the same $\forall n$. Therefore in view of the corollary in Section 2, it suffices to show that U_n, V_n have solvable word problems, and that the extended word problem for U_n mod V_n is solvable.

i) The proof for the case $d = 2$ will be exhibited; the other cases are handled identically.

ii) In G_2

$$U_n = \left\{\alpha_n, \alpha_{n+1}, \beta_n, \beta_{n+1};\ \alpha_n^2 = \alpha_{n+1}^2 = (\alpha_n\alpha_{n+1})^2 = 1,\ \beta_n^3 = \beta_{n=1}^3 = (\beta_{n+1}\beta_n^{-1})^3 = 1,\ \alpha_n\alpha_{n+1} = \beta_{n+1}\beta_n^{-1}\right\}$$

with V_n the subgroup generated by $\alpha_{n+1}, \beta_{n+1}$.

iii) So U_n $\{a,b,u,v;\ a^2=b^2=(ab)^2=1,\ u^3=v^3=(uv^{-1})^2=1,\ ba=uv^{-1}\}$, with V_n generated by b, u.

j) So U_n is isomorphic to the free product of $Z_2 \times Z_2$ and A_4 with a 2-cycle amalgamated.

k) Since both $Z_2 \times Z_2$ and A_4 are finite, all relevant word problems are solvable $\Longrightarrow$ each U_n has a solvable word problem.

ki) If we choose $(1, b)$ as coset representatives of the amalgamated 2-cycle in $Z_2 \times Z_2$ and $(1,u,u^{-1},v^{-1},u^{-1}v^{-1},v^{-1}u)$ as coset representatives of the amalgamated 2-cycle in A_4, then each $g \in U_n$ is represented as a unique sequence

$$(h, c_1, \ldots, c_k)$$

where h is 1 or a 2-cycle and the c_i are chosen from the coset representatives above with c_i and c_{i+1}, not both in the same factor [5].

kii) V_n is the subgroup of U_n, generated by b, u and as was shown it is their free product.

Claim. A word W in the generators of U_n represents an element of V_n iff in the sequence for W, $(h, c_1, \ldots, c_k)$, $h = 1$ and $c_i = (1, b)$ or $c_i = (1,u,u^{-1})$.

i) If W has the sequence $(1, c_1, \ldots, c_k)$ with c_i's above, then W defines the element $g = c'_1 \ldots c'_k$ under the obvious homomorphism.

j) Since each c_i is just a word in u or b, g defines an element in the free product of u and b, and so W represents an element of V_n.

ii) Conversely, say W defines an element of V_n. Then W is equivalent to a word

$$W_1 = u^{\varepsilon_1} b^{\varepsilon_2} u^{\varepsilon_3} \ldots u^{\varepsilon_n},\ \varepsilon_i = 0, \pm 1 .$$

j) So the sequence for W is the same as the sequence for W_1. But by using the process for determining this sequence found in [5] the sequence for W_1 has the desired form. Q.E.D.

Therefore the extended word problem for U_n mod V_n in G_2 is solvable, so it follows that the word problem in G_2 is solvable.

This can be extended to all of $PSL_2I(-2)$ by the following.

LEMMA. *Say* $N \lhd G$, *and* N *has a solvable word problem (in terms of the presentation given by the Reidemeister-Scheir method). Further* G/N *has a solvable word problem in terms of the factor group presentation and these presentations can be found constructively. Then* G *has a solvable word problem.*

Proof. Let $W = W(g_\nu)$ be a word in the generators of G.

i) Let $\phi : G \to G/N$ be the determined homomorphism. Consider $\phi(W)$. Since the presentation for G/N is obtained constructively we write $\phi(W)$ in the generators of G/N. Then we determine if $\phi(W)$ in the generators of G/N. Then we determine if $\phi(W)$ is 1 or not. If $\phi(W) \neq 1$ then $W \neq 1$.

ii) If $\phi(W) = 1$, $W \in N$. Since the Reidemeister-Scheir method is done constructively we can write W in the generators of N and determine if it is 1 or not. Q.E.D.

Now $PSL_2I(-2) = \Gamma$, satisfies the conditions of the lemma. $N = G_2$ has a solvable word problem while Γ/N is infinite cycle. This proves the theorem.

REFERENCES

[1]. Cohn, P. M., "On the Structure of the GL of a Ring," Publ. Math. IHES, No. 30 (Paris 1966), pp. 5-53.

[2]. ————, "A Presentation of SL for Euclidean Imaginary Quadratic Number Fields," Mathematika 15 (1968), pp. 156-163.

[3]. Coxeter, H. S. M., Moser, W. O., "Generators and Relations for Discrete Groups," Springer-Verlag, Berlin, 1957.

[4]. Fine, B., "The Structure of $PSL_2(R)$; R, the Ring of Integers in a Euclidean Quadratic Imaginary Number Field," Ph.D. thesis, New York University, 1973.

[5]. Magnus, W., Karrass, A., Solitar, D., "Combinatorial Group Theory," Interscience-Wiley, New York, 1966.

[6]. Mennicke, J., "A note on regular coverings of closed orientable surfaces," Proc. Glasgow Math. Assoc. 5 (1961), pp. 49-66.

[7]. Serre, J. P., to be published.

[8]. Swan, R. G., "Generators and Relations for Certain Special Linear Groups," Advances Math. 6 (1971), pp. 1-77.

[9]. Waldinger, H., "On the subgroups of the Picard group," Proc. Amer. Math. Soc. 16 (1965), pp. 1373-1378.

QUASI-CONFORMAL MAPPINGS AND LEBESGUE DENSITY

F. W. Gehring* and J. C. Kelly

§1. *Introduction*

Suppose that E is a Lebesgue measurable set in Euclidean n-space R^n. For each point $x \in R^n$ we call

$$\overline{d}(E,x) = \limsup_{r \to 0} \frac{m(E \cap B)}{m(B)}, \qquad \underline{d}(E,x) = \liminf_{r \to 0} \frac{m(E \cap B)}{m(B)}$$

the *upper* and *lower densities* of E at x, respectively, where B denotes the open n-ball $B(x,r)$ with center at x and radius r, and where m denotes the Lebesgue measure in R^n. When $\overline{d}(E,x) = \underline{d}(E,x)$, the common value $d(E,x)$ is called the *density* of E at x. A point x is said to be a *point of density* for E whenever $d(E,x) = 1$. It is well known that almost all $x \in E$ are points of density for E.

Suppose next that D and D' are domains in R^n where $n \geq 2$, that $f: D \to D'$ is a quasi-conformal mapping, and that $E \subset D$ is measurable. Since f maps sets of measure zero onto sets of measure zero, $f(E)$ is measurable and

$$d(f(E), f(x)) = 1 \text{ if and only if } d(E,x) = 1 \tag{1}$$

for all $x \in D \sim X$, where X is an exceptional set of measure zero.

The purpose of this note is to remark that (1) holds for *all* $x \in D$, and hence that points of density are preserved under quasi-conformal mappings. This fact is an immediate consequence of the following result.

* This research was supported in part by the U. S. National Science Foundation, Contract GP 28115.

THEOREM 1. *If* $f: D \to D'$ *is a* K*-quasi-conformal mapping and if* $E \subset D$ *is measurable, then*

$$\overline{d}(f(E), f(x)) \le b\,\overline{d}(E, x)^a, \qquad \underline{d}(f(E), f(x)) \ge 1 - b(1 - \underline{d}(E, x))^a \tag{2}$$

for all $x \in D$, *where* a *and* b *are positive constants which depend only on* K *and* n.

Theorem 1, in turn, follows from a still stronger uniform density property which actually characterizes quasi-conformal mappings.

§2. *Uniform density property*

Suppose that $f: D \to D'$ is a homeomorphism, and for each open n-ball $B_1 \subset D$ with center at x let B_2 denote the largest open n-ball in $f(B_1)$ with center at $f(x)$. We say that f has the *uniform density property* if there exists a continuous function $\phi: [0, \infty) \to [0, \infty)$ with $\phi(0) = 0$ such that for each $x \in D$ and each sufficiently small B_1 with center at x,

$$\frac{m(f(E) \cap B_2)}{m(B_2)} \le \phi\left(\frac{m(E \cap B_1)}{m(B_1)}\right) \tag{3}$$

for every measurable set $E \subset D$.

THEOREM 2. *If* $f: D \to D'$ *is a* K*-quasi-conformal mapping, then* f *has the uniform density property with* $\phi(t) = bt^a$, *where* a *and* b *are positive constants which depend only on* K *and* n.

Proof. Fix $x \in D$, choose $B_1 = B(x, r) \subset D$, and let $Q = Q(x, r)$ denote the smallest oriented n-cube containing B_1. By choosing r sufficiently small, we may insure that

$$\operatorname{dia}(f(Q)) < \operatorname{dist}(f(Q), \partial D') . \tag{4}$$

Let $J_f = J_f(x)$ denote the absolute value of the Jacobian of f at x, defined almost everywhere in D. By (4) and the proof of Theorem 1 in [3], there exists a positive constant c, which depends only on K and n, such that

$$\frac{1}{m(Q)} \int_Q J_f{}^p \, dm \leq \frac{c}{n+c-np} \left(K \frac{m(f(Q))}{m(Q)}\right)^p \tag{5}$$

for $p \in \left[1, 1+\frac{c}{n}\right)$.

Set $p = 1 + \frac{c}{2n}$ and let E be any measurable set in D. Hölder's inequality and (5) imply that

$$\begin{aligned} m(f(E) \cap B_2) &\leq \int_{E \cap B_1} J_f \, dm \leq \left(\int_Q J_f{}^p dm\right)^{\frac{1}{p}} m(E \cap B_1)^{\frac{p-1}{p}} \\ &\leq (2m(Q))^{\frac{1}{p}} K \frac{m(f(Q))}{m(Q)} m(E \cap B_1)^{\frac{p-1}{p}} , \end{aligned}$$

whence

$$\frac{m(f(E) \cap B_2)}{m(f(Q))} \leq 2K \left(\frac{m(E \cap B_1)}{m(Q)}\right)^{\frac{p-1}{p}} .$$

By geometry and Lemma 3 in [4], $m(B_1) = c_1 m(Q)$ and $m(f(Q)) \leq c_2 m(B_2)$ where c_1 and c_2 are positive constants which depend only on n and on K and n, respectively. Hence

$$\frac{m(f(E) \cap B_2)}{m(B_2)} \leq b \left(\frac{m(E \cap B_1)}{m(B_1)}\right)^a , \tag{6}$$

where $a = \frac{p-1}{p}$ and $b = 2Kc_1{}^a c_2$, and this completes the proof of Theorem 2.

Proof of Theorem 1. Choose $x \in D$. By Theorem 2 there exists a positive constant d such that (6) holds for each measurable set $E \subset D$ and each open n-ball B_1 with center at x and radius $r \in (0, d)$. If s denotes the radius of B_2, then s is continuous and increasing in r and $s \to 0$ as $r \to 0$. Thus r is continuous and increasing in s and (6) implies the first half of (2). The second half of (2) then follows from the relation

$$\underline{d}(E, x) + \overline{d}(D \sim E, x) = 1$$

and the first half of (2) applied to $D \sim E$.

§3. *Characterization of quasi-conformality*

We observe next that the uniform density property yields a new characterization for quasi-conformal mappings.

THEOREM 3. *A homeomorphism* $f : D \to D'$ *is quasi-conformal if and only if it has the uniform density property.*

Proof. The necessity follows from Theorem 2. For the sufficiency we show that f is quasi-conformal whenever f^{-1} has the uniform density property. For this, given $x \in D$, let

$$H_0(x) = \limsup_{r \to 0} \frac{m(B')}{m(f(B))},$$

where $B = B(x, r)$ and B' denotes the smallest open n-ball with center at $f(x)$ containing $f(B)$. Let $B_1 = B'$, $B_2 = B$ and $E = f(B)$. For sufficiently small r the uniform density property for f^{-1} implies that

$$1 \leq \phi\left(\frac{m(f(B))}{m(B')}\right),$$

and hence that

$$H_0(x) \leq K \tag{7}$$

for each $x \in D$, where K is a positive constant which depends only on ϕ.

Now by a trivial modification of a familiar argument, (7) implies that f is absolutely continuous on lines or ACL in D. (See, for example, pp. 9-12 in [2] or pp. 106-108 in [6].) Next (7) and the Lebesgue theorem applied to the additive set function $\mu(E) = m(f(E))$ imply that

$$L_f(x) = \limsup_{y \to x} \frac{|f(y) - f(x)|}{|y - x|} < \infty$$

almost everywhere in D. Hence by the Rademacher-Stepanoff theorem, f is differentiable almost everywhere in D. Finally from (7) it follows that

$$L_f(x)^n \leq K J_f(x)$$

at each point of differentiability $x \epsilon D$, and hence almost everywhere in D. Thus f is quasi-conformal with outer dilatation $K_0(f) \leq K$.

§4. *An application*

We conclude with an application of Theorem 1 to the behavior of the moduli of surfaces under quasi-conformal mappings.

Suppose that E is a measurable set in R^n. We call the set of points $x \epsilon R^n$ for which

$$\underline{d}(E, x) < 1 \quad \text{and} \quad \underline{d}(R^n \sim E, x) < 1$$

the *essential boundary* of E, denoted by $\sigma(E)$. Next we say that E has an *approximate tangent plane* T at x if T is an (n–1)-dimensional hyperplane which contains x and divides R^n into half spaces H_1 and H_2 such that

$$d(E \cap H_1, x) = 0 \quad \text{and} \quad d((R^n \sim E) \cap H_2, x) = 0 \ .$$

The set of all x at which E has an approximate tangent plane is called the *reduced boundary* of E, denoted by $\beta(E)$. It is not difficult to verify that $\sigma(E)$ and $\beta(E)$ are Borel sets and that

$$\beta(E) \subset \sigma(E) \subset \partial E \ .$$

If the Gauss-Green theorem as given in [1] holds for E, then

$$H^{n-1}(\sigma(E) \sim \beta(E)) = 0 \ , \tag{8}$$

where H^{n-1} denotes the normalized (n–1)-dimensional Hausdorff measure in R^n. By Theorem 4.5.11 in [1], a sufficient condition that this theorem holds for E is that $H^{n-1}(\sigma(E))$ be finite, in which case E has an approximate tangent plane T at H^{n-1} almost all points $x \epsilon \sigma(E)$.

Suppose next that D is a domain in R^n. We say that σ is a *boundary surface* in D if there exists a bounded set E with $\overline{E} \subset D$ such that $\sigma = \sigma(E)$. For each $p \in [1, \infty)$, the p-modulus of a family Σ of boundary surfaces σ is defined by

$$M_p(\Sigma) = \inf \int_{R^n} g^p dm \ ,$$

where the infimum is taken over all Borel functions $g : R^n \to [0, \infty]$ such that

$$\int_\sigma g \, d\mathcal{H}^{n-1} \geq 1$$

for all $\sigma \in \Sigma$.

Finally suppose that $f : D \to D'$ is a quasi-conformal mapping and that σ is a boundary surface in D. Then Theorem 1 implies that $f(\sigma)$ is a boundary surface in D'. The following result shows that f is $\mathcal{H}^{n-1}$ absolutely continuous on almost all such surfaces σ.

THEOREM 4. *If* $f : D \to D'$ *is a quasi-conformal mapping and if* Σ *is a family of boundary surfaces in* D, *then there exists a subfamily* $\Sigma_0 \subset \Sigma$ *with*

$$M_{n'}(\Sigma_0) = 0, \qquad n' = \frac{n}{n-1} \ , \tag{9}$$

such that for $\sigma \in \Sigma \sim \Sigma_0$, $\mathcal{H}^{n-1}(\sigma) < \infty$, f *is differentiable* $\mathcal{H}^{n-1}$ *almost everywhere in* σ, *and*

$$\int_\sigma (g \circ f) J_f^T d\mathcal{H}^{n-1} = \int_{f(\sigma)} g \, d\mathcal{H}^{n-1} \tag{10}$$

for each Borel function $g : D' \to [0, \infty]$.

In (10) if $\sigma = \sigma(E)$, then at $\mathcal{H}^{n-1}$ almost all $x \in \sigma$, E has an approximate tangent plane T, f has a differential $df : R^n \to R^n$, and $J_f^T = J_f^T(x)$ is the $(n-1)$-dimensional Jacobian of $df|T$, that is,

(11) $$J_f^{\,T}(x) = \frac{\mathcal{H}^{n-1}(df(T \cap B))}{\mathcal{H}^{n-1}(T \cap B)}$$

where $B = B(0, 1)$.

Proof. Suppose that $g : D' \to [0, \infty]$ is a Borel function. Since f is differentiable almost everywhere in D,

$$D = \bigcup_{j=0}^{\infty} D_j ,$$

where the D_j are disjoint Borel sets such that $m(D_0) = 0$ and such that for $j \geq 1$ f is differentiable at each $x \in D_j$ and $f|D_j$ is Lipschitzian. (Cf. Theorem 3.1.8 in [1].) Next Theorem 3.3.3 in [5], (8), and Theorem 1 imply that there exists a subfamily $\Sigma_0 \subset \Sigma$ satisfying (9) such that $\mathcal{H}^{n-1}(\sigma \cap D_0) = 0$, $\mathcal{H}^{n-1}(\sigma) < \infty$ and

(12) $$\mathcal{H}^{n-1}(f(\sigma)) \leq \int_\sigma J_f^{\,T} d\,\mathcal{H}^{n-1} < \infty$$

for $\sigma \in \Sigma \sim \Sigma_0$. Choose $\sigma \in \Sigma \sim \Sigma_0$. Then

(13) $$\int_{\sigma \cap D_j} (g \circ f) J_f^{\,T} d\,\mathcal{H}^{n-1} = \int_{f(\sigma \cap D_j)} g\, d\,\mathcal{H}^{n-1}$$

for $j \geq 1$ by Corollary 3.2.20 in [1], while (12) and (13) with $g = 1$ imply that

$$\mathcal{H}^{n-1}(f(\sigma \cap D_0)) = \mathcal{H}^{n-1}(f(\sigma)) - \sum_{j=1}^{\infty} \int_{f(\sigma \cap D_j)} d\,\mathcal{H}^{n-1} \leq \int_{\sigma \cap D_0} J_f^{\,T} d\,\mathcal{H}^{n-1} = 0 .$$

Hence (13) holds for $j = 0$, and we obtain (10) by summing on j.

COROLLARY. *If* $f : D \to D'$ *is a* K*-quasi-conformal mapping, then*

(14) $$K^{\frac{1}{1-n}} M_{n'}(\Sigma) \leq M_{n'}(f(\Sigma)) \leq K^{\frac{1}{n-1}} M_{n'}(\Sigma), \qquad n' = \frac{n}{n-1} ,$$

for each family Σ *of boundary surfaces* σ *in* D.

Proof. Choose a Borel function $g': R^n \to [0, \infty]$ such that

$$\int_{f(\sigma)} g' \, d\mathcal{H}^{n-1} \geq 1$$

for each $\sigma \in \Sigma$. Next let F denote the set of $x \in D$ at which f is differentiable and set

$$g(x) = \begin{cases} 0 & \text{if} \quad x \in R^n \sim F \ , \\ g' \circ f(x) \sup_T J_f^{\,T}(x) & \text{if} \quad x \in F \ , \end{cases}$$

where the supremum is taken over all $(n-1)$-dimensional hyperplanes T which contain x and $J_f^{\,T}(x)$ is as in (11). Then $g: R^n \to [0, \infty]$ is a Borel function, and by Theorem 4

$$\int_{\sigma} g \, d\mathcal{H}^{n-1} \geq \int_{\sigma} (g' \circ f) J_f^{\,T} d\mathcal{H}^{n-1} = \int_{f(\sigma)} g' \, d\mathcal{H}^{n-1} \geq 1$$

for $\sigma \in \Sigma \sim \Sigma_0$, where $M_{n'}(\Sigma_0) = 0$. Since f is K-quasi-conformal,

$$J_f^{\,T}(x)^{n'} \leq K^{\frac{1}{n-1}} J_f(x)$$

almost everywhere in D,

$$M_{n'}(\Sigma) \leq K^{\frac{1}{n-1}} \int_D (g' \circ f)^{n'} J_f \, dm \leq K^{\frac{1}{n-1}} \int_{R^n} (g')^{n'} dm \ ,$$

and we obtain the first half of (14). The second half then follows from the first half applied to f^{-1}.

UNIVERSITY OF MICHIGAN
ANN ARBOR, MICHIGAN

UNIVERSITY OF MISSOURI
COLUMBIA, MISSOURI

[1]. Federer, H., *Geometric measure theory*, Springer-Verlag, New York, 1969.

[2]. Gehring, F. W., *The definitions and exceptional sets for quasi-conformal mappings*, Ann. Acad. Sci. Fenn. 281 (1960), pp. 1-28.

[3]. ——————, *The* L^p*-integrability of the partial derivatives of a quasi-conformal mapping*, Acta Math. 130 (1973), pp. 265-277.

[4]. Gehring, F. W., and Väisälä, J., *Hausdorff dimension and quasi-conformal mappings*, J. London Math. Soc. 6 (1973), pp. 504-512.

[5]. Kelly, J. C., *Quasi-conformal mappings and sets of finite perimeter*, Trans. Amer. Math. Soc., (to appear).

[6]. Väisälä, J., *Lectures on* n*-dimensional quasi-conformal mappings*. Lecture notes in mathematics 229, Springer-Verlag, 1971.

ON THE MODULI OF COMPACT RIEMANN SURFACES WITH A FINITE NUMBER OF PUNCTURES

Jane Gilman

Introduction

Let S be a compact Riemann Surface of genus g and let X(g, 0) be the Riemann Space of S (the Space of Moduli). Baily has shown that X(g, 0) is a quasi-projective variety [2]. If S is a compact surface of genus g with k punctures or distinguished points, we let X(g, k) denote the Riemann Space of S. The purpose of this paper is to prove the following:

If $g \neq 0$ and $k = 6$ or $k \geq 8$, then X(g, k) is a quasi-projective variety.

In Section I we review the definitions of the Riemann Space of a Fuschian group representing a Riemann Surface and the relative Riemann Space of two such groups. We investigate the relationship between the algebraic structures on both spaces and conclude that if the Riemann Space is quasi-projective then the relative space is and conversely that if the relative space is quasi-projective and if a certain group of analytic automorphisms is algebraic, then the Riemann Space is also quasi-projective.

In Section II we choose two Fuschian groups G and G′ so that one represents a punctured surface of genus g with an even number k of punctures and the other a two-sheeted covering of the surface branched over those points. The covering surface will thus have a conformal involution. We use various facts about conformal involutions on compact surfaces and Baily's theorem for the compact case to conclude that the relative Riemann Space is quasi-projective. For the case of an odd number of punctures we choose G and G' in a slightly different manner but still obtain the result that the relative Riemann Space is quasi-projective.

In Section III we prove that when G and G' are chosen as in Section II, then the group of analytic automorphisms of the relative Space is regular for the given algebraic structure. Combining the results of the three sections we obtain the desired result.

§I. Throughout this whole paper, let G and G' be finitely generated Fuchsian groups with G' a normal subgroup of G of finite index, say n. We assume that G and G' are of the first kind. We use the terminology of [4]. In particular, let $T(G)$ and $\mathrm{Mod}(G)$ denote the Teichmüller space and the Teichmüller modular group of G. Then it is known that $T(G) \subset T(G')$, [4]. Let 1 be the group of fractional linear transformations of the upper half plane, U, into itself consisting of only the identity element. Then $\mathrm{Mod}(G)$ and $\mathrm{Mod}(G')$ are both subgroups of $\mathrm{Mod}(1)$, and we can, therefore, consider their intersection.

DEFINITION 1. Let $\mathrm{Mod}(G, G') = \mathrm{Mod}(G) \cap \mathrm{Mod}(G')$. $\mathrm{Mod}(G, G')$ is called the *relative modular group of* G *and* G'. $\mathrm{Mod}(G, G')$ can also be described as $\{g \in \mathrm{Mod}(G') \mid g(T(G)) = T(G)\}$.

DEFINITION 2. If G'' is also a normal subgroup of G, G' and G'' are said to be *related* if G/G' is isomorphic to G/G''.

Note: Let m' be the number of subgroups related to G'. Then m' is finite. Each such subgroup corresponds to the kernel of a homomorphism of G onto G/G'. Since G' is of finite index in G, G/G' is a finite group. If G is finitely generated, there are only a finite number of distinct homomorphism of G into any finite group.

DEFINITION 3. If G'' is a subgroup of G, then G' and G'' are said to be *strongly related* if there is an element h of $\mathrm{Mod}(G)$ with $hG'h^{-1} = G''$.

REMARK. If G'' is strongly related to G', then G'' is related to G', but it is easy to produce examples of groups that are related but not strongly related.

Let $G' = G_1, G_2, \ldots, G_m$ be all strongly related subgroups of G, m is finite since $m \leq m'$. Let $h_1 = \mathrm{id}$, $h_2, \ldots, h_m$ be such that $h_i G' h_i^{-1} = G_i$ with $h_i G h_i^{-1} = G$.

PROPOSITION 1. $[\mathrm{Mod}(G) : \mathrm{Mod}(G, G')] = m$.

Proof. If $g \in \mathrm{Mod}(G)$, then $gG'g^{-1}$ is strongly related to G'. Thus $gG'g^{-1} = G_i$ for some i. Let $\gamma_g = g^{-1} \circ h_i$. Then $\gamma_g \in \mathrm{Mod}(G)$ and $\gamma_g \in \mathrm{Mod}(G')$. Therefore, $\gamma_g \in \mathrm{Mod}(G, G')$. Since $g = h_i \circ \gamma_g^{-1}$ we are done.

Since G itself is a subgroup of $\mathrm{Mod}(G')$, we can look at the fixed point set of G in $T(G')$. This set will be denoted by F_G. It is known that $T(G) = F_G$ [9].

Recall that the Riemann Space of G, written $R(G)$, is equal to $T(G)/\mathrm{Mod}(G)$.

DEFINITION 4. Let $R(G, G') = T(G)/\mathrm{Mod}(G, G')$. $R(G, G')$ is called the *relative Riemann Space of* G *and* G'. [5].

As the quotient of a complex manifold by a properly discontinuous group of analytic automorphisms $R(G, G')$ is a normal analytic space.

PROPOSITION 2. *The natural map* $\phi, \phi : R(G, G') \to R(G)$ *is holomorphic. It makes* $R(G, G')$ *into an* m*-sheeted branched covering of* $R(G)$ *where* m *is the index of* $\mathrm{Mod}(G, G')$ *in* $\mathrm{Mod}(G)$.

Proof. Let pr_1 be the projection $T(G) \to R(G, G')$ and let pr_2 be the projection $pr_2 : T(G) \to R(G)$. Let $\phi : R(G, G') \to R(G)$ be defined as follows: if $x \in R(G, G')$, let y be in $T(G)$ with $pr_1(y) = x$; then set $\phi(x) = pr_2(y)$. It is clear that ϕ is well defined, and we have a commutative diagram:

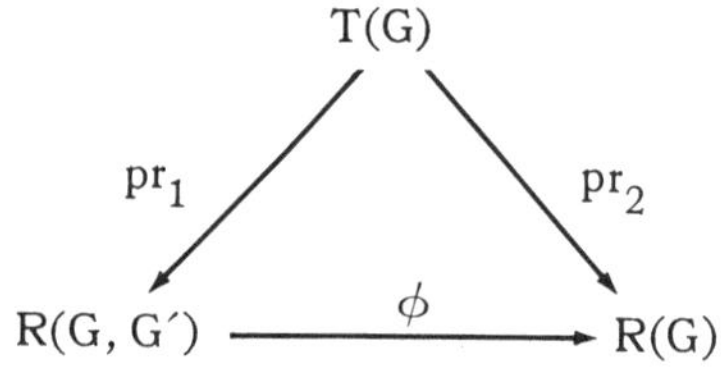

If 0 is an open set $R(G)$, $pr_2^{-1}(0)$ is an open set in $T(G)$; it is a $Mod(G)$ invariant open set and, therefore, a $Mod(G, G')$ invariant open set. But $pr_1(pr_2^{-1}(0))$ is open in $R(G, G')$ if $pr_2^{-1}(0)$ is a $Mod(G, G')$ invariant open set. Thus $\phi^{-1}(0) = pr_1(pr_2^{-1}(0))$ is open; ϕ is continuous.

To show that ϕ is analytic, we must show that if f is a holomorphic function in a neighborhood N of x in $R(G)$, then $f \circ \phi$ is a holomorphic function on $\phi^{-1}(N)$ in $R(G, G')$.

Let C be the complex numbers. Then $R(G)$ is locally isomorphic to C^s, for some integer s, factored out by the action of a finite linear group. Therefore, we may assume that N is isomorphic to $\tilde{N}/M_0$ where $\tilde{N}$ is a neighborhood of y in $T(G)$; M_0 is a finite subgroup of $Mod(G)$; and $pr_2(y) = x$. Then a holomorphic function f on N is a M_0 invariant holomorphic function on $\tilde{N}$; call this function $\tilde{f}$. Let $M_0(G, G') = M_0 \cap Mod(G, G')$. Then $\phi^{-1}(N)$ is isomorphic to $\tilde{N}/M_0(G, G')$, and a holomorphic function on $\phi^{-1}(N)$ is a $M_0(G, G')$ invariant function on $\tilde{N}$. $\tilde{f}$ is M_0 invariant implies that $\tilde{f}$ is $M_0(G, G')$ invariant, and therefore, it induces a holomorphic function f_0 on $\phi^{-1}(N)$. Clearly, $f_0 = f \circ \phi$. Thus $f \circ \phi$ is analytic on $\phi^{-1}(N)$ and ϕ is analytic.

Since $Mod(G, G')$ is a subgroup of $Mod(G)$ of index m, ϕ is in general an m-sheeted covering.

Let $G' = G_1, G_2, \ldots, G_m$ be all strongly related subgroups of G as before. Then for each i, we can consider $Mod(G, G_i)$. Since $Mod(G, G_i)$ is a subgroup of $Mod(G)$, the following definition makes sense:

DEFINITION 5. Let $Mod(G, G_1, \ldots, G_m) = \bigcap_{i=1}^{m} Mod(G, G_i)$. $Mod(G, G_1, \ldots, G_m)$ is called the *second relative modular group of* G *and* $G' = G_1$.

REMARK. $\mathrm{Mod}(G, G_1, \ldots, G_m)$ can be described as $\{g \in \mathrm{Mod}(G) \mid g \in \mathrm{Mod}(G_i)\ \forall\ i\}$.

PROPOSITION 3. $\mathrm{Mod}(G, G_1, \ldots, G_m)$ *is a normal subgroup of* $\mathrm{Mod}(G)$ *of finite index.*

Proof. We will define a homomorphism ψ of $\mathrm{Mod}(G)$ into the group of permutations on m elements. $\mathrm{Mod}(G, G_1, \ldots, G_m)$ will be the kernel of this homomorphism and as such is a normal subgroup of $\mathrm{Mod}(G)$ of finite index.

If $g \in \mathrm{Mod}(G)$, then g induces a permutation of $G_1, \ldots, G_m$ since for every i, $gG_ig^{-1} = G_j$ for some j. We let $\psi(g)$ be the permutation which g induces. Clearly $\psi(g \cdot h) = \psi(g) \cdot \psi(h)$ so that ψ is a group homomorphism.

DEFINITION 6. Let $R(G, G_1, \ldots, G_m) = T(G)/\mathrm{Mod}(G, G_1, \ldots, G_m)$. $R(G, G_1, \ldots, G_m)$ is called the *second relative Riemann Space of* G *and* $G' = G_1$.

As the quotient of a complex manifold by a properly discontinuous group of analytic automorphisms, $R(G, G_1, \ldots, G_m)$ is a normal analytic space.

Since $\mathrm{Mod}(G, G_1, \ldots, G_m)$ is a normal subgroup of $\mathrm{Mod}(G)$, we can form the factor group $\mathrm{Mod}(G)/\mathrm{Mod}(G, G_1, \ldots, G_m)$. Call this group $\mathfrak{M}(G, G_1, \ldots, G_m)$. Then $\mathfrak{M}(G, G_1, \ldots, G_m)$ is a finite group, (of order $\leq m!$) and $\mathfrak{M}(G, G_1, \ldots, G_m)$ acts as a group of analytic automorphisms of $R(G, G_1, \ldots, G_m)$. We have $R(G, G_1, \ldots, G_m)/\mathfrak{M}(G, G_1, \ldots, G_m) = R(G)$ and if θ is the projection

$$\theta : R(G, G_1, \ldots, G_m) \to R(G, G_1, \ldots, G_m)/\mathfrak{M}(G, G_1, \ldots, G_m) ,$$

θ is analytic.

Similarly, we have $\mathrm{Mod}(G, G_1, \ldots, G_m)$ is a normal subgroup of $\mathrm{Mod}(G, G_1)$ of finite index. Let $\mathfrak{M}(G, G_1) = \mathrm{Mod}(G, G_1)/\mathrm{Mod}(G, G_1, \ldots, G_m)$. Then $\mathfrak{M}(G, G_1)$ acts as a finite group of analytic automorphisms of

$R(G, G_1, \ldots, G_m)$. We have $R(G, G_1, \ldots, G_m)/\mathfrak{M}(G, G_1) = R(G, G_1)$ and if Ψ is the projection $\Psi : R(G, G_1, \ldots, G_m) \to R(G, G_1, \ldots, G_m)/\mathfrak{M}(G, G_1)$, then Ψ is analytic.

PROPOSITION 4. *There is a commutative diagram*

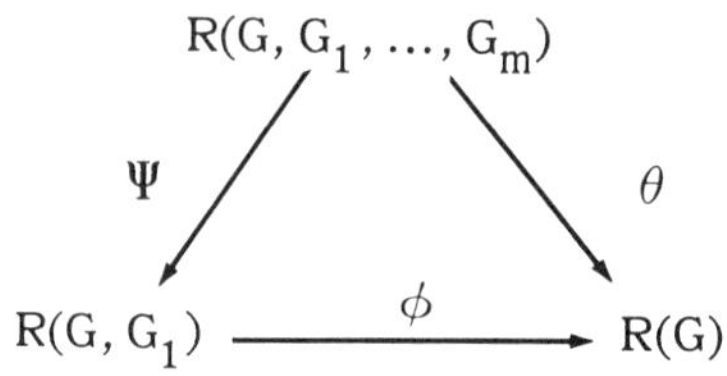

where Ψ, θ, ϕ *are all proper analytic with finite fibers and* θ *and* Ψ *are given by factoring out by the action of a finite group.*

Proof. The proof that these maps are all analytic and that the diagram commutes is straightforward. To show that these maps are all proper it suffices to show that factoring out by the action of a finite group is proper for then Ψ and θ are proper and since the diagram commutes ϕ must be proper too.

To simplify notation then, let M be a finite group with $R'/M = R$ and let p be the projection. Let W be a compact subset of R and $W' = p^{-1}(W)$. Let $\{U_a\}$ be an open cover for W'. We replace this open cover by another open cover as follows: for each x in R' and each U in the cover with $x \in U$ pick $U_1, \ldots, U_n$ in the open cover with $g_i(x) \in U_i$. Assume $M = \{g_1, \ldots, g_n\}$. Let $U_x = \bigcap_{i=1}^{n} g_i^{-1}(U_i)$. Let $U_i' = g_i(U_x)$. The new open cover is the set $\{U_a'\}$. This cover has the property that if U is in the set, so is $g_i(U)$ for all g_i in M. Finally, the original cover has the property that for each U' in the new cover there is a U in the old cover with $U' \subset U$.

Since p is factoring out by the action of a finite group, it is an open map. Thus $\{p(U_a')\}$ is an open cover for W. Thus a finite number cover W. Say, $p(U_1'), \ldots, p(U_n')$. Then the set $\{g_j(U_i')\}$ $i = 1, \ldots, r$ $j = 1, \ldots, n$

cover W′. But each open set here is contained in some open set of the original cover. Thus W′ is compact.

We are now in a position to prove:

THEOREM 1. *If* R(G) *is quasi-projective, then so are* $R(G, G_1, \ldots, G_m)$ *and* R(G, G′). *Conversely, if* R(G, G′) *is quasi-projective then so is* $R(G, G_1, \ldots, G_m)$. *If in addition* $\mathfrak{M}(G, G_1, \ldots, G_m)$ *is a group of regular maps, then* R(G) *is also quasi-projective.*

Note. Since R(G), R(G, G′) and $R(G, G_1, \ldots, G_m)$ are all manifolds factored out by the action of a properly discontinuous group of analytic automorphisms, they are all normal analytic spaces. It can, therefore, be shown that if these spaces are quasi-projective varieties, they are normal varieties. A proof of this fact is outlined in the Appendix.

Proof of theorem. Assume first that R(G) is quasi-projective. Then $R(G, G_1, \ldots, G_m)$ is a normal analytic space; $\theta : R(G, G_1, \ldots, G_m) \to R(G)$ is a proper holomorphic map with finite fibers; the branch locus, X, is a proper subvariety of R(G) with R(G)–X dense and $R(G, G_1, \ldots, G_m) - \theta^{-1}(X)$ dense. We can, therefore, apply the Generalized Riemann Existence theorem [13, p. 227 and the references given there] to conclude that $R(G, G_1, \ldots, G_m)$ is quasi-projective. Since θ is then a morphism, $\mathfrak{M}(G, G_1, \ldots, G_m)$ is a group of regular maps. The subgroup $\mathfrak{M}(G, G')$ is thus also regular. But the quotient of a quasi-projective variety by a finite group is a quasi-projective variety [18]. Thus R(G, G′) is quasi-projective.

To prove the converse, we can again apply the Generalized Riemann Existence theorem to $R(G, G_1, \ldots, G_m)$ and R(G, G′) to conclude that $R(G, G_1, \ldots, G_m)$ is quasi-projective if R(G, G′) is. Then if $\mathfrak{M}(G, G_1, \ldots, G_m)$ is a group of regular maps, R(G) as the quotient of a quasi-projective variety by a finite group will be quasi-projective.

§ II. The purpose of this section is to show that if G and G' are chosen appropriately, then $R(G, G')$ is in fact a quasi-projective variety.

In what follows let U be the upper half plane. If G is any Fuschian Group, let U_G be the upper half plane minus the elliptic fixed points of G. Then by the surface which G represents we shall mean the surface U_G/G.

We now fix G to represent a compact surface of genus g with k punctures. From now on assume that g and k are both positive. G is, therefore, generated by $2g$ hyperbolic elements $a_1, \ldots, a_g, b_1, \ldots, b_g$ and k elliptic elements $c_1, \ldots, c_k$ with relations $c_i^2 = 1$ for $i = 1, \ldots, k$ and $\prod_{i=1}^{g} [a_i, b_i] c_1 \ldots c_k = 1$. Let $S = U_G/G$. We next want to choose G'. We do this differently in the two cases that k is even or odd.

Assume that k is even. Let ϕ be a homomorphism of G into $\mathbb{Z}_2$, the integers mod 2, with $\phi(c_i) = 1$ for all i and $\phi(a_i)$ and $\phi(b_i)$ arbitrary. Let $G' = \ker \phi$. Then G' has no elliptic fixed points and it is a normal subgroup of G. Let $H = G/G'$. H is of order 2. Let $S' = U/G'$. Then H acts as a group of conformal self maps of S'; namely, the group consisting of a single involution j. The fixed points of j are precisely the images in S' of elliptic fixed points of G. Thus j has k fixed points. Let $<j>$ be the group generated by j. Then $(S' - \mathcal{F}_j)/<j> = S$ where $\mathcal{F}_j$ is the fixed point set of j. Throughout this section Q_1 will mean this group G'. In this case, $R(G) = X(g, k)$.

Assume that k is odd and at least 9. Let a and b be generators for $\mathbb{Z}_2 \times \mathbb{Z}_2$. Define a homomorphism ϕ from G onto $\mathbb{Z}_2 \times \mathbb{Z}_2$ by $\phi(c_1) = a$; $\phi(c_2) = b$; $\phi(c_3) = ab$; $\phi(c_4) = \phi(c_5) = b$; $\phi(c_6) = \phi(c_7) = \cdots = \phi(c_k) = a$. Let $\phi(a_i)$ and $\phi(b_i)$ be arbitrary. Let $Q_2 =$ Kernel of ϕ. Let $S' = U/Q_2$. Let $H = G/Q_2$. Then H is a group of conformal automorphisms of S' consisting of three involutions which all commute. If the involutions in H are hj, h, and j then one can compute that they all have different numbers of fixed points, namely 2, 6, and $2(k-5)+2$ respectively. Let $\mathcal{F}_H$ be the set of fixed points of H on S'. Then $S = (S' - \mathcal{F}_H)/H$ and in this case $R(G) = X(g, k)$.

REMARK. There are two diagrams at the end of this section which summarize the relationships between the various spaces and groups which are considered in this section.

We adopt the following notation:

$\mathcal{H}_n$ = the Siegel upper half plane of degree n.

Mod_n = the Siegel modular group of degree n.

C_r = the centralizer of r in Mod_n for r in Mod_n.

F_g = the fixed point set of g in whatever space g acts.

E_n = the set of points in $\mathcal{H}_n$ corresponding to period matrices of Riemann surfaces.

p_n is the projection $p_n : \mathcal{H}_n \to \mathcal{H}_n/\mathrm{Mod}_n$.

$J_{k,n}$ = the involution in Mod_n as defined in [6].

From now on we assume that $k = 6$ or $k \geq 8$.

For the rest of this section we adopt the language of Riemann Surfaces and identify T(G) with T(S), Mod(G) with Mod(S), Mod(G, G′) with Mod(S, S′), etc. Let $S_i = U/Q_i$ and $H_i = G/Q_i$. We will not distinguish in our notation between elements of H_i as conformal automorphisms on S_i and as elements of $\mathrm{Mod}(S_i)$.

Let $\sim$ denote homotopic and $\approx$ homologous. Let $C_g = \{h \in \mathrm{Mod}(S') \mid gh \sim hg\}$ for g in Mod(S′) and let $\underline{C}_g = \{h \in \mathrm{Mod}(S') \mid gh \approx hg\}$. Then $\mathrm{Mod}(S, S_1) = C_j$ and $\mathrm{Mod}(S, S_2) = C_j \cap C_h$. The latter follows from the fact that $\mathrm{Mod}(S, S_2)$ is the normalizer of H_2 in $\mathrm{Mod}(S_2)$ and j, jh, and h all commute and have different numbers of fixed points.

Let p be the mapping of $T(S_i)$ into $\mathcal{H}_n$ where n is the appropriate integer. If $N = \{g \in \mathrm{Mod}(S') \mid g \approx \pm 1\}$, then $\mathrm{Mod}(S')/N = \mathrm{Mod}_n$. (Note: $S' = U/G'$ so S′ is either S_1 or S_2.) To simplify the notation let p also denote the map from $\mathrm{Mod}(S_i)$ onto Mod_n. Let $\mathrm{Mod}_n(S, S_1) = C_{p(j)}$; $\mathrm{Mod}_n(S, S_2) = C_{p(j)} \cap C_{p(h)}$; and let $\mathit{Mod}(S, S_i) = p^{-1}(\mathrm{Mod}_n(S, S_i))$.

Let *pr* be the mapping $pr : T(S_i) \to T(S_i)/\mathit{Mod}(S, S_i)$. Let δ be the natural mapping $\delta : T(S_i)/\mathit{Mod}(S, S_i) \to \mathcal{H}_n/\mathrm{Mod}_n(S, S_i)$. Set $R'(S_i) = T(S_i)/\mathit{Mod}(S, S_i)$. Then $\delta \cdot pr$ factors through $\mathcal{H}_n$.

We know that $T(S) \subset T(S_i)$. Thus p maps $T(S)$ into $\mathcal{H}_n$ and one can ask whether $R(S, S_i)$ is contained in $R'(S_i)$ and if so whether the mapping $\delta \circ pr^*$ factors through $\mathcal{H}_n$ where pr^* is the natural projection $T(S) \to R(S, S_i)$. We prove:

LEMMA 1. a) *If* $x = (S', f, S'') \in T(S')$ *and* $x \in F_j$, *then we may assume that* fjf^{-1} *is conformal.*

b) *If* $tF_j \cap F_j \neq \emptyset$ *and* $t \in \underline{C}_j$, *then* $t \in C_j$ *and* $tF_j = F_j$.

c) $E_n \cap F_{p(j)} = \bigcup_{g \in \underline{C}_j} p(gF_j)$ *and this is a disjoint union.*

Proof. a) Let $x = (S', f, S'') \in F_j$. Then $fjf^{-1} \sim j''$ and j'' is conformal. If f is the extremal map in its homotopy class, then so are fj and $j''f$ in their homotopy classes since j is conformal. But $fj \sim j''f$. By the uniqueness of the extremal map, we must have $fj = j''f$.

b) $tF_j = F_{tjt^{-1}}$. Thus $(S', f, S'') \in tF_j$ implies $ftjt^{-1}f^{-1} \sim j''$ and j'' conformal. By (a) we may assume $fjf^{-1} = j'$ and j' conformal. Since $t \in \underline{C}_j$, $j'' \approx j'$. Since two homologous conformal maps are equal, $j'' = j'$. This implies $ftjt^{-1}f^{-1} \sim fjf^{-1}$ so $t \in C_j$. Thus $tF_j = F_j$.

c) If $p(x) \in F_{p(j)} \cap E_n$, then we may assume from the results of [6] that $p(j) = J_{k,n}$ and that $fjf^{-1} \approx j''$ where j'' is conformal with k fixed points. Since there is only one conjugacy class in Mod(S′) of involutions with a given number of fixed points [see 9], there is a q with $qjq^{-1} \sim j''$ and by (a), $qjq^{-1} = j''$. Because fjf^{-1} and qjq^{-1} are homolgous, $f^{-1}q \in \underline{C}_j$. Because $f(f^{-1}qjq^{-1}f)f^{-1} = qjq^{-1} = j''$, $x \in f^{-1}qF_j$. Thus $E_n \cap F_{p(j)} = \bigcup_{g \in \underline{C}_j} p(gF_j)$.

To show that this union is a disjoint union, assume that $x \in p(gF_j) \cap p(g'F_j)$. Then there is an $h \in N$ and an $\underline{x}$ with $p(\underline{x}) = x$ and $\underline{x} \in gF_j \cap hg'F_j$. Thus $g'^{-1}h^{-1}g \in C_j$ by part (b) so $gF_j = hg'F_j$ and $p(gF_j) = p(g'F_j)$.

COROLLARY 1. a) $R(S, S_1)$ *can be identified with the image of* F_j *in* $\mathcal{H}n/p(\underline{C}_j)$ *and this image is in fact equal to* $(E_n \cap F_{p(j)})/C_{p(j)}$.

b) $R(S, S_2)$ *can be identified with the image of* $F_j \cap F_h$ *in* $\mathcal{H}n/[C_{p(j)} \cap C_{p(h)}]$. *While the image of* $R(S, S_2)$ *can thus be thought of as lying in* $[E_n \cap F_{p(j)} \cap F_{p(h)}]/[C_{p(j)} \cap C_{p(h)}]$, *we do not have equality as in part* (a).

Proof. a) $T(S) = F_j$. $R(S, S_1) = F_j/C_j$. Assume that x and y are in F_j and $r(p(x)) = p(y)$ for some r in $p(\underline{C}_j)$. Because j has more than four fixed points it cannot be conjugate to an involution that is homologous to $-j$, [see 6]. Thus $p(\underline{C}_j) = C_{p(j)}$. If $p(\underline{r}) = r$, $r(p(x)) = p(\underline{r}(x)) = p(y)$. Thus $\underline{r}(x) = t(y)$ for some t in N. Apply Lemma 1b to conclude that $t^{-1}\underline{r} \in C_j$ so that $pr^*(x) = pr^*(y)$. Thus $\delta \circ pr^*$ factors through $\mathcal{H}_n$. In fact, since $F_{p(j)} \cap E_n = \bigcup_{g \in \underline{C}_j} p(gF_j) = \bigcup_{p(g) \in C_{p(j)}} p(g)p(F_j)$, we have $\delta \circ pr^*(T(S)) = (F_{p(j)} \cap E_n)/C_{p(j)}$.

b) As before, $C_{p(j)} = p(\underline{C}_j)$, $C_{p(h)} = p(\underline{C}_h)$, and $p(\underline{C}_j \cap \underline{C}_h) = C_{p(j)} \cap C_{p(h)}$. Repeat the argument used in (a) to conclude that $\delta \circ pr^*(T(S)) = p(T(S))/p(\underline{C}_j \cap \underline{C}_h)$.

From now on we identify $R(S, S_i)$ with its image in $\mathcal{H}n/\mathrm{Mod}_n(S, S_i)$. Since δ is a biholomorphic map between $R'(S_i)$ and its image in $\mathcal{H}n/\mathrm{Mod}_n(S, S_i)$, it will be clear that the analytic structure we assign to $\delta(R(S, S_i))$ is biholomorphically equivalent to that of $R(S, S_i)$. Thus the identification makes sense.

The proof that $R(S, S_i)$ is quasi-projective consists of solving two problems.

Problem 1. We want to find a biholomorphic projective embedding of $p(T(S))/\mathrm{Mod}_n(S, S_i) = T(S)/\mathrm{Mod}(S, S_i)$. We proceed as follows:

Let R be the subgroup of Mod_n consisting of matrices congruent to the identity mod 3. Then R is a normal subgroup of Mod_n of finite index containing no elements of finite order. Let M' be the smallest subgroup of Mod_n containing R and $\mathrm{Mod}_n(S, S_i)$. Then M' is a subgroup of Mod_n of finite index. The modular forms of high enough weight give a biholomorphic embedding of $\mathcal{H}_n/M'$ in some projective space [p. 17-10 of 17]. Let p' be the projection of $\mathcal{H}_n$ onto $\mathcal{H}_n/M'$ and let p'' be the projection of $\mathcal{H}_n$ onto $\mathcal{H}_n/\mathrm{Mod}_n(S, S_i)$. To show that $R(S, S_i)$ has a biholomorphic projective embedding it is enough to show that there is a holomorphic mapping $\hat{\phi}$ of $\mathcal{H}_n/\mathrm{Mod}_n(S, S_i)$ onto $\mathcal{H}_n/M'$ such that $\hat{\phi}$ restricted to $R(S, S_i)$ is a biholomorphic embedding.

We can define $\hat{\phi} : \mathcal{H}_n/\mathrm{Mod}_n(S, S_i) \to \mathcal{H}_n/M'$ by $\hat{\phi}(x) = p'(p'')^{-1}(x)$. $\hat{\phi}$ is well defined, and we have a commutative diagram:

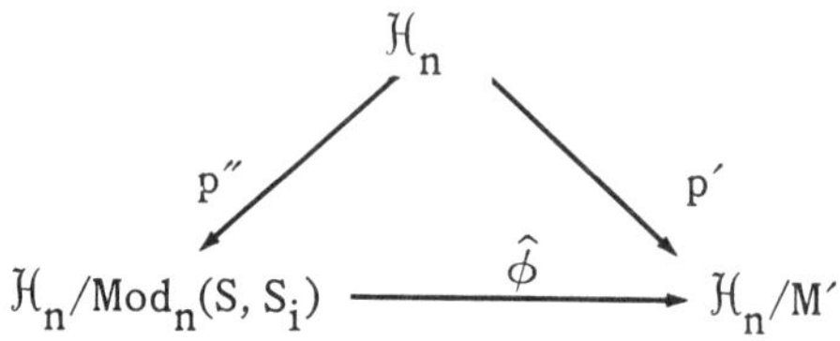

Let $F = F_{p(j)}$ or $F = F_{p(j)} \cap F_{p(h)}$. The context will make it clear which set we mean.

LEMMA 2. *If* $Z \in \mathcal{H}_n$ *with* $Z \in F$ *and* $g(Z) \in F$ *for some* g *in* M', *then we may assume that* g *is in* $\mathrm{Mod}_n(S, S_i)$.

Proof. Assume $i = 1$. $F_{p(j)} = F_{J_{k,n}}$ and $\mathrm{Mod}_n(S, S_1) = C_{J_{k,n}}$. Set $J = J_{k,n}$. Since R is a normal subgroup of Mod_n, every element g in M' can be written as $g = hs$ where $h \in R$ and $s \in C_J$. Let $JhJ = \underline{h}$. Then $\underline{h} \in R$ since R is normal and $J = J^{-1}$. We have:

$$J(Z) = Z$$

$$J(hs(Z)) = hs(Z)$$

$$J(hs(Z)) = \underline{h}Js(Z) = \underline{h}sJ(Z) = \underline{h}s(Z) .$$

Thus

$$hs(Z) = \underline{h}s(Z)$$

or

$$(\underline{h})^{-1}h(s(Z)) = s(Z) .$$

Thus $(\underline{h})^{-1}h$ has a fixed point in $\mathcal{H}_n$. $(\underline{h})^{-1}h \in R$. All of the elements of R are of infinite order. Since Mod_n acts properly discontinuously on $\mathcal{H}_n$, no element of R other than the identity has a fixed point. Thus $\underline{h}^{-1}h = \text{id}$. Thus $JhJ = h$ so that $h \in C_J$.

Assume $i = 2$. $\text{Mod}(S, S_2) = C_j \cap C_h$. j and h are chosen so that $p(j) = J_{m,n}$ where $m = 2(k-3) - 2$ and $p(h) = J_{6,n}$. Set $J = J_{m,n}$ and $J' = J_{6,n}$. Then $p(\underline{C}_j) = C_J$ and $p(\underline{C}_h) = C_{J'}$. This is true because both J and J' have more than 4 fixed points. $\text{Mod}_n(S, S_2) = C_J \cap C_{J'}$.

If g in M' we can write $g = ts$ where $t \in R$ and $s \in C_J \cap C_{J'}$. Repeat the previous argument twice to conclude that $t \in C_J \cap C_{J'}$.

We fix the following notation for the rest of this section: If V is any complex manifold, B a properly discontinuous group of analytic automorphisms of V, and $b : V \to V/B$ the projection, then a base for the topology of V/B is given by the following distinguished neighborhoods, N: For $\underline{x} \in V$ with $b(\underline{x}) = x$ there is a neighborhood $\underline{N}$ of $\underline{x}$ with $g(\underline{N}) \cap \underline{N} = \emptyset$ for all g in B unless $g \in B_{\underline{x}}$, the stability subgroup of B at $\underline{x}$, in which case $g(\underline{N}) = \underline{N}$. Let $N = b(\underline{N})$.

REMARK. Assume that D is a subgroup of B with $D_{\underline{x}} = B_{\underline{x}}$ for some $\underline{x}$. Let $q : V \to V/D$. Let N be a distinguished neighborhood of $q(\underline{x})$ and W a distinguished neighborhood of $b(\underline{x})$. Let $\underline{W}' = \underline{N} \cap \underline{W}$. Then $W' = q(\underline{W}')$ is a distinguished neighborhood of $q(\underline{x})$ and $W' \subset N$. If we replace such an N by W' we still have a base for the topology of V/D.

LEMMA 3. *Let* g *and* g' *be any two elements of* B. *Then* $q(F_g)$ *and* $q(F_g \cap F_{g'})$ *are analytic subvarieties of* V/D.

Proof. Let $N = \underline{N}/D_{\underline{x}}$. Since $D_{\underline{x}}$ is a finite group, the restriction of q to $\underline{N}$ is proper. Thus the image of an analytic subvariety of $\underline{N}$ is an analytic subvariety of N. In general then, the image of an analytic subvariety K of V is the union of local subvarieties of V/D.

We say that a subvariety K of V satisfies * if there is a subvariety K′ of V such that:

(*) If $q(y) \in q(K) \cap N$ and $y \in K$ with $t(y)$ in $\underline{N}$ for some t in D, then $t(y) \in K' \cap \underline{N}$.

If K satisfies *, then $q(K) = q(K')$ and $q(K')$ is an analytic subvariety of V.

Let $F = F_g$. Then F satisfies *:

Proof: Let $F' = \bigcup_{r \in B} F_{rgr^{-1}}$. $F' \cap \underline{N} = \bigcup_{\substack{r \in B \\ rgr^{-1} \in B_{\underline{x}}}} F_{rgr^{-1}}$

and this is a finite union. Thus F′ is a subvariety of V. Suppose $q(y) \in q(F) \cap N$. Then we may assume $y \in F$ and $h(y) \in \underline{N}$ for some h in D. Then $hgh^{-1}(h(y)) = h(y)$ so $hgh^{-1} \in B_{\underline{x}}$. Thus $h(y) \in F' \cap \underline{N}$.

Let $F = F_g \cap F_{g'}$. Then F satisfies condition *.

Proof: Let $F' = \bigcup_{r \in B} (rF_g \cap rF_{g'})$ and proceed as above.

PROPOSITION 5. *$\hat{\phi} : p''(F) \to p'(F)$ is biholomorphic.*

Proof. Lemma 2 says that the restriction of $\hat{\phi}$ to $p''(F)$ is $1-1$. Lemma 2 also says that for x in F, $M'_x = [\mathrm{Mod}_n(S, S_i)]_x$. If we combine these facts plus Lemma 3 and its preceding remark with the V-manifold structures of $\mathcal{H}n/\mathrm{Mod}_n(S, S_i)$ and $\mathcal{H}n/M'$, it is not difficult to see that the restriction of $\hat{\phi}$ to $p''(F)$ is biholomorphic.

As an immediate consequence we have:

COROLLARY 2. *The restriction of $\hat{\phi}$ to $R(S, S_i)$ is $1-1$. If $\hat{\phi}(R(S, S_i))$ is quasi-projective, then $R(S, S_i)$ is, too. Moreover, in that case $R(S, S_i)$ and $\hat{\phi}(R(S, S_i))$ are biholomorphically equivalent.*

Let $W' = \mathcal{H}n/M'$ and W'^* be its Satake-Baily compactification [16, 17].

Problem 2. To show that $R(S, S_i)$ is quasi-projective we must show that the closure of $p'(p(T(S)))$ in W'^* is an analytic subvariety and that there is an analytic subvariety T in W'^* such that $p'(p(T(S))) = \overline{p'(p(T(S)))} - T$. (The bar over a set denotes its topological closure.) Once $p'(p(T(S)))$ is the difference of two analytic subvarieties of projective space we can apply Chow's theorem to obtain the desired result.

We will use repeated applications of the following:

LEMMA 4. *Let* X *be an analytic space and* Y *an analytic subvariety of* X. *Assume* $X - Y = \cup Z_i$ *where the* Z_i *are closed disjoint subvarieties of* $X - Y$. *Then for each irreducible* Z_i *the closure of* Z_i *in* X *is an analytic subvariety of* X *and* $Z_i = \overline{Z}_i - Y$.

Proof. We may assume that each Z_i is irreducible. If not, replace it by the union of its irreducible components.

Let X_i be the smallest subvariety of X containing Z_i. Set $Y_i = X_i \cap Y$. Since Z_i is irreducible we may assume that X_i is also. Thus $\dim X_i > \dim Y_i$ so that $\dim X_i = \dim X_i - Y_i$.

Fix i and let $\{T_j\}$ be the irreducible components of $X_i - Y_i$. There is one T_j with $\dim T_j = \dim X_i$. Apply the Remmert-Stein Theorem [7, p. 169] to conclude that the closure of T_j in X_i is analytic. Since X_i and $\overline{T}_j$ are both irreducible of the same dimension they must be equal.

Using the fact that the $\{Z_i\}$ are disjoint closed subvarieties one can see that $T_j \subset Z_k$ for some k. Since $X_i = \overline{T}_j$ and $X_i \supset Z_i$, we must have $k = i$ and $T_j = Z_i$.

The Case of an Even Number of Punctures

We are now in a position to prove that $R(S, S_1)$ is quasi-projective.

Let $W = \mathcal{H}n/R$ and $W' = \mathcal{H}n/M'$. Let W^* and W'^* be their Satake-Baily compactifications. Let r be the natural map $r : \mathcal{H}n \to W$ and s the map $s : W^* \to W'^*$. s is a proper holomorphic map. Let $\underline{J}$ be the analytic automorphism of W^* induced by J. ($\underline{J}$ acts on W and extends to W^* in a natural manner.)

LEMMA 5. *Let* $M = \{g \in R \mid JgJ = g^{-1}\}$. *Then* $W \cap F_{\underline{J}} = \bigcup_{g \in M} r(F_{gJ})$.

Proof. $r(Z) \in F_{\underline{J}}$ implies $J(Z) = g(Z)$ for some g in R. Thus $Jg(Z) = Z$. Hence, $JgJg(Z) = Z$. Let $\underline{g} = JgJ$. Since $J = J^{-1}$ and R is a normal subgroup of Mod_n, $\underline{g} \in R$. Thus $\underline{g}g(Z) = Z$. No element of R distinct from the identity has a fixed point in $\mathcal{H}n$. $\underline{g}g$ in R. Thus $\underline{g}g = \mathrm{id}$ and $\underline{g} = g^{-1}$.

LEMMA 6. *Let* g *and* g' *be any two elements of* R. *Then either* $r(F_{gJ}) \cap r(F_{g'J}) = \emptyset$ *or* $r(F_{gJ}) = r(F_{g'J})$.

Proof. Assume that $Z \in F_{gJ}$ and that $h(Z) \in F_{g'J}$ for some h in R. Then it suffices to show that $h(F_{gJ}) = F_{g'J}$. But $h(F_{gJ}) = F_{hgJh^{-1}}$. Let $h(Z) = Z'$. Then $Z' \in F_{g'J} \cap F_{hgJh^{-1}}$. Thus $g'JhgJh^{-1}(Z') = Z'$. Since R is a normal subgroup, $JhgJ$ is in R. Thus $g'JhgJh^{-1}$ is in R so it cannot have a fixed point unless it is the identity. Thus $g'J = (hgJh^{-1})^{-1} = hgJh^{-1}$.

THEOREM 2a). $R(S, S_1) = sr(p(T(S)))$ *is quasi-projective.*

Proof.

$$s(F_{\underline{J}}) \cap W' = \bigcup_{g \in M} sr(F_{gJ}) .$$

Since s is proper, this is a union of analytic subvarieties of W'. We can show that the union is disjoint: Assume $s(x) \in sr(F_{gJ}) \cap sr(F_{g'J})$. Then $x \in r(F_{gJ})$ and $\underline{t}(x) \in r(F_{g'J})$ where $\underline{t}$ is the analytic automorphism of W^* induced by some t in C_J. Thus $x \in r(tF_{g'J})$. If $g'' = tg't^{-1}$, then $tF_{g'J} = F_{g''J}$ since $t \in C_J$. Thus $r(F_{g''J}) = r(F_{gJ})$ by the proof of Lemma 6. Since $r(F_{g''J}) = \underline{t}r(F_{g'J})$, $sr(F_{g''J}) = sr(F_{g'J}) = sr(F_{gJ})$.

Apply Lemma 4 to this union as follows: Baily's theorem [2] says that $sr(E_n) = p'(p(T(S_1))) = X_1 - X_2$ where X_i is an analytic subvariety of W'^*. We let $X = X_1 \cap sr(F_{\underline{J}})$. Let $Y = X_2$ (i.e., $X_2 \supset (W'^* - W') \cap X_1$). Let $Z_g = sr(F_{gJ}) \cap (X - X_2)$ where g is in M. Then $Z_g = sr(F_{gJ} \cap E_n)$.

In particular, $Z_1 = \mathrm{sr}(F_J \cap E_n) = R(S, S_1)$ and Z_1 is irreducible. Thus by Lemma 4, $\overline{Z}_1$ is an analytic subset of X and we are done.

The Case of an Odd Number of Punctures

We need additional notation. In this case $M' = R(C_{p(j)} \cap C_{p(h)})$ and $p': \mathcal{H}n \to \mathcal{H}n/M'$ as before. Let $W' = \mathcal{H}n/M'$, let P' be the projection $P': \mathcal{H}n \to \mathcal{H}n/RC_{p(j)}$, and let $W'' = \mathcal{H}n/RC_{p(j)}$. Let W'^* and W''^* be the corresponding compactifications and let $\underline{p}$ be the proper holomorphic mapping $\underline{p}: W'^* \to W''^*$.

LEMMA 7. $p'(p(F_j)) = T_1 - T_2$ *and* $p'(p(F_h)) = R_1 - R_2$ *where* R_i *and* T_i *are analytic subvarieties of* W'^*. *Further* $T_1 = \overline{p'p(F_j)}$ *and* $R_1 = \overline{p'p(F_h)}$.

Proof. We give the proof for F_j. The proof for F_h is the same.

By Theorem 2a), $P'(p(F_j)) = V_1 - V_2$. Let $X = \underline{p}^{-1}(V_1)$ and let $Y = \underline{p}^{-1}(V_2)$. We may assume that if $p'(E_n) = L_1 - L_2$, then $L_1 \supset X$ and $Y \supset L_2$. Thus

$$X \cap p'(E_n) = X - Y$$

and

$$X \cap p'(E_n) = \bigcup_{t \in \underline{C}_j - \underline{C}_h} p'(p(tF_j)) \; .$$

It is not difficult to verify that this is a disjoint union. To see that these sets $p'(p(tF_j))$ are analytic subvarieties of $X - Y$, apply Lemma 3 to conclude that $p'(p(tF_j)) = p'(p(F_{tjt}-1))$ is an analytic subvariety of $p'(p(T(S_2))) = p'(E_n)$.

Let $Z_t = p'(p(tF_j))$. Then $Z_1 = p'p(F_j)$ which is irreducible. Apply Lemma 4 to conclude that the closure T_1 of Z_1 is analytic and that $Z_1 = T_1 - T_2$ where T_2 is also an analytic subvariety of W'^*.

THEOREM 2b). $R(S, S_2) = p'(p(F_j \cap F_h))$ *is quasi-projective.*

Proof. $p'(p(F_j \cap F_h)) \subset p'p(F_j) \cap p'p(F_h)$ so that $p'(p(F_j \cap F_h)) \subset R_1 \cap T_1 - L_2$. $R_1 \cap p'(E_n) = p'p(F_j)$ and $T_1 \cap p'(E_n) = p'p(F_h)$. $R_1 \cap T_1 - L_2 \subset R_1 \cap p'(E_n) \cap T_1 \cap p'(E_n)$ so that $R_1 \cap T_1 - L_2 \subset p'p(F_j) \cap p'p(F_h)$. Let $\tilde{R}$ be the normal subgroup of $\mathrm{Mod}(S_2)$ defined by $p(\tilde{R}) = R$.

$$\begin{aligned} p'p(F_j) \cap p'p(F_h) &= \bigcup_{t \in M'} p'[p(F_j) \cap tp(F_h)] \\ &= \bigcup_{\underline{t} \in \tilde{R}(\underline{C}_j \cap \underline{C}_h)} p'[p(F_j) \cap p(\underline{t}F_h)] \\ &= \bigcup_{\substack{\underline{t} \in \tilde{R}(\underline{C}_j \cap \underline{C}_h) \\ q \in N}} p'[p(F_j \cap q\underline{t}F_h)] \\ &= \bigcup_{\underline{t} \in \tilde{R}(\underline{C}_j \cap \underline{C}_h)} p'[p(F_j \cap \underline{t}F_h)] . \end{aligned}$$

By Lemma 3 each of the components in this union is an analytic subvariety of $p'p(T(S_2)) = p'(E_n) = L_1 - L_2$. To see that the components of this union are disjoint use Lemma 2.

Apply Lemma 4 as follows: Let $X = R_1 \cap T_1$ and let $Y = L_2$. Let $Z_t = p'[p(F_j \cap \underline{t}F_h)] \cap (X - Y)$. Since $p'[p(F_j \cap F_h)] \subset X - Y$, $Z_1 = R(S, S_2)$. From Lemma 4 we can conclude that the closure of Z_1 is an analytic subvariety of W'^*. This proves the theorem.

To summarize the above results: Let G and Q_i be the finitely generated Fuschian groups defined at the beginning of this section. We have shown:

THEOREM 2′. *If* $k = 6$ *or* $k \geq 8$, $R(G, Q_i)$ *is a quasi-projective variety.*

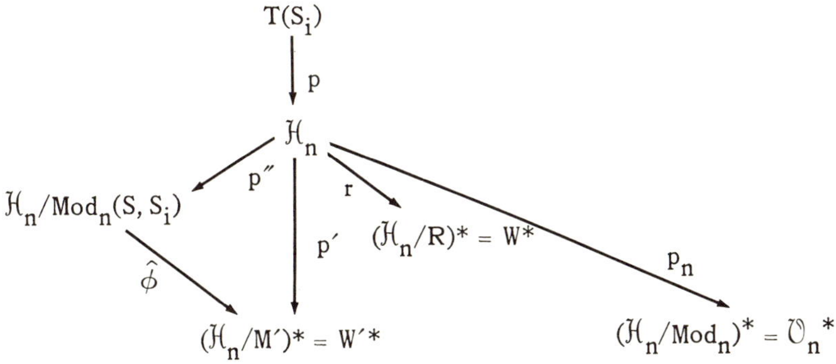

II

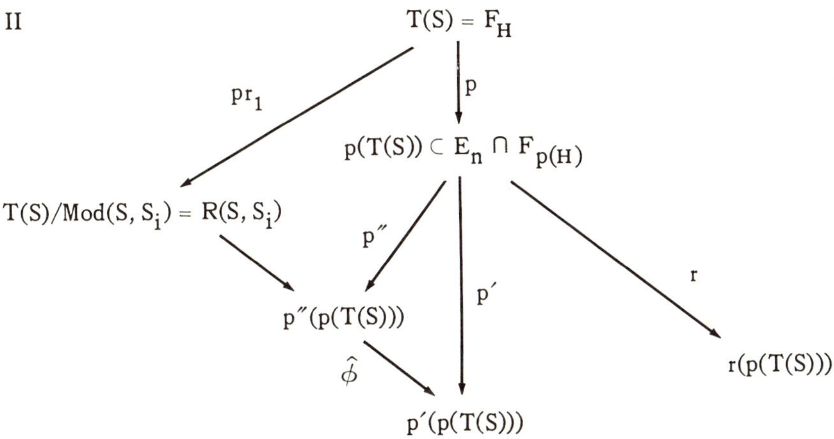

The above diagrams may help to summarize all of the various maps and spaces. Diagram II is a "subset" of diagram I in the sense that the sets in diagram II are subsets of the corresponding spaces in diagram I.

§III. Let G and G' be as in Section II. The purpose of this section is to show that $\mathfrak{M}(G, G_1, \ldots, G_m)$ is a group of algebraic morphisms of $R(G, G_1, \ldots, G_m)$. Assume that as a subset of $(\mathcal{H}_n/M')^*$ that $R(G, G') = V_0 - A_0$ where V_0 and A_0 are projective varieties and that $R(G, G_1, \ldots, G_m) = V - A$ where V and A are projective varieties. Recall that $\Psi : V \to V_0$.

LEMMA 8. *Let* $\underline{V}_0 = V_0 \cap \mathcal{H}_n/M'$. *Then* $\underline{V}_0$ *is a hyperbolic complex space and it is hyperbolically imbedded in* V_0. (For the definitions of these terms see p. 99 of [12].)

Proof. If $\mathcal{H}_n/M'$ is a hyperbolic complex space which is hyperbolically imbedded in its compactification, (see p. 237 of [11] and the references given there) then $\underline{V}_0$ as a closed analytic subvariety of a hyperbolic space is hyperbolic. Let d be the distance on $\mathcal{H}_n/M'$ and d' the distance on $\underline{V}_0$. Then $d'(p,q) \geq d(p,q)$ for all p, q in $\underline{V}_0$. To show that $\underline{V}_0$ is hyperbolically imbedded in V_0 it suffices to show that if p and q are in $V_0 - \underline{V}_0$ and $\{p_n\}$ and $\{q_n\}$ are sequences in $\underline{V}_0$ with $p_n \to p$ and $q_n \to q$ and $d'(p_n, q_n) \to 0$, then $p = q$. (This is condition 3′ on p. 237 of [11].) But $d'(p_n, q_n) \geq d(p_n, q_n)$. Thus $d(p_n, q_n) \to 0$. But 3′ holds for $\mathcal{H}_n/M'$ and its compactification. Thus $p = q$.

LEMMA 9. $\mathfrak{M}(G, G_1, \ldots, G_m)$ *is algebraic.*

Proof. Let $h \in \mathfrak{M}(G, G_1, \ldots, G_m)$. Then $h: V-A \to V-A$. To show that h is algebraic it suffices to show that Graph h is an algebraic subvariety of $(V-A) \times (V-A)$ [see Proposition 8 of 19]. To do this it suffices to produce an analytic subvariety of $V \times V$ of the same dimension as Graph h, for then by Proposition 4 of [14] using the fact that Graph h is irreducible we can conclude that the closure of Graph h in $V \times V$ is analytic. Since $V \times V$ is a projective variety, we can use Chow's theorem to conclude that this closure is an algebraic subvariety.

Let $\underline{h} = \Psi \circ h$. Then $\underline{h}: V-A \to \underline{V}_0$. By Theorem 3 of [10] we can conclude that $\underline{h}$ extends to a meromorphic mapping of V into V_0. This means among other things that the graph of $\underline{h}$ extends to an analytic subvariety of $V \times V_0$ which we will also call Graph $\underline{h}$ to simplify notation:

By Theorem 3, Graph h is a complex subspace of $V \times V_0$. To see that it is a closed complex subspace we note that if (Z, B, ϕ) is a resolution of A in V as in the proof of Theorem 3 of [10] (e.g., the Hironaka desingularization [8]), and f is the extension of $h \circ \phi$, then Graph f is a subvariety of $Z \times V_0$. Since ϕ is proper and Z and V are projective varieties, $(\phi \times \mathrm{id}) : Z \times V_0 \to V \times V_0$ is proper. Thus $(\phi \times \mathrm{id})$ Graph f = Graph $\underline{h}$ is a subvariety of $V \times V_0$.

The map $\mathrm{idx}\Psi : V \times V \to V \times V_0$ is holomorphic so that if $F = (\mathrm{idx}\Psi)^{-1}$ (Graph h), then F is an analytic subvariety of $V \times V$. Clearly Graph h is contained in F. Let B be the branch locus of Ψ. B' its inverse image in V. Let $A'' = A \cup B'$ and let $A' = A'' \times V \cup V \times A''$. We may assume F is irreducible. If it is not, we merely replace it by its irreducible component containing Graph h. Then $\dim F = \dim F - A'$. $F - A' = F \cap (V - A'') \times (V - A'') = \bigcup_{g \in \mathfrak{M}(G,G')}$ Graph $g \circ h$. Since we have removed the branch locus this is a disjoint union of closed subvarieties. Thus $\dim F = \dim$ Graph h.

REMARK. Since M' has fixed points, the metric $d_{W'}$ on W' has to be replaced by the metric $d'_{W'}$ [see p. 103 of 12], in order to say that W' is hyperbolic. Kiernan's extension theorem, therefore, may not be strictly applicable. We can get around this difficulty as follows:

Let R be as in Section II. Then the methods of Section II yield the result that the image of $p(T(S))$ in $(\mathcal{H}n/R)^*$ is quasi-projective. Assume that this image is $V_1 - A_1$. Then since R has no fixed points, we have $V_1 - A_1$ is hyperbolically imbedded in V_1.

Let $\underline{R}_1$ be the normal subgroup of $\mathrm{Mod}(G, Q_i)$ defined by $\underline{R}_1 = \{g \text{ in } \mathrm{Mod}(G, Q_i) \mid p(g) \in R\}$. (Note: we have changed back to the language of groups so $Q_i = G_1 = G'$ and p now maps $\mathrm{Mod}(Q_i) = \mathrm{Mod}(G')$ into Mod_n.) Let $h_i G' h_i^{-1} = G_i$. Let $\underline{R}_i = h_i \underline{R}_1 h_i^{-1}$. We have $h_i \mathrm{Mod}(G, G_1) h_i^{-1} = \mathrm{Mod}(G, G_i)$. Let $RR = \bigcap \underline{R}_i$. Then RR is a normal subgroup of finite index in both $\underline{R}_i$ and Mod(G). Let $H_i = \underline{R}_i / RR$ and let $H = \mathrm{Mod}(G)/RR$. There is a commutative diagram:

$$\begin{array}{ccc} & T(G)/RR & \\ {\scriptstyle p_1} \swarrow & & \searrow {\scriptstyle p_2} \\ ((T(G))/RR)/H_i = T(G)/\underline{R}_i & \xrightarrow[\;p_3\;]{} & T(G)/\mathrm{Mod}(G) = (T(G)/RR)/H \end{array}$$

We have:

Theorem 1 is true when RR replaces $\mathrm{Mod}(G, G_1, \ldots, G_m)$ and $\underline{R}_i$ replaces $\mathrm{Mod}(G, G_i)$.

Theorem 2′ is true when $\underline{R}_1$ replaces $\mathrm{Mod}(G, G_1)$.

Lemmas 8 and 9 are true when $\underline{R}_1$ replaces $\mathrm{Mod}(G, G_1)$.

Combining these result we can conclude:

THEOREM 3. *If* $g \neq 0$ *and* $k = 6$ *or* $k \geq 8$, *then* $X(g, k)$ *is a quasi-projective variety.*

Appendix. In what follows let V represent $R(G, G')$ or $R(G, G_1, \ldots, G_m)$ where G and G′ are as before. V is a normal analytic space. We write ${}_V\theta$ to denote the sheaf of germs of holomorphic functions on V. We want to prove that if V is a quasi-projective variety, then it is a normal quasi-projective variety. The proof outlined below is due to Michael Engber. The numbers in the proof (e.g. (18.4)) refer to theorems, lemmas, and definitions in Nagata [15].

Let $\mathbb{C}$ be the complex numbers; $\mathbb{C}[z_1, \ldots, z_n]$ the polynomial ring in n variables over $\mathbb{C}$; $\mathbb{C}\{z_1, \ldots, z_n\}$ the ring of convergent power series in n variables; and $\mathbb{C}[[z_1, \ldots, z_n]]$ the ring of formal power series in n variables over $\mathbb{C}$. If R is a local ring and m its maximal ideal and if $\cap\, m^p = 0$, we let $\hat{R}$ denote the m-adic completion of R.

Let z be a point of V and N a neighborhood of z with $(N, {}_V\theta|N)$ isomorphic to $(Y, {}_Y\theta)$ where Y is a subvariety of a domain in $\mathbb{C}^n$ for some n. If $\mathcal{J}$ is the ideal sheaf of Y, ${}_Y\theta = {}_n\theta/\mathcal{J}|Y$ where ${}_n\theta$ is the sheaf of germs of holomorphic functions on $\mathbb{C}^n$. Let $R = {}_Y\theta = \mathbb{C}\{z_1, \ldots, z_n\}/\mathcal{J}$. Then the statement that V is a normal analytic space means that R is normal.

We let $\underline{z}$ be the image of z under the isomorphism of $(N, {}_V\theta|N)$ and $(Y, {}_Y\theta)$. Let p be the ideal of $\mathbb{C}[z_1, \ldots, z_n]/\mathcal{J}$ of functions that are zero at $\underline{z}$. Note that since V is quasi-projective we may assume that $\mathcal{J}$ corresponds to an ideal of $\mathbb{C}[z_1, \ldots, z_n]$. Let $P = \mathbb{C}[z_1, \ldots, z_n]/\mathcal{J}$ and let

$\theta = P_p$ be P localized at p. Then V is normal at z if and only if θ is normal (i.e., integrally closed in its quotient field). To show that θ is normal we will show:

(i) R normal implies that $\hat{R}$ is normal

(ii) $\hat{\theta} = \hat{R}$.

(iii) $\hat{\theta}$ normal implies θ normal.

To prove (i): We use (37.8) of [15] to show that $\hat{R}$ is normal. We need to show that

a) there is an analytically normal ring S such that R is of finite type over S and R/S is separable, and

b) R is local and pseudo-geometric.

We use (45.5) of [15] to show that there exist $y_1, \ldots, y_s$ which are analytically independent such that R is a finite module over $\mathbb{C}\{y_1, \ldots, y_s\}$. Let $S = \mathbb{C}\{y_1, \ldots, y_s\}$. Then by (45.5) again $\hat{S} = \mathbb{C}[[y_1, \ldots, y_s]]$. $\hat{S}$ is, therefore, a regular local ring and thus a unique factorization domain. Thus $\hat{S}$ is normal. S is analytically normal, for by definition S analytically normal if $\hat{S}$ is normal. R/S is separable since we are over a field of characteristic zero. Thus a) is proved.

To show b) we remark that by (45.5) and (45.6) R is Weierstrass. But Weierstrass by definition (45.1) implies pseudo-geometric. R is local since $\mathbb{C}\{z_1, \ldots, z_n\}$ is a local ring and the image of its maximal ideal in R is a maximal ideal of R. Thus b) is proved so that we have (i).

(ii) is merely Proposition 3 of [19].

To prove (iii): We remark that $\hat{\theta}$ and θ satisfy the conditions of (18.3) and thus the conditions of (18.4) which says essentially that $\hat{\theta}$ normal implies θ normal.

RUTGERS UNIVERSITY
NEWARK, NEW JERSEY

REFERENCES

[1]. Accola, Robert D., "Automorphisms of Riemann Surfaces," *Journal D'Analyse Mathematique*, vol. 18 (1967), pp. 1-5.

[2]. Baily, Walter L., Jr., "On the Moduli of Jacobian Varieties," *Annals of Mathematics*, vol. 71, ser. 2 (1960), pp. 303-314.

[3]. Bers, L., *Riemann Surfaces*, New York University Lecture Notes, 1957-58.

[4]. Bers, L., and Greenberg, L., "Isomorphisms between Teichmüller Spaces," *Advances in the Theory of Riemann Surfaces*, Princeton University Press, 1971, pp. 53-80.

[5]. Earle, Clifford J., "On the Moduli of Closed Riemann Surfaces with Symmetries," *Advances in the Theory of Riemann Surfaces*, Princeton University Press, 1971, pp. 119-130.

[6]. Gilman, Jane, "Compact Riemann Surfaces with Conformal Involutions," *Proceedings of the AMS*, Vol. 37 (1973), pp. 105-107.

[7]. Gunning, R. C., and Rossi, H., *Analytic Functions of Several Variables*, Prentice-Hall, Inc., Englewood Cliffs, N. J., 1965.

[8]. Hironaka, H., "Resolution of Singularities of an Algebraic Variety over a Field of Characteristic Zero: I and II," *Annals of Math.*, Vol. 79 (1964), pp. 109-326.

[9]. Harvey, W. J., "On Branch Loci in Teichmüller Space," *Transactions of the AMS*, vol. 153 (1971), pp. 387-399.

[10]. Kiernan, P. J., "Extension of holomorphic maps." *Trans. AMS*, (1972), pp. 347-355.

[11]. ———— and Kobayashi, "Satake Compactification and Extension of Holomorphic Mappings," *Inventiones Math*, 16 (1972), pp. 237-248.

[12]. Kobayashi, *Hyperbolic Manifolds and Holomorphic Mappings*, Marcel Dekker, 1970.

[13]. Mumford, D., "Abelian Quotients of the Teichmüller Modular Group," *Journal D'Analyse Mathematique*, vol. 18 (1965), pp. 227-244.

[14]. Narasimhan, Raghavan, *Introduction to the Theory of Analytic Spaces*, Springer-Verlag 1966.

[15]. Nagata, Masayoshi, *Local Rings*, Interscience Publishers, London, 1962.

[16]. Satake, I., "On the compactification of the Siegel Space," *J. Indian Math. Soc.*, (1956), pp. 259-281.

[17]. Seminaire de H. Cartan, 1957-58.

[18]. Serre, Jean Pierre, *Groups Algebrique et Corps de Class*, Herman, Paris, 1959.

[19]. ————, "Geometrie Algebrique et Geometrie Analytique, *Annales Inst. Fourier*, 6, 1955-56, pp. 1-42.

MAXIMAL GROUPS AND SIGNATURES

Leon Greenberg

§1. *Maximal signatures of Fuchsian groups*

Let Γ be a Fuchsian group whose quotient surface has finite area. Γ is *maximal* if it is not a subgroup of any other Fuchsian group. A signature $\sigma = (g; \nu_1, \ldots, \nu_k; t)$ will be called *maximal* if every group of signature σ is maximal. We shall be interested in obtaining a qualitative and quantitative estimate on the number of maximal signatures. For this purpose, we introduce the following notation:

(1) $L^k = \{(\nu_1, \ldots, \nu_k) \in Z^k : \nu_i \geq 2,\ 1 \leq i \leq k\}$.

(2) For non-negative integers g, k, t, we define $M(g, k, t)$ to be the set of all points $\nu = (\nu_1, \ldots, \nu_k) \in L^k$, such that $\sigma = (g; \nu_1, \ldots, \nu_k; t)$ is a maximal signature.

We regard L^k as a subset of $\mathbb{Q}^k$, in which we have the usual notion of "affine subspace," namely the set of solutions of a system of linear (not necessarily homogeneous) equations with rational coefficients.

THEOREM 1. (1) $M(g, k, t)$ *is empty if and only if*

(a) *For* $g > 0$, $k \leq 2g+2$ (t *even*) *or* $k \leq 2g+1$ (t *odd*).

(b) *For* $g = 0$, $k \leq 2$ *and* $t > 1$.

(2) *Except for the above cases,* $M(g, k, t)$ *is the complement in* L^k *of a finite set of proper affine subspaces.*

As an illustration of the theorem, consider the case where $g = t = 0$, $k = 3$. In this case we are dealing with triangle groups (ν_1, ν_2, ν_3). There is (up to conjugation) only one group per signature here. Thus the theorem asserts that the non-maximal triangle groups fall into a finite number of families, in which the integers ν_i are related by linear equations. This is confirmed in [1] and [2], where it is shown that the non-maximal triangle groups are (up to permutations): (a, a, b), $(2, a, 2a)$ and $(3, a, 3a)$.

We shall make Theorem 1 more quantitative by finding a rough estimate on the exceptional subspaces.

THEOREM 2. *Let* $k > 2g + 2$, $N = 42(2(g-1) + k + t)$ *and* $\Sigma = \{a/b : a, b \in \mathbf{Z}, 1 \le a, b \le N-1\}$. *The signature* $\sigma = (g; \nu_1, \ldots, \nu_k; t)$ *is maximal if* $\nu_i \notin \Sigma$ *and* $\nu_i/\nu_j \notin \Sigma$ *(for* $1 \le i, j \le k$, $i \ne j$).

By using this theorem, one can construct any number of maximal signatures. For example (taking $t = 0$) the following signatures are maximal: $(0; p_1, p_2, p_3, p_4)$, where the p_i are different primes > 84; $(1; q_1, q_2, q_3, q_4, q$ where the q_i are different primes > 210; $(2; r_1, \ldots, r_7)$, where the r_i are different primes > 378. Note that these statements are purely group theoretical. The abstract groups of the above signatures are not isomorphic to a subgroup of any other Fuchsian group. (A similar statement may be made in the non-compact case, when $t > 0$. However, in this case, one must consider only isomorphisms which preserve parabolic elements.)

We shall prove several lemmas which will lead to proofs of Theorems 1 and 2. First we shall itemize some standard notation and terminology which will be used in Section 1.

(i) D will always denote the unit disc $\{z \in \mathbf{C} : |z| < 1\}$.

(ii) S and T will always denote Riemann surfaces of finite type. Recall that a surface of finite type (g, t) is conformally equivalent to a closed surface of genus g, minus t points.

(iii) Γ and Δ will denote Fuchsian groups whose quotients have finite area.

(iv) Recall that a ramified covering $\phi : A \to B$ of Riemann surfaces is called *normal*, if the group G of covering transformations is transitive on each ϕ-fiber. In this case $B \approx A/G$.

LEMMA 1. *Let* $\phi: D \to S$ *be a ramified covering. Then* ϕ *is normal* $\Longleftrightarrow$ *the branching orders are constant along each fiber.*

Proof. ($\Longrightarrow$) This is clear, since the covering group preserves branching orders.

($\Longleftarrow$) Let $B = \{b_1, b_2, \ldots\}$ denote the set of branch points. Let $D_0 = D - B$, $S_0 = S - \phi(B)$. $\phi: D_0 \to S_0$ is an unramified covering, which is normal $\Longleftrightarrow$ $\phi: S \to D$ is normal. Choose a point $z_0 \in D_0$, and let $s_0 = \phi(z_0)$. From covering space theory, ϕ is normal $\Longleftrightarrow$ $\phi^*\pi_1(D_0, z_0)$ is a normal subgroup of $\pi_1(S_0, s_0)$.

Choose a small circle γ_i around each branch point b_i, such that the γ_i which correspond to branch points on the same fiber project down onto the same curve in S_0. ϕ wraps γ_i $\nu(b_i)$-times around $\phi(b_i)$ (where $\nu(b_i)$ is the branching order of b_i).

There exist curves β_i connecting z_0 to γ_i, such that the homotopy classes of $\alpha_i = \beta_i \gamma_i \beta_i^{-1}$ generate $\pi_1(D_0, z_0)$. To show that $\phi^*\pi_1(D_0, z_0)$ is normal, it suffices to show that for each loop δ at s_0 in S_0, the homotopy class $[\delta(\phi \circ \alpha_i)\delta^{-1}] \in \phi^*\pi_1(D_0, z_0)$. This will be true $\Longleftrightarrow$ $\delta(\phi \circ \alpha_i)\delta^{-1}$ lifts to a loop at z_0 in D_0.

The curve δ lifts to a curve $\tilde{\delta}$, connecting z_0 to a point z_1 on the same fiber. $\phi \circ \beta_i$ lifts to a curve β which connects z_1 to γ_j (a circle going around the branch point b_j, which lies on the same fiber as b_i). Now we come to the essential point of our argument: Since $\nu(b_i) = \nu(b_j) = \nu$, both circles γ_i and γ_j are wrapped ν-times around $\phi(b_i) = \phi(b_j)$. Therefore, the projection $\phi \circ \gamma_i$ lifts (near b_j) to the closed curve γ_j. This means that the curve $\delta(\phi \circ \alpha)\delta^{-1}$ lifts to $\tilde{\delta}\beta\gamma_j\beta^{-1}\tilde{\delta}^{-1}$, which is a loop at z_0. Q.E.D.

DEFINITION. A *signature* of order k on S is a function $\nu: S \to \mathbf{Z}$, such that

(1) $\nu(s) \geq 1$ for all $s \in S$,

(2) $\nu(s) > 1$ for exactly k points $s \in S$.

DEFINITION. Let $\phi : D \to S$ be a ramified covering, and ν a signature on S. Then ϕ is a *uniformization of* S *with signature* ν, if for each $s \in S$, the branching orders in $\phi^{-1}(s)$ are all equal to $\nu(s)$. By Lemma 1, ϕ is necessarily normal.

Suppose that S has type (g, t). Let $\phi : D \to S$ be a uniformization with signature ν, and Γ the group of covering transformations. Then Γ is a Fuchsian group of signature $\sigma = (g; \nu_1, \ldots, \nu_k; t)$, where the ν_i are the values $\nu(s) > 1$. (It is well-known that if the above σ is an admissible signature for Fuchsian groups, then S may be uniformized with signature ν.) Conversely, if Γ is a Fuchsian group, then $S = D/\Gamma$ is a Riemann surface with a naturally defined signature ν, such that the projection $\phi : D \to S$ is a uniformization with signature ν.

DEFINITION. Let $\phi : S \to T$ be a ramified covering. Let ν be a signature on S and for each $s \in S$, let $\beta(s)$ be the branching order of ϕ at s. Then ν and ϕ are called *compatible* if $\nu(s)\beta(s)$ is constant along each ϕ-fiber.

The basic idea of our proof is contained in the following two lemmas.

LEMMA 2. *Let* Γ, Δ *be Fuchsian groups, such that* $\Gamma \subset \Delta$. *Let* $S = D/\Gamma$ *(with the signature* ν *inherited from* Γ*),* $T = D/\Delta$ *and* $\phi : S \to T$ *the covering induced by the group inclusion. Then* ϕ *and* ν *are compatible.*

Proof. Let $\psi : D \to S$ be the quotient map (mod Γ). *Then* $\phi \circ \psi : D \to T$ is the quotient map (mod Δ). We use the notation β_ϕ, β_ψ and $\beta_{\phi\circ\psi}$ to denote branching orders with respect to the different coverings.

Let $t \in T$, $s_1, s_2 \in \phi^{-1}(t)$, and choose $z_1, z_2 \in D$ such that $\psi(z_i) = s_i$. Then $\beta_{\phi\circ\psi}(z_i) = \beta_\psi(z_i)\beta_\phi(\phi(z_i)) = \nu(s_i)\beta_\phi(s_i)$. But $z_1, z_2 \in (\phi\circ\psi)^{-1}(t)$, while $\phi\circ\psi$ is a normal covering. Therefore $\beta_{\phi\circ\psi}(z_1) = \beta_{\phi\circ\psi}(z_2)$ and $\nu(s_1)\beta_\phi(s_1) = \nu(s_2)\beta_\phi(s_2)$. Q.E.D.

LEMMA 3. *Let* $\phi : S \to T$ *be a ramified covering and* ν *a signature on* S, *compatible with* ϕ. *Let* $\psi : D \to S$ *be a uniformization of* S *with signature* ν, *and let* Γ, Δ *be the groups of covering transformations of* ψ *and* $\phi \circ \psi$, *respectively. Then* ψ *and* $\phi \circ \psi$ *are normal coverings,* $S = D/\Gamma$, $T = D/\Delta$, $\Gamma \subset \Delta$, *and* ϕ *is the covering induced by the group inclusion.*

Proof. All statements follow easily from the normality of ψ and $\phi \circ \psi$, and this is implied by Lemma 1. Q.E.D.

The point of Lemma 3 is that whenever we choose a signature ν, compatible with a covering ϕ, we are led to a group inclusion $\Gamma \subset \Delta$. Conversely, Lemma 2 says that any group inclusion is obtained in this way.

LEMMA 4. *Let* $\phi : S \to T$ *be an* n*-sheeted covering. Suppose that* S *has type* (g, t) *and* b *branch points of order* n. *Then*

(1) $b \leq 2g + 2$,

(2) *If* t *is odd and* $g > 0$, *then* $b \leq 2g + 1$.

(3) *If* $t = 1$ *and* $g = 0$, *then* $b \leq 1$.

Proof. (1) Let $h = \text{genus}(T)$ and let $\bar{S}$ and $\bar{T}$ denote the closed surfaces containing S and T. ϕ extends to a ramified covering $\bar{\phi} : \bar{S} \to \bar{T}$. If r denotes the total ramification of $\bar{\phi}$, then by the Hurwitz relation:

$$2(g-1) = 2n(h-1) + r \geq -2n + (n-1)b \ .$$

Therefore

$$b \leq \frac{2(g-1)+2n}{n-1} = \frac{2g}{n-1} + 2 \leq 2g + 2 \ .$$

(2) If $b = 2g + 2$, then all of the above inequalities are equalities. If $g > 0$, then $n = 2$ (and $h = 0$). Since $\bar{\phi} : \bar{S} \to \bar{T}$ has at most $2g + 2$ branch points, and $\phi : S \to T$ has $2g + 2$ branch points, all branch points in $\bar{S}$ are contained in S. Therefore the ϕ-fibers in $\bar{S} - S$ each have two points, and $t = \text{card}(\bar{S} - S)$ is even.

(3) Suppose $t = 1$ and $g = 0$. Then $S = \overline{S} - \{p\}$. The point p must be the only point on its fiber. Since $\overline{S}$ has at most 2 branch points of order n, and p is one of these, it follows that S can have at most one point of order n. Q.E.D.

Proof of Theorem 1. Let (g, k, t) be a fixed triple, $\phi : S \to T$ a ramified covering and $p \in P = \{(p_1, \ldots, p_k) \in S^k : p_i \neq p_j \text{ for } i \neq j\}$. We shall say that the pair (ϕ, p) is *admissible* if

(1) S has type (g, t).

(2) There exists a signature ν of order k on S, such that ν is compatible with ϕ and $\nu(p_i) > 1$ $(1 \leq i \leq k)$.

(3) The signature $\sigma = (g; \nu_1, \ldots, \nu_k, t)$ (where $\nu_i = \nu(p_i)$) is the signature of a Fuchsian group.

We shall call ϕ *admissible* if (ϕ, p) is admissible for some $p \in P$.

For a given admissible pair (ϕ, p), let $A(\phi, p)$ denote the set of all signatures ν which satisfy (2). Let $\nu \in A(\phi, p)$, $\nu_i = \nu(p_i)$ and $\overline{\nu} = (\nu_1, \ldots, \nu_k)$. Lemma 3 says that $\nu \in L^k - M(g, k, t)$. Put $\overline{A}(\phi, p) = \{\overline{\nu} : \nu \in A(\phi, p)\}$. Lemmas 2 and 3 tell us that $L^k - M(g, k, t) = \bigcup \{\overline{A}(\phi, p) : (\phi, p) \text{ admissible}\}$.

We claim that $\overline{A}(\phi, p)$ is (the intersection with L^k of) an affine subspace, and as (ϕ, p) runs through the (possibly infinite) set of admissible pairs, we obtain only finitely many subspaces.

Let $\beta(s)$ be the branching order of ϕ at s. The signature ν belongs to $A(\phi, p)$ if and only if $\nu(p_i) > 1$ $(1 \leq i \leq k)$ and, for each ϕ-fiber $s_1, s_2, \ldots, s_h$,

$$(*) \qquad \beta(s_1)\nu(s_1) = \beta(s_2)\nu(s_2) = \ldots = \beta(s_h)\nu(s_h) .$$

When we restrict condition $(*)$ to the fibers containing the points p_i, we obtain a system of linear equations in the variables $\nu_i = \nu(p_i)$, which define the set $\overline{A}(\phi, p)$.

We must now show that there are only a finite number of the above subspaces. Let $|\phi|$ denote the number of sheets of the covering $\phi : S \to T$. We shall first show that $|\phi|$ is bounded.

For each $\bar{\nu} = (\nu_1, \ldots, \nu_k) \epsilon \bar{A}(\phi, p)$ there exist Fuchsian groups $\Gamma \subset \Delta$, such that Γ has signature $\sigma = (g; \nu_1, \ldots, \nu_k; t)$. Furthermore $|\phi| = [\Delta : \Gamma] = A(\Gamma)/A(\Delta)$, where A denotes the area of a fundamental polygon. Thus

$$\frac{\pi}{21}|\phi| \leq A(\Delta)|\phi| = A(\Gamma)$$
$$= 2\pi\left(2(g-1) + t + \sum_{i=1}^{k}\left(1 - \frac{1}{\nu_i}\right)\right) \leq 2\pi(g-1) + t + k) .$$

Therefore

$$|\phi| \leq 42(2(g-1) + k + t) = N .$$

Next we note that the admissible coverings $\phi : S \to T$ give rise to only a finite number of branching schemes. Let h be the genus of T, and r the total ramification of ϕ. We have the Hurwitz relation:

$$2(g-1) = 2|\phi|(h-1) + r \geq -2|\phi| + r ,$$

so that

$$r \leq 2(g-1) + 2|\phi| \leq 2(g-1+N) .$$

Therefore the number of branch points and the branching orders are bounded, and there are only finitely many branching schemes.

The system of equations (*) which defines $\bar{A}(\phi, p)$ depends only on the branching scheme and the location of the points p_i. For each branching scheme, there are only finitely many non-isomorphic fibers. Therefore there are only finitely many different systems of equations (*) as the points p_i vary on S. Since there are only finitely many possible branching schemes, we see that the number of subspaces $\bar{A}(\phi, p)$ is finite.

We have now shown that $M(g, k, t)$ is the complement in L^k of finitely many affine subspaces. Therefore, $M(g, k, t)$ can be empty only when one of these subspaces is the entire space. This is possible only if the equations (*) (when restricted to the fibers containing the p_i) are empty equations, in the sense that each p_i is the only point on its fiber. By Lemma 4, this implies that for $g > 0$, $k \leq 2g+2$ (t even) or $k \leq 2g+1$ (t odd), while for $g = 0$, $k \leq 2$ and $t > 1$. (The cases $g = 0$, $t \leq 1$,

$k \leq 1$ do not give rise to Fuchsian signatures.) Conversely, if k satisfies these conditions, then the existence of hyperelliptic surfaces guarantees that $M(g, k, t)$ is empty if $g > 0$. The case $g = 0$ is taken care of by the functions z^n on the sphere. Q.E.D.

Proof of Theorem 2. In the proof of Theorem 1, it was shown that for any admissible covering $\phi : S \to T$, the number of sheets $|\phi| \leq N = 42(2(g-1) + k + t)$. Therefore all branching orders are $\leq N$.

Now suppose that $\bar{\nu} = (\nu_1, \ldots, \nu_k) \in \bar{A}(\phi, p)$, while $\nu_i \notin \Sigma$ and $\nu_i/\nu_j \notin \Sigma$. Since $k > 2g + 2$, Lemma 4 guarantees that one of the points p_i is not the only point in its fiber. All branch orders $\beta(s)$ on this fiber are $\leq N-1$. The corresponding equation (*) implies that either $\nu_i \in \Sigma$ or $\nu_i/\nu_j \in \Sigma$ (for some ν_j, with $i \neq j$). Q.E.D.

§2. *Maximal signatures of Kleinian groups*

If Γ is a Kleinian group, there always exists a Kleinian group Δ, such that the isometric circles of each group are exterior to those of the other. The group generated by Γ and Δ will then be Kleinian. Thus no Kleinian group is maximal. We shall define a restricted notion of maximality, which is quite analogous to the notion we have been using for Fuchsian groups.

DEFINITION. A Kleinian group is *finitely maximal* if it is not a subgroup of finite index in any other Kleinian group.

Let Γ be a finitely generated, non-elementary Kleinian group, with region of discontinuity Ω. The Ahlfors finiteness theorem asserts that Ω/Γ is a finite union of surfaces $S_i (1 \leq i \leq n)$ of finite type, which have only a finite number of branch images. Suppose that S_i has type (g_i, t_i) and that the projection $\Omega \to \Omega/\Gamma$ has branching orders $\nu_{i1}, \ldots, \nu_{ik_i}$ over S_i. Let $\sigma_i = (g_i; \nu_{i1}, \ldots, \nu_{ik_i}; t_i)$ and $\sigma = (\sigma_1, \ldots, \sigma_n)$. We shall call σ the signature of Γ. Note that each σ_i is the signature of a Fuchsian group, since Γ is non-elementary.

DEFINITION. The signature σ is *finitely maximal* if every Kleinian group of signature σ is finitely maximal.

THEOREM 3. *If* $\sigma_1, \sigma_2, \ldots, \sigma_n$ *are distinct maximal signatures of Fuchsian groups, then* $\sigma = (\sigma_1, \ldots, \sigma_n)$ *is finitely maximal.*

Proof. Let Γ be a Kleinian group of signature σ, and suppose $\Gamma \subset \Delta$, where $[\Delta : \Gamma] < \infty$. Then Γ and Δ have the same region of discontinuity Ω. Let Ω_0 be a connected component of Ω, and let Γ_0 and Δ_0 be the subgroups of Γ and Δ which preserve Ω_0. Let $S = \Omega_0/\Gamma_0$, $T = \Omega_0/\Delta_0$ and $\phi : S \to T$ the covering induced by the group inclusion $\Gamma_0 \subset \Delta_0$. Let ν be the signature on S inherited from Γ (in the same sense as in Section 1). The proof of Lemma 2 shows that ϕ is compatible with ν. Let $\psi : D \to S$ be a uniformization of S with signature ν, and F and G the groups of covering transformations of ψ and $\phi \circ \psi$. Lemma 3 tells us that $F \subset G$ and $\phi : S \to T$ is the covering induced by the group inclusion. But the signature of F is σ_i, for some i. Therefore F is maximal, $F = G$ and $\Gamma_0 = \Delta_0$.

Now let $\delta \in \Delta$. $\Omega_1 = \delta(\Omega_0)$ is another component of Ω. Let Γ_1 and Δ_1 be the subgroups of Γ and Δ which preserve Ω_1. By our previous argument, $\Gamma_1 = \Delta_1$. Clearly $\delta\Delta_0\delta^{-1} = \Delta_1$ and therefore $\delta\Gamma_0\delta^{-1} = \Gamma_1$. Let $p : \Omega \to \Omega/\Gamma$ be the natural projection, and $S_i = p(\Omega_i)$ (for $i = 0, 1$). Then δ induces a map $\bar{\delta} : S_0 \to S_1$, such that the diagram

$$\begin{array}{ccc} \Omega_0 & \xrightarrow{\ \delta\ } & \Omega_1 \\ \downarrow p & & \downarrow p \\ S_0 & \xrightarrow{\ \bar{\delta}\ } & S_1 \end{array}$$

commutes. This implies that S_0 and S_1 give rise to the same signature.

Since the signatures σ_i $(1 \le i \le n)$ are distinct, this means that $S_0 = S_1$. Therefore, there exists $\gamma \epsilon \Gamma$ such that $\gamma(\Omega_0) = \Omega_1$. It now follows that $\gamma^{-1}\delta \epsilon \Delta_0 = \Gamma_0$, and therefore $\delta \epsilon \Gamma$. Q.E.D.

It is important to note that we make no claim that a group Γ of signature $\sigma = (\sigma_1, \ldots, \sigma_n)$ exists. The theorem merely states that if one exists, then it is finitely maximal. However, by using the results of Section 1, one can easily construct Kleinian groups which have finitely maximal signatures. For example, let p_i $(1 \le i \le 5)$ and q_i $(1 \le i \le 5)$ be ten different prime numbers > 504. Let $\sigma_1 = (1; p_1, \ldots, p_5)$, $\sigma_2 = (1; q_1, \ldots, q_5)$, $\sigma_3 = (2; p_1, \ldots, p_5, q_1, \ldots, q_5)$ and $\sigma = (\sigma_1, \sigma_2, \sigma_3)$. Theorem 2 implies that the σ_i $(1 \le i \le 3)$ are all maximal signatures, so σ is finitely maximal. Let Γ_1, Γ_2 be Fuchsian groups of signatures σ_1, σ_2 and with invariant discs D_1, D_2. Choose D_1 and D_2 far enough apart, so that the isometric circles of each group are exterior to those of the other. Then the group Γ, generated by Γ_1 and Γ_2 is Kleinian, and has signature σ.

§3. *Uniformization by maximal groups*

The results in Section 1 are purely group theoretic. In this section we shall prove a result which involves conformal structure. More precisely, it is the location of the branch images on the quotient surface which will play a crucial role.

Let S be a closed Riemann surface of genus g, and let $\nu_1, \ldots, \nu_k$ be integers ≥ 2. We ask the following question: Does there exist a maximal Fuchsian group Γ of signature $\sigma = (g; \nu_1, \ldots, \nu_k)$, such that $D/\Gamma \approx S$? Theorem 4 answers this affirmatively, provided k is sufficiently large, relative to g. This result differs qualitatively from the results in Section 1 in two respects. In the first place, we are finding a maximal group, not a maximal signature. However, we don't lose too much, because we can still cover all conformal structures. In the second place, Theorems 1 and 2 have restrictions on k and the integers ν_i. Theorem 4 does not restrict the ν_i. This will be important for an application in Section 4.

Let Γ be a Fuchsian group with compact quotient D/Γ, and let $\phi : D \to D/\Gamma$ be the projection map. We shall use the following notation:

(i) $g(\Gamma)$ is the genus of Γ.

(ii) For an integer $\nu \geq 2$, $k_\nu(\Gamma)$ is the number of points in D/Γ, such that the points in $\phi^{-1}(s)$ have branching order ν.

(iii) $k(\Gamma) = \sum_{\nu=2}^{\infty} k_\nu(\Gamma)$.

(iv) $A(\Gamma) = \frac{1}{2\pi}$ (area of D/Γ).

(v) $T(\Gamma)$ is the Teichmüller space.

(vi) $d(\Gamma) = \dim_{\mathbf{C}} T(\Gamma)$.

When there can be no confusion, we shall suppress the symbol Γ in the above notation. The following formulas are well known:

(α) $A = 2(g-1) + \sum_{\nu=2}^{\infty} k_\nu\left(1 - \frac{1}{\nu}\right)$.

(β) $d = 3(g-1) + k$.

The following formula, which follows from (α) and (β), will be useful.

(γ) $3A = 2d + \sum_{\nu=2}^{\infty} k_\nu\left(1 - \frac{3}{\nu}\right)$.

LEMMA 5. *If* $\Gamma \subset \Delta$, *and* $[\Delta : \Gamma] = n$, *then*

$$\sum_{\nu=2}^{\infty} k_\nu(\Gamma)\left(1 - \frac{3}{\nu}\right) \leq n \sum_{\nu=3}^{\infty} k_\nu(\Delta)\left(1 - \frac{3}{\nu}\right) .$$

Proof. Let $S = D/\Gamma$, $T = D/\Delta$. We have projections $\psi : D \to S$, $\theta : D \to T$ and a ramified covering $\phi : S \to T$ (induced by the group inclusion) such that the following diagram commutes:

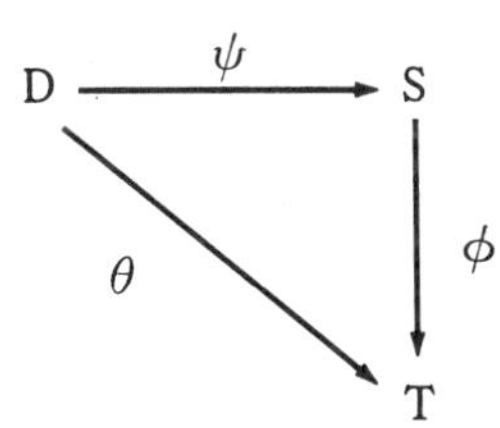

Let $t \in T$, such that $\theta^{-1}(t)$ has branching order $\nu > 2$. Let $\phi^{-1}(t) = \{r_1, r_2, \ldots, r_p, s_1, \ldots, s_q\}$ where $\psi^{-1}(r_i)$ is unbranched $(1 \le i \le p)$ and $\psi^{-1}(s_i)$ has branching order $\nu_i > 1$ $(1 \le i \le q)$. For $s \in S$, let $\beta(s)$ denote the ϕ-branching order of s. Since branching orders multiply when maps are composed, we have the following:

(a) $\beta(r_i) = \nu \ (1 \le i \le p)$

(b) $\beta(s_i) = \mu_i$, where $\mu_i \nu_i = \nu \ (1 \le i \le q)$.

Since ϕ is an n-sheeted covering,

(c) $p\nu + \sum_{i=1}^{q} \mu_i = n$.

Using these formulas, a short computation shows:

(d) $\sum_{i=1}^{q} \left(1 - \frac{3}{\nu_i}\right) = 3p + q - \frac{3n}{\nu}$.

Since we have assumed $\nu \ge 3$, formula (c) implies:

(e) $3p + q \le n$.

Now (d) implies:

(f) $\sum_{i=1}^{q} \left(1 - \frac{3}{\nu_i}\right) \le n\left(1 - \frac{3}{\nu}\right)$.

For each $t \in T$ which is a θ-branch image of branching order $\nu > 2$, we obtain an inequality (f). If we add these, we obtain

(g) $\sum_{\nu=3}^{\infty} k_\nu(\Gamma)\left(1 - \frac{3}{\nu}\right) + k'_2(\Gamma)\left(1 - \frac{3}{2}\right) \le n \sum_{\nu=3}^{\infty} k_\nu(\Delta)\left(1 - \frac{3}{\nu}\right)$,

where $k'_2(\Gamma)$ is the number of points $s \in S$ which are ψ-branch images of branching order 2 and are mapped by ϕ to one of the above points $t \in T$. Since $k'_2(\Gamma) \le k_2(\Gamma)$, we have

$$\sum_{\nu=2}^{\infty} k_\nu(\Gamma)\left(1 - \frac{3}{\nu}\right) \le n \sum_{\nu=3}^{\infty} k_\nu(\Delta)\left(1 - \frac{3}{\nu}\right). \qquad \text{Q.E.D.}$$

LEMMA 6. *Let* $k = k(\Gamma)$, $g = g(\Gamma)$ *and suppose that* $\Gamma \subset \Delta$.

(1) *If* $k > 6g-3$, *then* $d(\Delta) < k$.

(2) *If* $g = 1$, $k > 6$, *then* $d(\Delta) < k-1$.

(3) *If* $g = 0$, $k > 6$, *then* $d(\Delta) < k-3$.

Proof. If $[\Delta:\Gamma] = n$, then $A(\Gamma) = nA(\Delta)$. Using formulas (α), (β), (γ) and Lemma 5, we have:

$$\begin{aligned}
6(g(\Gamma)-1) + 3 \sum_{\nu=2}^{\infty} k_\nu(\Gamma)\left(1-\tfrac{1}{\nu}\right) &= 3A(\Gamma) = 3nA(\Delta) \\
&= n\left[2d(\Delta) + \sum_{\nu=2}^{\infty} k_\nu(\Delta)\left(1-\tfrac{3}{\nu}\right)\right] \\
&\geq 2nd(\Delta) + \sum_{\nu=2}^{\infty} k_\nu(\Gamma)\left(1-\tfrac{3}{\nu}\right) - \tfrac{n}{2}\, k_2(\Delta)\ .
\end{aligned}$$

Therefore:

(a) $d(\Delta) \leq \frac{1}{n}[3(g-1)+k] + \frac{1}{4}\, k_2(\Delta)$,

where $g = g(\Gamma)$, $k = k(\Gamma)$.

We also have:

$$d(\Delta) = 3(g(\Delta)-1) + k_2(\Delta) + \sum_{\nu=3}^{\infty} k_\nu(\Delta) \geq k_2(\Delta) - 3\ ,$$

so that:

(b) $k_2(\Delta) \leq d(\Delta) + 3$.

The inequalities (a) and (b) imply:

(c) $d(\Delta) \leq \frac{2}{3}[3(g-1)+k] + 1$.

The inequality (c) implies (1), (2), and (3). Q.E.D.

DEFINITION. Let $C(S)$ denote the group of conformal homeomorphisms of a Riemann surface S, and let $S_1, \ldots, S_k$ be a collection of pairwise disjoint subsets of S. The sets S_i will be called *dependent* mod $C(S)$ if there exist points $z_i \in S_i$, an element $f \neq 1$ in $C(S)$, and a permutation π of $\{1, 2, \ldots, k\}$ such that $f(z_i) \in S_{\pi(i)}$. Otherwise, the sets S_i are *independent* mod $C(S)$.

Note that if S is a sphere or torus, then $C(S)$ is a continuous group. Therefore, any collection of open subsets $S_1, \ldots, S_k$ is dependent mod $C(S)$.

LEMMA 7. *Let* S *be a closed Riemann surface of genus* g, *and let* k *be a positive integer.*

(1) *If* $g \geq 2$, *then there exist* k *open subsets* $S_1, \ldots, S_k$ *which are independent* mod $C(S)$.

(2) *If* $g = 1$, *let* $k > 1$, $p \in S$ *and* $S_0 = \{p\}$. *There exist* k *open subsets* $S_1, \ldots, S_k$ *such that the sets* $S_0, S_1, \ldots, S_k$ *are independent* mod $C(S)$.

(3) *If* $g = 0$, *let* $k > 1$, *let* p_1, p_2, p_3 *be different points in* S, *and* $S_{-i} = \{p_i\}$ $(1 \leq i \leq 3)$. *There exist* k *open subsets* $S_1, \ldots, S_k$ *such that the sets* $S_{-3}, S_{-2}, S_{-1}, S_1, \ldots, S_k$ *are independent* mod $C(S)$.

Proof. (1) If $g \geq 2$, then the group $C(S)$ has finite order: $|C(S)| = n < \infty$. We have an n-sheeted ramified covering $\phi : S \to S/C(S)$. Choose disjoint open sets $U_1, \ldots, U_k$ in $S/C(S)$, which do not contain any branch images, and such that (for $1 \leq i \leq k$) $\phi^{-1}(U_i)$ is the union of n disjoint, open subsets of S. Note that $C(S)$ permutes the connected components of $\phi^{-1}(U_i)$. For each i, let S_i be one of these connected components in $\phi^{-1}(U_i)$. Then the sets $S_1, \ldots, S_k$ are independent mod $C(S)$. In fact, they are even disjoint mod $C(S)$, in the sense that if $f \in C(S)$, $f \neq 1$, then $f(S_i) \cap S_j = \emptyset$ (for $1 \leq i, j \leq k$).

(2) If $g = 1$, then $C(S)$ is the isometry group of the metric on S, inherited from the Euclidean plane. Choose a disc D (with center p), which contains the shortest geodesic (in S) between any two of its points. Choose a geodesic σ in D, which begins at p, and let $p_1, \ldots, p_k$ be k points on σ, with the following properties:

(a) The points $p, p_1, \ldots, p_k$ are listed in consecutive order along σ.

(b) The distance $d(p, p_1) = m\varepsilon$ and $d(p_i, p_{i+1}) = 3\varepsilon$ (where m and ε are determined below).

Let S_i be the open disc of radius ε and center p_i. The disc S_1 subtends an angle $\theta = 2 \arcsin\left(\frac{1}{m}\right)$ at p. Choose m large enough so that $m \geq 7$ and $\theta < \frac{\pi}{6}$. (Actually, $m = 7$ satisfies these conditions.) Now choose ε small enough, so that the entire configuration is contained in the original disc D.

We claim that the sets $S_0 = \{p\}$, $S_1, \ldots, S_k$ are independent mod C(S). Suppose there are points $z_i \in S_i$ $(0 \leq i \leq k)$, $f \in C(S)$ and a permutation π of $\{0, 1, \ldots, k\}$ such that $f(z_i) \in S_{\pi(i)}$. We shall show that $f = 1$. Since p is the only point in S_0, $f(z_i) = p$ for some i. If $z_i \neq p$, then there is another point z_j ($j = i-1$ or $i+1$) in a neighboring disc, such that $d(z_i, z_j) < 5\varepsilon$. Therefore $d(p, f(z_j)) = d(f(z_i), f(z_j)) = d(z_i, z_j) < 5\varepsilon$. But for any point $z \in \bigcup_{1 \leq i \leq k} S_i$, $d(p, z) > 6\varepsilon$. From this contradiction, we see that $f(p) = p$. (Note that we used the condition $k > 1$, to find the neighboring point z_j. If $k = 1$, the statement is false when C(S) contains involutions.) Thus f is a rotation about p. The smallest possible angle of rotation on a torus is $\pi/6$. Since all S_i are contained in a sector at p of angle $\theta < \pi/6$, we see that $f = 1$.

(3) We now consider the case $g = 0$. Let $A = \{-3, -2, -1, 1, \ldots, k\}$. For each subset $\{a, b, c, d\} \subset A$, of order 4, we choose six orderings which give rise to the six generically different cross-ratios. (I.e., we choose only one of the orderings (a, b, c, d), (c, d, a, b), (b, a, d, c), (d, c, b, a).) Let B denote the set of these chosen ordered subsets.

Let $z = (z_1, \ldots, z_k) \in \mathbb{C}^k$. For $i \in A$, let $i(z) = p_{-i}$ (if $i < 0$) and $i(z) = z_i$ (if $i > 0$). For $\alpha = (a, b, c, d) \in B$, let $\alpha(z) = (a(z), b(z), c(z), d(z))$ and $r_\alpha(z) = r(\alpha(z))$ be the cross-ratio. If $\alpha, \beta \in B$, $\alpha \neq \beta$, the equation $r_\alpha(z) = r_\beta(z)$ defines a proper algebraic subset $V_{\alpha,\beta}$ in $\mathbb{C}^k$. Since B is a finite set, the set $W = \mathbb{C}^k - \bigcup_{\alpha \neq \beta} V_{\alpha,\beta}$ cannot be empty.

Let $z = (z_1, \ldots, z_k) \in W$. For $\alpha, \beta \in B$, $\alpha \neq \beta$, we have $r_\alpha(z) \neq r_\beta(z)$. If z', z'' are close enough to z, then $r_\alpha(z') \neq r_\beta(z'')$. Choose discs S_i, centered at z_i, which are small enough so that the above inequalities are satisfied in the polydisc $P = \Pi S_i$. We now claim that the sets $S_{-1} = \{p_1\}$, $S_{-2} = \{p_2\}$, $S_{-3} = \{p_3\}$, $S_1, \ldots, S_k$ must be independent mod C(S). For suppose that $z'_i \in S_i$, $f \in C(S)$ and π is a permutation of $\{-3, -2, -1, 1, \ldots, k\}$ so that $z''_i = f(z'_i) \in S_{\pi(i)}$. If $\pi \neq \text{id}$, then π cannot leave invariant every subset of order 4. (We are using the condition $k > 1$

here.) Since f preserves cross-ratios, we would obtain an equation $r_\alpha(z') = r_\beta(z'')$ for some $\alpha, \beta \in B$, $\alpha \neq \beta$. This contradicts our choice of the polydisc $P = \Pi S_i$. Therefore $\pi = \mathrm{id}$, $f(p_i) = p_i$ $(1 \leq i \leq 3)$ and $f = 1$.

Q.E.D.

THEOREM 4. *Let S be a closed surface of genus* g. *Suppose that* $k > 6g-3$ *(if* $g \geq 2$*) and* $k > 6$ *(if* $g \leq 1$*). Let* $\nu_1, \ldots, \nu_k$ *be integers such that* $\nu_i \geq 2$ $(1 \leq i \leq k)$. *Then there exists a maximal Fuchsian group* Γ *of signature* $(g; \nu_1, \ldots, \nu_k)$, *such that* $D/\Gamma \approx S$.

Proof. If $g \geq 2$, Lemma 7 implies that there exist k (pairwise disjoint) open subsets $D_1, \ldots, D_k$ in S, which are independent mod C(S). By choosing smaller sets, if necessary, we may assume that each D_j is conformally equivalent to a disc, that the $\bar{D}_j$ are pairwise disjoint, and $\gamma_j = \partial D_j$ is a simple, closed curve. Let $S_0 = S - \bigcup_{1 \leq j \leq k} D_j$.

If $g = 1$ and $p \in S$, there exist open sets $D_2, \ldots, D_k$, such that $\{p\}, D_2, \ldots, D_k$ are independent mod C(S). As before, we may assume that each D_j is equivalent to a disc, the $\bar{D}_j$ are pairwise disjoint, and $\gamma_j = \partial D_j$ is a simple closed curve. Let $S_0 = S - \bigcup_{2 \leq j \leq k} D_j - \{p\}$. Let γ_1 be a simple closed curve in S_0 which goes around p.

If $g = 0$ and $p_1, p_2, p_3 \in S$, there exist open sets $D_4, \ldots, D_k$, such that $\{p_1\}, \{p_2\}, \{p_3\}, D_4, \ldots, D_k$ are independent mod C(S). We may assume that the D_j have the same properties as in the previous cases. Let $S_0 = S - \bigcup_{4 \leq j \leq k} D_j - \{p_1, p_2, p_3\}$ and (for $1 \leq i \leq 3$) let γ_i be a simple closed curve in S_0 which goes around p_i.

In all of the above cases, choose a base point $s_0 \in S_0$ and (if $g > 0$) closed loops α_i, β_i (based at s_0 and lying in S_0) which go around the handles in the usual manner. For $1 \leq j \leq k$, let δ_j be a curve in S_0, which connects s_0 to γ_j, and let $\varepsilon_j = \delta_j \gamma_j \delta_j^{-1}$. The curves δ_j may be chosen so that the homotopy classes $a_i = [\alpha_i]$, $b_i = [\beta_i]$, $e_j = [\varepsilon_j]$ $(1 \leq i \leq g,\ 1 \leq j \leq k)$ generate $\pi_1(S_0, s_0)$ with a single defining relation:

$\left(\prod_{1 \le i \le g} a_i b_i a_i^{-1} b_i^{-1}\right) e_1 \dots e_k = 1$. Let N be the smallest normal subgroup in $\pi_1(S_0, s_0)$ which contains the elements $e_i^{\nu_i}$ $(1 \le i \le k)$, and let Γ_0 be the quotient group. Γ_0 is isomorphic to a Fuchsian group of signature $(g; \nu_1, \dots, \nu_k)$.

Now let $z_i \in D_i$ ($z_1 = p$ if $g = 1$, $z_i = p_i$ $(1 \le i \le 3)$ if $g = 0$) and $z = (z_1, \dots, z_k)$. Let $\psi_z : D \to S$ be a uniformization of S, which has branching order ν_i over z_i, and let Γ_z be the group of covering transformations. Let $\hat{D} = D - \{\text{fixed points of } \Gamma_z\}$, $\hat{S} = S - \{z_1, \dots, z_k\}$. $\psi_z : \hat{D} \to \hat{S}$ is an unramified normal covering. Since $\hat{S}$ retracts to S_0, we may identify their fundamental groups. Then, by lifting paths, we obtain a homomorphism $\pi_1(S_0, s_0) \to \Gamma_z$ with kernel N, and so we obtain an isomorphism $i_z : \Gamma_0 \to \Gamma_z$. Γ_z and i_z are determined by z up to conjugation. (By normalizing Γ_z in some manner, we could determine i_z completely, if we wish.)

Let $\mathcal{L}(D)$ denote the linear fractional group which preserves the disc D. We regard the Teichmüller space $T(\Gamma_0)$ to be the set of injections $i : \Gamma_0 \to \mathcal{L}(D)$ (such that $i(\Gamma_0)$ is discrete), modulo the relation of conjugacy. The complex structure comes from the complex space of Beltrami differentials, via the quasi-conformal mappings which induce isomorphisms. Let $U = \prod_{h \le j \le k} D_j$, where $h = 1, 2, 4$, if $g \ge 2$, $g = 1$ or $g = 0$ (respectively). In the previous paragraph, we obtained an injection $i_z : \Gamma_0 \to \Gamma_z \subset \mathcal{L}(D)$, corresponding to each $z \in U$. This induces a map $\phi : U \to T(\Gamma_0)$. Since the complex structure in $T(\Gamma_0)$ is known to be compatible with variation of branch points, the map ϕ is holomorphic. However, we shall indicate why this is true in a more explicit manner.

Let $|z| < \frac{1}{2}$, and consider the map $w : D \to D$, defined by: $w(\zeta) = \zeta + (1 - |\zeta|^2)z$. It is easily seen that w is a diffeomorphism of D, which leaves ∂D pointwise fixed and maps 0 to z. Furthermore, $\mu = w_{\bar{\zeta}}/w_{\zeta} = \frac{\zeta z}{\bar{\zeta} z - 1}$. Thus μ depends holomorphically on z, and $|\mu| \le \frac{|z|}{1-|z|} < 1$, so w is quasi-conformal. If we replace w by fwf^{-1} and z by $f^{-1}(z)$ (where $f \in \mathcal{L}(D)$), then $\mu(\zeta, z)$ is replaced by $f'(\zeta)\overline{f'(\zeta)}^{-1} \mu(f^{-1}(\zeta), f^{-1}(z))$.

The new mapping has the same properties as the old, except that now z is allowed to vary inside a non-Euclidean circle of radius $\frac{1}{2}$, centered at $z^0 = f(0)$.

We now transfer the above mappings to the Riemann surface S. Let ζ_j be a parameter in D_j, and $z_j^0 \in D_j$. There exists a neighborhood N_j of z_j^0 in D_j, and a mapping $w_j : \overline{D}_j \to \overline{D}_j$, which depends on $z_j \in N_j$. w_j is continuous on $\overline{D}_j$, a quasi-conformal diffeomorphism of D_j, leaves ∂D_j pointwise fixed, maps z_j^0 to $z_j \in N_j$, and gives rise to a Beltrami differential $\mu_j(\zeta_j, z_j)$, which depends holomorphically on z_j. We define: $\mu = \mu_j$ in D_j, $\mu = 0$ in $S - \bigcup D_j$, $w = w_j$ in D_j, $w = \mathrm{id}$ in $S - \bigcup D_j$. The Beltrami differential μ defines a new conformal structure on S, which we denote S_μ. w is a conformal map from S_μ to S, which moves the branch images z_i^0 to z_i. When w is lifted to D, it conjugates Γ_{z^0} to Γ_z (when these groups are suitably normalized) and i_{z^0} to i_z. The most important property is that μ depends holomorphically on $z = (z_1, \ldots, z_k)$, and therefore the mapping $\phi : U \to T(\Gamma_0)$ is holomorphic.

We shall now show that the mapping ϕ is injective. Suppose that $\phi(z) = \phi(z')$. Then there exists $f \in \mathfrak{L}(D)$, such that $\Gamma_{z'} = f\Gamma_z f^{-1}$. The map $f : D \to D$ maps Γ_z-orbits to $\Gamma_{z'}$-orbits, and therefore induces a conformal transformation $\hat{f} : S \to S$, which maps the branch images $\{z_1, \ldots, z_k\}$ to $\{z'_1, \ldots, z'_k\}$. This contradicts the independence mod C(S) of the sets $D_1, \ldots, D_k$. (Here we are using the notation $D_1 = \{p\}$ if $g = 1$, and $D_i = \{p_i\}$ (for $1 \leq i \leq 3$) if $g = 0$.) Therefore $\phi : U \to T(\Gamma_0)$ is a holomorphic, injective map. The set $P = \phi(U)$ is holomorphically equivalent to a polydisc of dimension k (if $g \geq 2$), $k - 1$ (if $g = 1$) or $k - 3$ (if $g = 0$).

If Δ is a Fuchsian group, and $i : \Gamma_0 \to \Delta$ is an injection, there is a holomorphic map $T(\Delta) \to T(\Gamma_0)$, induced by $\alpha \to \alpha \circ i$ (where $\alpha : \Delta \to \mathfrak{L}(D)$ is an injection). By Lemma 6, $\dim T(\Delta) < \dim P$ so the image of $T(\Delta)$ in $T(\Gamma_0)$ cannot contain P.

There are only a finite number of signatures σ, such that a group Γ_0 of signature σ can contain a group isomorphic to Γ_0. (This follows easily from the area formula.) For each such σ, there are only a countable number of injections $i: \Gamma_0 \to \Gamma_\sigma$. In this way, we obtain a countable number of images I_n of spaces of dimension $< \dim P$. $I = \bigcup I_n$ cannot contain P. $T(\Gamma_0) - I$ is precisely the set of points in $T(\Gamma_0)$, corresponding to maximal groups. Since $P - I \neq \emptyset$, and since the points in P correspond to groups which uniformize S, we are finished. Q.E.D.

§4. *Conformal groups and Galois groups*

Let S be a closed Riemann surface, and $\mathfrak{M}(S)$ the field of meromorphic functions on S. A ramified covering $\phi: S \to T$ corresponds to an injection $\phi^0: \mathfrak{M}(T) \hookrightarrow \mathfrak{M}(S)$. Furthermore ϕ is normal if and only if $\mathfrak{M}(S)$ is a Galois extension of $\phi^0\mathfrak{M}(T)$. In this case the Galois group is isomorphic to the group of covering transformations. Using this dictionary, it is easy to see that the following two theorems are equivalent:

THEOREM 5. *Let* K *be a function field in one variable* (*over* $\mathbb{C}$) *and* G *a finite group. Then there exists a Galois extension* $L \supset K$, *whose Galois group* $\mathrm{Aut}_K(L)$ *is isomorphic to* G.

THEOREM 5′. *Let* S *be a closed Riemann surface, and* G *a finite group. Then there exists a normal covering* $\phi: T \to S$, *whose group of covering transformations is isomorphic to* G.

This theorem is quite easy to prove. If we puncture the surface S at n points, and denote the punctured surface by S_0, then $\pi_1(S_0)$ is a free group of rank $\geq n-1$. Since G is the homomorphic image of some free group, we can use covering space theory to construct the surface T.

The following two theorems are also easily equivalent. However I am not aware of an easy proof.

THEOREM 6. *Let* K *be a function field in one variable* (*over* $\mathbb{C}$) *and* G *a non-trivial finite group. Then there exists a Galois extension* $L \supset K$, *whose Galois group* $\mathrm{Aut}_K(L)$ *is isomorphic to* G, *and* $\mathrm{Aut}_K(L) = \mathrm{Aut}_{\mathbb{C}}(L)$.

THEOREM 6′. *Let* S *be a closed Riemann surface, and* G *a non-trivial finite group. Then there exists a normal covering* $\phi : T \to S$, *whose group of covering transformations is isomorphic to* G, *and is the full group of conformal homeomorphisms of* T.

Proof of Theorem 6′. Let g be the genus of S. Choose a set of generators $x_1, \ldots, x_k$ for G, such that

(a) x_i has order $\nu_i > 1$,

(b) $x_1 x_2 \cdots x_k = 1$,

(c) $k > 6g-3$. ($k > 6$ if $g \leq 1$).

This can always be arranged by repeating generators more than once, and choosing x_k to satisfy (b).

By Theorem 4, there exists a maximal Fuchsian group Γ of signature $(g; \nu_1, \ldots, \nu_k)$, such that $S \approx D/\Gamma$. Γ has generators a_i, b_i, e_j $(1 \leq i \leq g, 1 \leq j \leq k)$, with defining relations: $\left(\prod_{1 \leq i \leq g} a_i b_i a_i^{-1} b_i^{-1}\right) e_1 \cdots e_k = 1$, $e_i^{\nu_i} = 1$ $(1 \leq i \leq k)$. Therefore the map $a_i \to 1$, $b_i \to 1$, $e_j \to x_j$, extends to a homomorphism $\theta : \Gamma \to G$. Let $\Delta = \mathrm{Ker}(\theta)$. Since e_j and x_j have the same order, Δ contains no elements of finite order, and $\Gamma/\Delta \approx G$. Let $T = D/\Delta$, let C(T) be the conformal group of T, and let N be the normalizer of Δ in $\mathfrak{L}(D)$. Then $C(T) \approx N/\Delta$. Since Δ is normal in Γ, and Γ is maximal, $N = \Gamma$. Thus $C(T) \approx \Gamma/\Delta \approx G$. The group inclusion $\Delta \subset \Gamma$ induces the normal covering $T \to S$. Q.E.D.

UNIVERSITY OF MARYLAND
COLLEGE PARK, MARYLAND

REFERENCES

[1]. Greenberg, L., Maximal Fuchsian groups, Bull. A.M.S. vol. 4 (1963), pp. 569-573.

[2]. Singerman, D., Finitely maximal Fuchsian groups, J. London Math. Soc. (2) vol. 6 (1972), pp. 29-38.

COMMENSURABLE GROUPS OF MOEBIUS TRANSFORMATIONS

Leon Greenberg

§1. *The Moebius group* $\mathfrak{M}_n$

Let $\mathfrak{M}_n$ be the group of Moebius transformations of $\overline{R^n} = R^n \cup \{\infty\}$ (i.e., the group generated by reflections in (n−1)-dimensional planes and spheres). It is well known that $\mathfrak{M}_n$ is isomorphic to a subgroup of index 2 in the orthogonal group $O(n+1, 1)$. Furthermore, if we consider R^n to be a subspace of R^{n+1}, then the action of $\mathfrak{M}_n$ extends to $\overline{R^{n+1}}$, leaving each component of $R^{n+1} - R^n$ invariant. Let one of these components be denoted H^{n+1}. There exists a Riemannian metric in H^{n+1}, such that H^{n+1} is isometric to an (n+1)-dimensional hyperbolic space, and its isometry group is $\mathfrak{M}_n$. We shall outline some basic facts concerning $\mathfrak{M}_n$ and H^{n+1}. These can be found in greater detail in [4].

$\mathfrak{M}_n$ acts transitively in H^{n+1}. The isotropy group at a point $p \in H^{n+1}$ is isomorphic to $O(n+1)$. The geodesics in H^{n+1} are semi-circles and half-lines orthogonal to R^n. (We shall call these h-lines.) The totally geodesic subspaces are hemispheres and half-planes orthogonal to R^n. (We shall call them h-planes.)

As in the 2-dimensional case, the elements in $\mathfrak{M}_n$ are classified into elliptic, parabolic and loxodromic types. An element $g \in \mathfrak{M}_n$ is a continuous transformation of $\overline{H^{n+1}} = H^{n+1} \cup \overline{R^n}$, which is homeomorphic to an (n+1)-dimensional, closed ball. Therefore g has a fixed point in $\overline{H^{n+1}}$. If g has a fixed point in H^{n+1}, it is called elliptic. In this case it is conjugate to an element in $O(n+1)$. (Here we have identified $\mathfrak{M}_n$ with a subgroup of index 2 in $O(n+1, 1)$.) If g has a single fixed point in $\overline{R^n}$, it is called parabolic. By moving the fixed point to ∞, g can be conjugated to an element (without fixed points in R^n) in the group of Euclidean isometries of R^n. If g has exactly 2 fixed points in $\overline{H^{n+1}}$ (which necessarily lie in $\overline{R^n}$) it is called loxodromic. If we move the fixed points to $\{0, \infty\}$, then g is conjugated into the form: $g(x) = tA(x)$, where $0 < t \neq 1$, and $A \in 0(n)$. If g has 3 fixed points in $\overline{R^n}$, then it preserves the circle C through them, and the 2-dimensional h-plane π, whose boundary is C. π is isometric to the non-Euclidean plane, and g is an isometry of π, with 3 fixed points on the boundary. Therefore $g|\pi = \mathrm{id}$, so g is elliptic.

If π is an h-plane (or h-line) let $K(\pi)$ be the subgroup of $\mathfrak{M}_n$, which leaves π pointwise fixed, and let $I(\pi)$ be the isometry group of π. $I(\pi)$ is naturally imbedded in $\mathfrak{M}_n$ (as the subgroup generated by reflections in n-dimensional h-planes orthogonal to π). The subgroup of $\mathfrak{M}_n$ which leaves π invariant is $K(\pi) \times I(\pi)$. We shall denote the identity components of $K(\pi)$, $I(\pi)$, $\mathfrak{M}_n$ by $K_0(\pi)$, $I_0(\pi)$, $\mathfrak{M}_n^0$.

The following theorem, which classifies the connected Lie subgroups of $\mathfrak{M}_n$ can be found in [4].

THEOREM 1. *Let* G *be a connected Lie subgroup of* $\mathfrak{M}_n$. *Then one of the following is true.*

(a) *The elements of* G *have a common fixed point in* H^{n+1}, *and* G *is conjugate to a subgroup of the orthogonal group* $SO(n+1)$.

(b) *The elements in* G *have a common fixed point in* $\overline{R^n}$, *and* G *is conjugate to a subgroup of the isogonal group of* R^n.

(c) *There is an* h-*line* λ, *so that* $G = K \times L$, *where* $K \subset K_0(\lambda)$ *and* L *is a* 1-*parameter loxodromic group, which leaves* λ *invariant.*

(d) *There is an* h-*plane* π, *so that* $G = K \times I_0(\pi)$, *where* $K \subset K_0(\pi)$.

(e) $G = \mathfrak{M}_n^0$.

This theorem can be used to show that certain subgroups are discrete, and it will be applied in this context below.

If G is a subgroup of $\mathfrak{M}_n$, the limit set $L(G)$ may be defined as in the 2-dimensional case: $L(G)$ is the intersection with $\overline{R^n}$ of the set of accumulation points of an orbit Gx, where $x \in H^{n+1}$. (If G is discrete, then Gx can accumulate only in $\overline{R^n}$.) The definition is independent of $x \in H^{n+1}$, because G preserves distances in H^{n+1}, while the (non-Euclidean) distance between two points in $\overline{R^n}$ is infinite. We shall also use the following notation: $S(G)$ is the intersection of all spheres and planes (of any dimension) in $\overline{R^n}$, which contain $L(G)$. ($\overline{R^n}$ is considered to be an n-dimensional sphere.) $L(G)$ and $S(G)$ are invariant under G. The following result is proved in [4].

LEMMA 1. *If* S *is a closed subset of* $\overline{R^n}$, *which contains more than one point and is invariant under* G, *then* $S \supset L(G)$.

LEMMA 2. *Let* G *and* H *be subgroups of* $\mathfrak{M}_n$. *Suppose that* $hGh^{-1} = G$ *for all* $h \in H$. *Then*:

(1) $L(G)$ *is invariant under* H.

(2) *If* $L(G)$ *contains more than one point, then* $L(H) \subset L(G)$.

Proof. (1) Let $z \in L(G)$, and let $\{g_n\}$ be a sequence in G, such that $\lim g_n(x) = z$ (for all $x \in H^{n+1}$). Then for $h \in H$, $\lim hg_nh^{-1}(x) = h \lim g_n(h^{-1}x) = h(z)$. Since $hg_nh^{-1} \in G$, $h(z) \in L(G)$.

(2) This follows from (1) and Lemma 1. Q.E.D.

We shall conclude this section with some results on discrete groups.

LEMMA 3. *Let* Γ *be a discrete subgroup of* $\mathfrak{M}_n$, *such that every element in* Γ *leaves* ∞ *fixed. Then either the elements in* Γ *have another common fixed point, or* Γ *is a subgroup of the Euclidean group* E_n.

Proof. If $\Gamma \not\subset E_n$, then Γ contains loxodromic elements. Suppose that this is true, and let Γ' be the commutator subgroup of Γ. In any case $\Gamma' \subset E_n$.

Suppose that $\Gamma' = \{1\}$. Then Γ is commutative. Let f be a loxodromic element in Γ, with fixed points $\{p, \infty\}$. Let $x \in \{p, \infty\}$ and $g \in \Gamma$.

Then $g(x) = gfg^{-1}(g(x)) = f(g(x))$. Thus g leaves $\{p, \infty\}$ invariant. Since $g(\infty) = \infty$, $g(p) = p$.

We now consider the general case, where Γ' need not be trivial. Let $g \in \Gamma'$ and let f be a loxodromic element in Γ, with fixed points $\{p, \infty\}$. Let $K = \{k \in E_n : k(p) = p\}$. Then $K \approx O(n)$. There exist $A, B \in K$, $t \in R^+$ and $a \in R^n$ so that $g(x) = Ax + a$, $f(x) = tBx$. We may suppose that $0 < t < 1$ (otherwise replace f by f^{-1}). Note that $a = 0$ precisely when $g(p) = p$. We have: $f^n gf^{-n}(x) = B^n AB^{-n}x + t^n B^n a$. Since K is compact, the transformations $f^n gf^{-n}$ must accumulate in E_n, if $a \neq 0$. Therefore $a = 0$ and $g \in K$. Since Γ is discrete, g has finite order. Thus every element in Γ' has finite order.

Now let f_1, f_2 be loxodromic elements in Γ, and let Δ be the subgroup generated by f_1 and f_2. Since Δ is finitely generated, a theorem of Selberg [7] implies that Δ has a subgroup Δ_1 of finite index, which has no elements of finite order. Therefore the commutator group $\Delta'_1 = \{1\}$, so (by the argument of the previous paragraph) f_1 and f_2 have the same fixed points. Thus all loxodromic elements in Γ have the same fixed points $\{p, \infty\}$. If f is a loxodromic element in Γ, $g \in \Gamma$ and $x \in \{p, \infty\}$, then $gfg^{-1}(g(x)) = g(x)$. Since gfg^{-1} is loxodromic, $g(x) \in \{p, \infty\}$. Thus g leaves $\{p, \infty\}$ invariant, and since $g(\infty) = \infty$, $g(p) = p$. Q.E.D.

Note that the lemma implies that a discrete group cannot have a pair of loxodromic elements with a single common fixed point or parabolic and loxodromic elements with a common fixed point. For the subgroup generated by these elements would violate the lemma.

The following theorem of Bieberbach can be found in [2].

THEOREM (Bieberbach). *Let Γ be a discrete subgroup of the Euclidean group E_n.*

(1) *If R^n/Γ is compact, then the subgroup $T(\Gamma)$ of translations in Γ has finite index and rank* n.

(2) *Suppose that* Γ *is infinite, but* R^n/Γ *is not compact. There exists an orthogonal decomposition* $R^n = V \times W$ *(where* V *and* W *are Euclidean subspaces) such that* $\Gamma \subset K \times I$, *where* K *is the orthogonal group which leaves* W *pointwise fixed and* I *is the Euclidean group of* W. *Let* ϕ *denote the projection into* I. *Then* $\Gamma' = \phi(\Gamma)$ *is discrete, and* W/Γ' *is compact.*

§2. *Commensurable groups*

Let Γ and Δ be subgroups of a group G. Γ and Δ are *commensurable* $(\Gamma \sim \Delta)$ if $\Gamma \cap \Delta$ is of finite index in Γ and Δ.

We shall use the following notation:

(i) $B(\Gamma)$ is the group generated by all subgroups of G which are commensurable to Γ.

(ii) $C(\Gamma) = \{g \in G : g\Gamma g^{-1} \sim \Gamma\}$.

$C(\Gamma)$ has sometimes been called "the commensurability group of Γ." It is easy to see that $B(\Gamma) \subset C(\Gamma)$, and $C(\Gamma)$ permutes the groups commensurable to Γ. Therefore, $B(\Gamma)$ is a normal subgroup of $C(\Gamma)$. We raise the following question: For which discrete subgroups $\Gamma \subset \mathfrak{M}_n$ is $B(\Gamma)$ or $C(\Gamma)$ discrete? We shall give a partial answer to this question.

EXAMPLES. (1) In the case $n = 1$, the component of the identity $\mathfrak{M}_1^0 \approx PSL(2, R)$. For $\Gamma = PSL(2, Z)$, it is easily seen that $B(\Gamma) = C(\Gamma) \supset PSL(2, Q)$. Thus Γ is discrete, while $B(\Gamma)$ is dense.

(2) Let Γ_q be the qth Hecke group in $PSL(2, R)$. Γ_q is generated by $f(z) = -1/z$ and $g(z) = z + 2\cos\frac{\pi}{q}$ $(q = 3, 4, 5, \ldots)$. Leutbecher [6] has shown that for $q \geq 5$ $(q \neq 6)$, $C(\Gamma_q) = \Gamma_q$.

The discrete subgroups $\Gamma \subset \mathfrak{M}_n$ for which volume $(\mathfrak{M}_n/\Gamma) < \infty$ are of special interest. The groups in the above examples have this property. It follows from Theorem 1 that, in this case, either $C(\Gamma)$ is discrete and the index $[C(\Gamma):\Gamma] < \infty$, or $B(\Gamma)$ is dense in $\mathfrak{M}_n$. If Γ is an arithmetic subgroup (cf. Borel [3]) then $B(\Gamma)$ is dense. This raises the question: Can $B(\Gamma)$ be dense, if Γ is a non-arithmetic discrete subgroup with volume $(\mathfrak{M}_n/\Gamma) < \infty$? I have not been able to answer this question. However, the following theorem provides some answers for the case of infinite volume.

THEOREM 2. *Let* Γ *be a discrete subgroup of* $\mathfrak{M}_n$.

(1) *If* $\dim S(\Gamma) < n-2$, *then there is a homomorphism* $\phi : \Gamma \to \Gamma'$, *such that*

(a) $\mathrm{Ker}(\phi)$ *is finite,*

(b) Γ' *is a discrete subgroup of* $\mathfrak{M}_n$, *with* $L(\Gamma') = L(\Gamma)$,

(c) $B(\Gamma')$ *is not discrete.*

(2) *If* $\dim S(\Gamma) = n-2$, *then* $B(\Gamma)$ *is not discrete.*

(3) *If* $\dim S(\Gamma) \geq n-1$, $n \geq 2$ *and* $L(\Gamma) \neq S(\Gamma)$, *then* $C(\Gamma)$ *is discrete.*

(4) *In the case* $n = 2$, *if* Γ *is a finitely generated, non-elementary Kleinian group, whose limit set* $L(\Gamma)$ *is different from a circle (or line), then the index* $[C(\Gamma) : \Gamma] < \infty$.

Proof. (1) Suppose that $L(\Gamma) = \emptyset$. Then each Γ-orbit in H^{n+1} is finite. This implies that Γ is finite. Therefore $\Gamma \sim \{1\}$. For $p \in H^{n+1}$, let $K_p = \{g \in \mathfrak{M}_n : g(p) = p\}$. Then $K_p \approx O(n+1)$. Since the elements of finite order are dense in $O(n+1)$, we see that $B(\Gamma)$ contains a dense subset in each K_p, and so it is not discrete. (In fact $B(\Gamma)$ is dense in $\mathfrak{M}_n$.)

Next consider the case $L(\Gamma) = \{p\}$. By conjugating the group if necessary, we may assume $p = \infty$. Γ cannot contain loxodromic elements, since the two fixed points would be limit points. Therefore Γ is a subgroup of the Euclidean group E_n. We now invoke Bieberbach's theorem.

Suppose first, that R^n/Γ is compact. Then the translation subgroup $T(\Gamma)$ in Γ has finite index and rank n. By adjoining k^{th} roots of translations in $T(\Gamma)$, we can construct translation groups which contain $T(\Gamma)$ as a subgroup of finite index. Thus, if we identify $T(\Gamma)$ with Z^n (by choosing a basis) then $B(\Gamma) \supset Q^n$, and is not discrete.

Now consider the case where R^n/Γ is not compact. Γ is infinite, since $L(\Gamma) \neq \emptyset$. The second part of Bieberbach's theorem says that there is an orthogonal decomposition $R^n = V \times W$, such that $\Gamma \subset K \times I$, where K leaves W pointwise fixed and I is the Euclidean group of W. Furthermore, if ϕ denotes the projection into I, then $\Gamma' = \phi(\Gamma)$ is discrete and W/Γ' is compact. By the argument of the previous paragraph, $B(\Gamma')$ is not discrete. $\mathrm{Ker}(\phi) = \Gamma \cap K$ is finite. Since Γ' contains translations, $L(\Gamma') = \{\infty\} = L(\Gamma)$.

We now consider the case where $S(\Gamma)$ contains more than one point, while $\dim S(\Gamma) = k < n-2$. (In this case $S(\Gamma)$ is a sphere, possibly of 0 dimension.) Let π be the (k+1)-dimensional h-plane, whose boundary is $S(\Gamma)$. Γ leaves $L(\Gamma)$ invariant, and therefore also preserves $S(\Gamma)$ and π. Thus $\Gamma \subset K(\pi) \times I(\pi)$. Let ϕ be the projection into $I(\pi)$ and $\Gamma' = \phi(\Gamma)$. For each $\gamma \in \Gamma$, γ and $\phi(\gamma)$ have the same (elliptic, parabolic or loxodromic) type. If γ is parabolic or loxodromic, then γ and $\phi(\gamma)$ have the same fixed points. Since the limit set is the closure of the parabolic or loxodromic fixed points, $L(\Gamma') = L(\Gamma)$. Since Γ is discrete and $K(\pi)$ compact, Γ' is discrete, and $\mathrm{Ker}(\phi) = K(\pi) \cap \Gamma$ is finite. If A is any finite subgroup of $K(\pi)$, $A \times \Gamma' \sim \Gamma'$, so that $A \subset B(\Gamma')$. Thus $B(\Gamma')$ contains all elements of finite order in $K(\pi)$ (which is isomorphic to $O(n-k)$) and therefore $B(\Gamma')$ is not discrete.

(2) If $\dim S(\Gamma) = n-2$, then (using the same notation as in the previous paragraph) $K(\pi) \approx O(2)$, and the identity component $K_0(\pi) \approx SO(2)$. Thus $K_0(\pi)$ is commutative, it commutes with $I(\pi)$, and therefore it commutes with Γ. If A is a finite subgroup of $K_0(\pi)$, then ΓA is a group, and $\Gamma A \sim \Gamma$. Thus $B(\Gamma)$ contains all elements of finite order in $K_0(\pi)$, and is therefore non-discrete.

(3) Suppose that $n \geq 2$, $\dim S(\Gamma) \geq n-1$ and $L(\Gamma) \neq S(\Gamma)$. Let $G = \{g \in \mathfrak{M}_n : gL(\Gamma) = L(\Gamma)\}$. G is a closed subgroup of $\mathfrak{M}_n$ which contains $C(\Gamma)$. We shall prove that G is discrete by using Theorem 1 to show that the identity component G_0 is trivial.

First note that $L(\Gamma)$ contains more than two points, since $\dim S(\Gamma) \geq n-1$ and $n \geq 2$. Since $G_0 \subset G$, G_0 leaves $L(\Gamma)$ invariant. Lemma 1 implies that $L(G_0) \subset L(\Gamma)$. On the other hand, $\Gamma \subset G$, and so Γ normalizes G_0. Lemma 2 implies that Γ leaves $L(G_0)$ invariant, and if $L(G_0)$ contains more than one point, then $L(\Gamma) \subset L(G_0)$. In that case, we would have $L(\Gamma) = L(G_0)$.

We shall now examine the possibilities (a)-(e) in Theorem 1.

(a) Suppose that the elements of G_0 have a common fixed point in H^{n+1}. Let $\pi = \{x \in H^{n+1} : g(x) = x,$ for all $g \in G_0\}$. If $\pi = H^{n+1}$, then $G_0 = \{1\}$, which is what we wish to prove. We shall show that all other possibilities lead to contradictions.

If $\pi \neq H^{n+1}$, then π is a proper h-plane. Since G_0 is normal in G, it follows that G leaves π invariant. Let S be the boundary sphere (in $\overline{R^n}$) of π. Then S is invariant under G, and therefore under $\Gamma \subset G$. By Lemma 1, $S \supset L(\Gamma)$, and so $S \supset S(\Gamma)$. Since $\dim S(\Gamma) \geq n-1$, we see that $\dim S \geq n-1$ and $\dim \pi \geq n$. Since $\pi \neq H^{n+1}$, $\dim \pi = n$. Thus $G_0 \subset K_0(\pi) \approx SO(1) = \{1\}$, and $\pi = H^{n+1}$.

(b) Suppose the elements in G_0 have a common fixed point $p \in \overline{R^n}$. Since Γ normalizes G_0, it leaves invariant the set $F = \{x \in \overline{R^n} : g(x) = x, g \in G_0\}$. If F contains more than two points, then G_0 has a fixed point in H^{n+1}, so we are back in case (a). If F contains two points, then Lemma 1 implies that $F \supset L(\Gamma)$. Therefore $\dim S(\Gamma) = 0 < n-1$ (since $n \geq 2$). This shows that F contains a single point, which we may assume to be ∞. Since Γ leaves ∞ fixed we may invoke Lemma 3, which implies that $L(\Gamma)$ contains ≤ 2 points. This again contradicts $\dim S(\Gamma) \geq n-1$ (and $n \geq 2$).

In the remaining cases (c)-(e), $L(G_0)$ contains more than one point. As we have observed above, this implies $L(\Gamma) = L(G_0)$.

(c) In this case $L(\Gamma) = L(G_0)$ consists of two points, which violates the conditions $\dim S(\Gamma) \geq n-1$, $n \geq 2$.

(d) and (e) In these cases $L(\Gamma) = L(G_0)$, (so $S(\Gamma) = S(G_0)$) and $L(G_0) = S(G_0)$. This contradicts our assumption that $L(\Gamma) \neq S(\Gamma)$.

We can now conclude that $G_0 = \{1\}$. Therefore G and its subgroup $C(\Gamma)$ are discrete.

(4) By (3), $C(\Gamma)$ is discrete. Since $C(\Gamma)$ leaves $L(\Gamma)$ invariant, Γ and $C(\Gamma)$ have the same limit set, and therefore the same region of discontinuity Ω. The Ahlfors finiteness theorem [1] implies that the area $A(\Omega/\Gamma) < \infty$. If ω is a fundamental region for $C(\Gamma)$, and $\{g_i\}$ is a set of coset representatives for $C(\Gamma) \bmod \Gamma$, then $\bigcup g_i(\omega)$ is a fundamental region for Γ. Since $A(\Omega/\Gamma) < \infty$, $\{g_i\}$ must be a finite set, and $[C(\Gamma):\Gamma] < \infty$. Q.E.D.

EXAMPLE. We shall give an example of a discrete subgroup $\Gamma \subset \mathfrak{M}_4$, such that $\dim S(\Gamma) = 1$, while $C(\Gamma)$ is discrete. Thus the difference between parts (1) and (2) in Theorem 2 is not caused by a defect in the method of proof.

Let π be a 2-dimensional h-plane in H^5. Let F be a Fuchsian group of the 2nd kind, which is a free group of rank 2, and which operates in π (i.e., $F \subset I(\pi)$). In this case $K_0(\pi) \approx SO(3)$, and $K_0(\pi)$ leaves invariant each 3-dimensional h-plane π', orthogonal to π. Choose one of these planes π', and let Σ^2 be the sphere of radius 1 in π', centered at $p = \pi \cap \pi'$. (The action of $K_0(\pi)$ on Σ^2 is isomorphic to the action of $SO(3)$ on a standard 2-sphere.) Let k_1 and k_2 be two elements in $K_0(\pi)$, which have infinite order, and no common fixed point in Σ^2. Let K denote the group generated by k_1 and k_2. K is dense in $K_0(\pi)$. (This follows easily from the fact that the only connected Lie subgroups of $SO(3)$ are 1-parameter groups.) Since K is a 2-generator group, and F is free of rank 2, there is a surjective homomorphism $\psi: F \to K$. Let $k_f = \psi(f)$ and $\Gamma = \{fk_f : f \in F\}$. Since $f = 1$ implies $k_f = 1$, $\Gamma \cap K_0(\pi) = \{1\}$, and Γ is discrete.

Note that $L(\Gamma) = L(F)$, so that $S(\Gamma) = S(F) = \partial\pi$ has dimension 1. Since $C(\Gamma)$ leaves $S(\Gamma)$ invariant, it also leaves π invariant, and

$C(\Gamma) \subset K(\pi) \times I(\pi)$. The subgroup $C_0(\Gamma) = C(\Gamma) \cap (K_0(\pi) \times I_0(\pi))$ has finite index in $C(\Gamma)$, and $C_0(\Gamma) \subset K_0(\pi) \times I_0(\pi)$. Let ϕ denote the projection of $C_0(\Gamma)$ into $I_0(\pi)$. Since commensurability is preserved by homomorphisms, $\phi C_0(\Gamma) \subset G = C(F, I_0(\pi))$ (the commensurability group of F in $I_0(\pi)$). Recall that Theorem 2 implies $[G:F] < \infty$.

We assert that $\mathrm{Ker}(\phi) = C_0(\Gamma) \cap K_0(\pi)$ is trivial. For suppose that $k \in C_0(\Gamma) \cap K_0(\pi)$. There is a positive integer n, such that $k\gamma^n k^{-1} \in \Gamma$, for all $\gamma \in \Gamma$. Each $\gamma \in \Gamma$ is of the form $\gamma = fk_f$ (for $f \in F$). Thus $\gamma^n = f^n k_f^n \in \Gamma$, and $k\gamma^n k^{-1} = f^n k k_f^n k^{-1} \in \Gamma$. Therefore $kk_f^n k^{-1} k_f^{-n} = k\gamma^n k^{-1} \gamma^{-n} \in \Gamma$. Since $K_0(\pi) \cap \Gamma = \{1\}$, $kk_f^n k^{-1} k_f^{-n} = 1$. If $k \neq 1$ and k_f has infinite order, then either k has the same fixed points as k_f in Σ^2, or k is an involution, which interchanges the fixed points of k_f. Now using the properties of $K = \{k_f : f \in F\}$ (namely K is dense in $K_0(\pi)$, and has elements of infinite order, so that the fixed points of these elements are dense in Σ^2) we see that $k = 1$ and $\mathrm{Ker}(\phi) = K_0(\pi) \cap C_0(\Gamma) = \{1\}$.

We are now ready to show that $C(\Gamma)$ is discrete. It suffices to show that the index $[C_0(\Gamma):\Gamma] < \infty$. Recall that $\phi C_0(\Gamma) \subset G$, and $[G:F] < \infty$. Since ϕ is an isomorphism, $[C_0(\Gamma):\Gamma] = [\phi C_0(\Gamma):\phi\Gamma] \leq [G:F] < \infty$.

§3. *Maximal groups*

In [5] a Kleinian group Γ is called *finitely maximal* if Γ is not a subgroup of finite index in any other Kleinian group. Let Γ be a finitely generated, non-elementary, Kleinian group whose region of discontinuity is Ω. The Ahlfors finiteness theorem says that $\Omega/\Gamma = S_1 \cup \ldots \cup S_n$, where the S_i are surfaces of finite type, with finitely many branch images. This permits us to attach a signature $\sigma = (\sigma_1, \ldots, \sigma_n)$ to Γ. (We shall say that σ has n components.) σ is called *finitely maximal*, if every Kleinian group of signature σ is finitely maximal. In [5] it is shown that many such signatures exist.

Two Kleinian groups Γ, Δ will be called *weakly commensurable* if there exists $g \in \mathfrak{M}_2^0$, such that $\Gamma \sim g\Delta g^{-1}$. The following theorem follows immediately from Theorem 2 (part (4)).

THEOREM 3. (1) *Let* Γ *be a finitely generated, non-elementary Kleinian group whose limit set* $L(\Gamma)$ *is different from a circle. If* Γ *is finitely maximal, then* $C(\Gamma) = \Gamma$. *Thus,* Γ *contains all of its commensurable groups.*

(2) *Let* σ *be a finitely maximal signature with more than two components, and let* Γ *and* Δ *have signature* σ. *Then* Γ *is weakly commensurable to* $\Delta \Longleftrightarrow \Gamma$ *is conjugate to* Δ.

(3) *Let* σ, τ *be different finitely maximal signatures with more than two components. Then no group of signature* σ *is weakly commensurable to a group of signature* τ.

UNIVERSITY OF MARYLAND
COLLEGE PARK, MARYLAND

REFERENCES

[1]. Ahlfors, L., *Finitely generated Kleinian groups*, Amer. J. Math. 86 (1964), pp. 413-429.

[2]. Bieberbach, L., *Über die Bewegungsgruppen der Euklidischen Räume I*, Math. Ann. 70 (1911), pp. 297-336.

[3]. Borel, A., Introduction aux groupes arithmétiques, Hermann (1969).

[4]. Greenberg, L., *Discrete subgroups of the Lorentz group*, Math. Scand. 10 (1962), pp. 85-107.

[5]. ———, *Maximal groups and signatures*, these proceedings.

[6]. Leutbecher, A., *Über die Heckeschen Gruppen* $G(\lambda)$, Abh. Math. Sem. Univ. Hamburg, 31 (1967), pp. 199-205.

[7]. Selberg, A., *On discontinuous groups in higher-dimensional symmetric spaces*, Contributions to function theory (Internat. Colloq. Function Theory, Bombay, 1960), 147-164.

CHABAUTY SPACES OF DISCRETE GROUPS

William Harvey

Introduction

The purpose of this talk is to demonstrate the importance, for the study of moduli of a Fuchsian or Kleinian group, of the space of all discrete subgroups with the Chabauty topology, which was first introduced in the context of lattices in R^n. Here we indicate how the space may be used to construct a compactification of the Riemann space of a Fuchsian group which is finitely generated and of the first kind. The compactification turns out to be of significance for related problems concerning the boundary of the Teichmüller space. Finally we point out the connection between the notion of stability and the natural question of determining the local structure of Chabauty space.

§1. *The space of Fuchsian groups*

For the present, G will denote the group PSL(2, R), and μ will denote the invariant measure on G. Let S(G) denote the collection of all discrete subgroups of G and $S_0(G)$ the subspace consisting of those subgroups Γ such that G/Γ is compact. The topology on S(G) is most briefly described thus: if Γ_n, $n = 1, 2, \ldots,$ is a sequence of groups we say $\Gamma_n \to \Gamma$ as $n \to \infty$ if

(i) each $\gamma \in \Gamma$ is a limit of elements $\gamma_n \in \Gamma_n$, $n = 1, 2, \ldots,$

(ii) if $\gamma_n \in \Gamma_n$, $n = 1, 2, \ldots$ and $\gamma_n \to \gamma$ as $n \to \infty$, then $\gamma \in \Gamma$.

Essentially this means that groups are close to each other if their intersections with compact sets in G are close. For more details on this topology see Macbeath [7]. A fundamental theorem proved there is that for $\Gamma_0 \in S_0(G)$ there is a neighborhood of Γ_0 in S(G) consisting entirely of groups isomorphic to Γ_0. It is natural then to study the open subspace $S(\Gamma_0, G) \subseteq S_0(G)$ of groups isomorphic to Γ_0. Its importance is emphasized by the fact that Riemann's space of moduli $X(\Gamma_0)$ sits underneath $S(\Gamma_0, G)$, via the quotient by the action of G by conjugation. One would like to know about the closure of $S(\Gamma_0, G)$ in $S(G)$. The first point of interest is a result of Chabauty [3].

THEOREM. *Let* Γ_n, $n = 1, 2, \ldots$, *be a sequence of discrete groups with the property that, for some open neighborhood* V *of* id *in* $G, \Gamma_n \cap V = \{\mathrm{id}\}$ *for all* n.[1] *Then some subsequence* $\{\Gamma'_n\}$ *converges to a discrete group* $\overline{\Gamma}$, *and if in addition* $\mu(G/\Gamma_n) \leq K$ *for all* n *then* $\mu(G/\overline{\Gamma}) \leq \liminf \mu(G/\Gamma'_n)$.

In the present situation $\mu(G/\Gamma_n) = \mu(G/\Gamma_0)$, so $\mu(G/\overline{\Gamma}) \leq \mu(G/\Gamma_0)$. It was observed by Mumford [10] that if one imposes the restriction on $\{\Gamma_n\}$ that all hyperbolic elements have trace bounded away from ± 2, then necessarily $\overline{\Gamma} \in S(\Gamma_0, G)$. Therefore all honest Fuchsian groups on the boundary of $S(\Gamma_0, G)$ possess parabolic elements and are not isomorphic to Γ_0. It is relatively simple to construct for any such group Γ, with $\mu(G/\Gamma) \leq \mu(G/\Gamma_0)$ and having its collection of primitive periods (nonconjugate) contained in that of Γ_0, a sequence of group in $S(\Gamma_0, G)$ tending to Γ; and no group Γ having $\mu(G/\Gamma_0) < \mu(G/\Gamma)$ can be so approximated. Let $\mathcal{F}(\Gamma_0)$ denote the family (finite) of isomorphism classes of groups of the former type. One then has the result which follows.

THEOREM 1. [5]
$$\partial S(\Gamma_0, G) = \bigcup_{\Gamma \in \mathcal{F}(\Gamma_0)} S(\Gamma, G)\ .$$

§2. *Compactifying moduli space* $X(\Gamma_0)$

The idea of making the space of curves of genus $g \geq 2$ into a compact variety by adding suitable singular curves goes back to Severi (and perhaps

[1] Such a set of groups is said to be *uniformly* discrete.

further). We shall show now how to associate to a sequence of Riemann surfaces of genus $g \geq 2$ a limit "surface" which is composed of a collection of punctured surfaces with a $(2:1)$ identification on the set of all punctures – in other words a singular surface with ordinary double points in the sense of Bers [2], or the complex points of a stable curve of genus g in the sense of Deligne-Mumford [4].

Let $\{\Gamma_n\}$ be a sequence of groups in $S(\Gamma_0, G)$, with Γ_0 a surface group of genus $g \geq 2$. As far as $X(\Gamma_0)$, the space of moduli, is concerned we are at liberty to conjugate each Γ_n by an element $t_n \in G$. This turns out, with the help of the next lemma, to enable us to reconstruct a collection of conjugacy classes of groups in $\partial S(\Gamma_0, G)$ which represent all possible surfaces arising as limits (in the sense of Chabauty for the uniformizing groups) of some fixed subsequence from the original sequence of surfaces.

What is needed is a method of adjusting by conjugation to make the sequence $\{\Gamma_n\}$ uniformly discrete and this is provided by a result from the non-Euclidean geometry of the unit disc Δ, the symmetric space associated to G.

LEMMA. *Let* $k > 0$. *There is a* $\delta > 0$ *such that any convex polygon in* Δ, *having invariant measure at least* k, *contains a (non-Euclidean) disc of radius* δ.

This was originally proved by the author under a restriction on the number of sides, and has since been proved independently by Shinnar and Sturm [10] and by Marden. It is closely related to a deep theorem of Kazdan and Margulis on semi-simple Lie groups ([6], see also H. C. Wang's paper [12]).

Let $z_n \in \Delta$ be the center of a non-Euclidean disc of radius δ contained in a Dirichlet fundamental polygon for Γ_n, $n = 1, 2, \ldots$, and let $t_n \in G$ be chosen so that $t_n(z_n) = 0$. Then the sequence of groups $\{t_n \Gamma_n t_n^{-1}\}$ is uniformly discrete since their Dirichlet regions contain a disc about 0. By Chabauty's theorem we may choose a subsequence $\{\Gamma_n^*\}$ which converges to a discrete group Γ with $\mu(G/\Gamma) \leq \mu(G/\Gamma_0)$. If equality holds that is the end of the process, but if the inequality is strict the situation becomes more interesting. There is in any case a mapping $\theta_n : \Gamma \to \Gamma_n^*$ defined for all sufficiently large n, by setting $\theta_n(\gamma)$ to be the element $\gamma_n \in \Gamma_n^*$ from a sequence of elements tending to γ as $n \to \infty$. For large n this is unambiguous and can be shown to be a monomorphism. The hyperbolic elements γ_n tending to a parabolic generator of Γ may be taken to correspond to curves in the surfaces $S_n = \Delta/\Gamma_n^*$ which are either all dividing cycles or all non-dividing cycles. Therefore the subgroups $\theta_n(\Gamma) \subset \Gamma_n^*$ define isomorphic pieces of well defined decompositions of these groups into either a free product with amalgamations or a product where two non-conjugate infinite cyclic subgroups become identified by introduction of a new element which conjugates the one into the other, or indeed any combination of a finite number of each.

For example, if $g = 2$ the possible boundary groups Γ are of signature $(0; n)$,[2] with $n = 3$ or 4, and $(1; n)$ with $n = 1$ or 2. The diagrams below illustrate the cases $(1; 1)$ and $(0; 3)$.

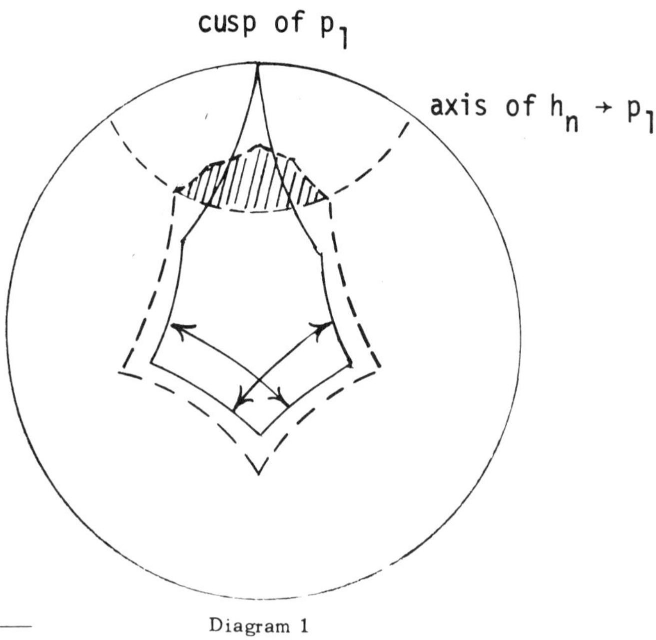

Diagram 1

[2] $(g; n)$ refers to a group with genus g and n boundary elements.

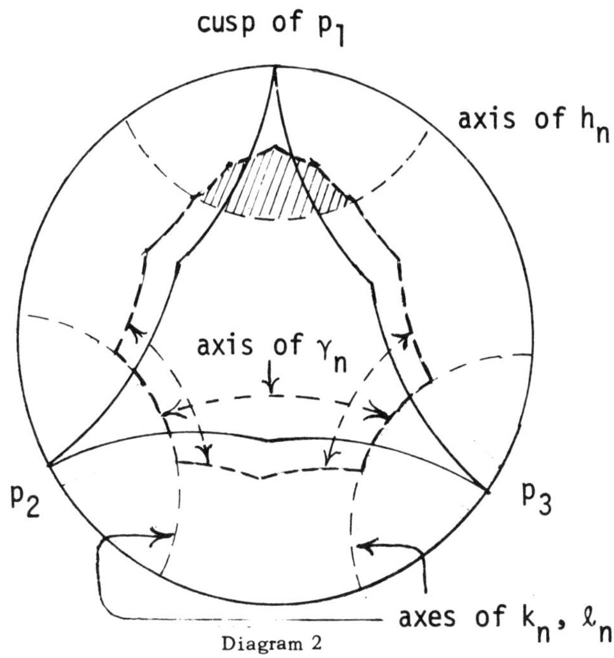

Diagram 2

In the first case $\mu(G/\Gamma) < \mu(G/\Gamma_0)$. This is because half of the fundamental region for the group Γ_n^* lies beyond the axis of the hyperbolic element h_n which approximates to the parabolic generator $p \in \Gamma$. Consequently as $n \to \infty$ half of the fundamental region is washed away to infinity down the waste pipe at the cusp. In the second case this is happening too, but in addition two hyperbolic elements k_n, ℓ_n are permuted by an element $\gamma_n \in \Gamma_n^*$; there is thus an implicit pairing of the parabolic elements $p_2 = \lim_{n\to\infty} k_n$ and $p_3 = \lim_{n\to\infty} \ell_n$ of Γ.

One recovers the second piece by employing the lemma to move the disappearing sub-polygons to positions containing a disc about 0. Thus we obtain a new sequence of conjugates $\{u_n \Gamma_n u_n^{-1}\}$ which is uniformly discrete and so has a subsequence Γ_n^{**} which converges to another piece of the limit. In fact this is the only other piece in this case and one would obtain from this subsequence a limit group with signature either (1; 1) or (0; 3). One has now for the second subsequence a decomposition of the approximating groups Γ_n^{**} into either the free product of two (1; 1) groups with the subgroup generated by h_n amalgamated, or a group obtained from the free product of a (1; 1) group with a (0; 3) group with amalgamation of $\langle h_n \rangle$ and subsequent adjunction of the element which permutes the remaining two boundary curves.

The situation in general may be viewed as a jigsaw puzzle where the linking of two pieces is accomplished by a process analogous to the first case and the self-linkings of a piece come by the second one. We can now state the general result. Let Γ_0 be any Fuchsian group with $\mu({}^G/\Gamma_0) < \infty$.

THEOREM 2. *Let* Γ_n, $n = 1, 2, \ldots,$ *be a sequence of groups in* $S(\Gamma_0, G)$. *There is a subsequence (again denoted by* $\{\Gamma_n\}$*) and a collection of groups* $\Gamma^1, \Gamma^2, \ldots, \Gamma^m$ *whose isomorphism classes belong to* $\mathcal{F}(\Gamma_0)$, *such that each* Γ_n *has a decomposition into a free product of* m *groups* $H_{n1}, \ldots, H_{nm}$ *isomorphic to* $\Gamma^1, \ldots, \Gamma^m$ *with certain amalgamations, followed by adjunction of a number of elements which permute pairs of hyperbolic generators of the* $\{H_{ni}\}$. *The decomposition structure is independent of* n, *that is the patterns of amalgamation and linking are the same, and as* $n \to \infty$ *suitably chosen conjugates of* $\{H_{ni}\}$ *converge to* Γ^i, *for* $i = 1, 2, \ldots, m$. *These represent all possible (discrete) limits of* $\{\Gamma_n\}$.

One can now strengthen the approximation result mentioned in Section 1, using the methods of Fricke, Nielsen and Fenchel, and Keen, to show that any such collection of groups $\Gamma^1, \ldots, \Gamma^m$ can be simultaneously approximated in any appropriate $S(\Gamma_0, G)$ – there must of course be an equality between the union of the periods of $\{\Gamma^i\}$ and Γ_0 and one must have $\sum_{i=1}^{m} \mu({}^G/_{\Gamma^i}) = \mu({}^G/_{\Gamma_0})$. In short the groups must represent a singular surface of the type of $\Delta/_{\Gamma_0}$. The purely combinatorial conditions necessary and sufficient for this have been described by Maskit [9].

Full proofs of Theorem 2 and the consequent compactification theorem for $X(\Gamma_0)$ will appear in [5].

§3. *Implications and speculations*

It should be pointed out that the approximation process described above is tailored for interpretation in terms of the Teichmüller spaces of the groups concerned, since to each collection $\Gamma^1, \ldots, \Gamma^m$ representing a limit of groups $\{\Gamma_n\}$ in $S(\Gamma_0, G)$ one can choose an underlying marking on Γ_0 such that the isomorphisms $r_n : \Gamma_0 \to \Gamma_n$ tend to a specific point of the boundary of $T(\Gamma_0)$. If one takes the representation of $T(\Gamma_0)$ inside the space of Schwarzian derivatives of Schlicht functions on the disc Δ (see [2a]), it can be shown that the point in question is represented by a regular b-group belonging to Γ_0. Conversely, given a regular b-group one can construct a sequence of points in $T(\Gamma_0)$ which tends to it, a result proved originally by Marden [8a] using 3-manifold methods. One thence obtains an imbedding of a fibred product of the spaces $T(\Gamma^i)$ into the boundary of $T(\Gamma_0)$, analogous to the adjunction of a rational point on the real axis to the upper half plane in the classical case of elliptic curves. This situation arises in the work of Abikoff [1], where these boundary components are obtained from the direct study of $T(\Gamma_0)$ and explicit construction of certain neighborhoods of the boundary in terms of quasiconformal mappings.

In order to capture all of the well-behaved points of $\partial T(\Gamma_0)$ one must now allow the Teichmüller modular group to act, in the manner described by Earle in his Maryland talk. The analogy with the case of elliptic curves is again striking. One would hope that this approach might yield a proof that the compactified moduli space $\overline{X(\Gamma_0)}$ is a projective variety along the lines of Bailey's theorem. Another speculation might be that more information can be gleaned about the mysterious degenerate b-groups. The Fuchsian approximation process can be applied to find such groups representing any given collection of surfaces satisfying the conditions detailed by Maskit in [9] as a limit of points of $T(\Gamma_0)$, but one knows too little in general about the moduli of such groups and no constructive information appears from the above method.

A final observation is that the study of the Chabauty spaces for Kleinian groups should yield information about the stability of such groups. Making the obvious analogy with the Fuchsian situation, it seems to be insufficient to study only the space of representations of a fixed group as there may well be non-isomorphic groups which are close to each other in Chabauty's sense.

UNIVERSITY OF LONDON,
KING'S COLLEGE

REFERENCES

[1]. Abikoff, W., Maryland lecture.

[2]. Bers, L., "Spaces of degenerating Riemann surfaces," Maryland lecture.

[2a]. ———, "On boundaries of Teichmüller spaces and on Kleinian groups I," Ann. of Math. 91 (1970), pp. 570-600.

[3]. Chabauty, C., "Limites d'ensembles et geometrie des nombres," Bull. Soc. Math. de France 78 (1950), pp. 143-151.

[4]. Deligne, P., Mumford, D., "The irreducibility of the space of curves of given genus," Publ. Math. I.H.E.S. 36 (1969).

[5]. Harvey, W., "Limits of Fuchsian groups and the compactification of moduli space," Manuscript in preparation.

[6]. Kazdan, D. A., Margulis, G. A., "A proof of Selberg's conjecture," Math. U.S.S.R. (Sbornik), 4 (1968), pp. 147-152.

[7]. Macbeath, A. M., "Groups of homeomorphisms of a simply connected space," Ann. of Math. 79 (1964), pp. 473-488.

[8]. Marden, A., "The geometry of finitely generated Kleinian groups," Ann. Math. (to appear).

[8a]. ———, preprint.

[9]. Maskit, B., "On boundaries of Teichmüller spaces and on Kleinian groups II," Ann. of Math. 91 (1970), pp. 607-639.

[10]. Mumford, D., "A remark on Mahler's compactness theorem," Proc. Amer. Math. Soc. 28 (1971), pp. 289-294.

[11]. Shinnar, M., Sturm, J., Maryland talk.

[12]. Wang, H. C., "On discrete nilpotent subgroups of Lie groups," J. Diff. Geom.

MONODROMY GROUPS AND LINEARLY POLYMORPHIC FUNCTIONS

Dennis A. Hejhal

§1. *Introduction*

Around 1880-1890, Klein and Poincaré studied monodromy groups for linear differential equations on compact Riemann surfaces. See Klein [4] and Poincaré [9, 10]. Roughly stated, there were two fundamental problems:

(A) given the DE, to calculate the monodromy group;

(B) to determine all DE [of a certain type] having a given monodromy group.

Neither of these problems was completely solved, by any means.

We are particularly concerned with problem (B) for monodromy groups which arise from Schwarzian differential equations of the form:

$$\{z, t\} = q(t), \qquad F = U/\mathcal{G} \; . \tag{1}$$

We assume here that F is a compact RS of genus $g \geqq 2$, U is the unit disk, $\mathcal{G}$ is the automorphic group, and that $q(t)$ is a regular [i.e., pole-free] quadratic differential mod $\mathcal{G}$. There is, of course, a well-known equivalence between (1) and the linear DE

$$u'' + \frac{1}{2} q(t)u = 0 \; . \tag{2}$$

This particular equivalence can be seen by forming quotients of linearly independent solutions of (2).

Using the Poincaré index theorem (for vector fields) and a certain cutting-and-pasting technique, we have been able to gain new insight into

Some abbreviations: (i) DE = differential equation; (ii) LF = linear fractional; (iii) RS = Riemann surface; (iv) S = (Riemann) sphere.

the nature of problem (B). In this paper, we present a survey of our results and sketch several of the proofs. A more detailed treatment of these results will appear elsewhere.

§2. *Preliminaries*

Let $\mathfrak{L}$ denote the group of LF transformations and consider the Schwarzian DE (1). It is easy to check that there is a homomorphism $L \to \tilde{L}$ taking $\mathcal{G} \to \mathfrak{L}$ such that

$$z(Lt) = \tilde{L}z(t) \; . \tag{3}$$

Thus, $z = z(t)$ transforms linear fractionally under the action of $\mathcal{G}$. For this reason, $z(t)$ is called a linearly polymorphic function. The group $\{\tilde{L}\}$ is called the monodromy group of $z(t)$.

Because $q(t)$ is a regular quadratic differential, it also follows that $z(t)$ is a locally schlicht meromorphic function on U.

Suppose next that F is a marked compact surface. Thus, F comes equipped with a canonical homotopy basis $\{\mathcal{A}_k, \mathcal{B}_k\}_{k=1}^{g}$. We may suppose that the corresponding commutator relation reads

$$\prod_{k=1}^{g} (\mathcal{A}_k, \mathcal{B}_k) = I \; . \tag{4}$$

In this case, we have (modulo an auxiliary LF map) a corresponding marking on $\mathcal{G}$:

$$\left\{ \begin{matrix} \mathcal{A}_k \to A_k, \qquad \mathcal{B}_k \to B_k \\ \\ \mathcal{G} = [A_1, \dots, A_g; B_1, \dots, B_g] \end{matrix} \right\} . \tag{5}$$

Corresponding to (4), we now have

$$\prod_{k=1}^{g} (A_k, B_k) = I \; . \tag{6}$$

To avoid any ambiguity, one can of course normalize three fixpoints of A_1 and A_2. In this way, the marked Fuchsian group $\mathcal{G}$ corresponding to the marked surface F is uniquely determined. Note: the fixpoints of A_1 and A_2 are mutually distinct, since otherwise (A_1, A_2) would be parabolic.

It is now clear that the monodromy group is generated by $\tilde{A}_1, \ldots, \tilde{B}_g$. We may therefore define the marked monodromy group $\mathfrak{M}[z]$ of $z(t)$ to be:

$$\mathfrak{M}[z] = [\tilde{A}_1, \ldots, \tilde{A}_g; \tilde{B}_1, \ldots, \tilde{B}_g] . \tag{7}$$

By equation (6),

$$\prod_{k=1}^{g} (\tilde{A}_k, \tilde{B}_k) = I .$$

In an obvious way, we may also write $\mathfrak{M}[z] \in \mathfrak{L}^{2g}$.

Let R denote the conglomeration of all marked monodromy groups – over all possible $z(t)$, $q(t)$, and F. Let $N = \{(X_1, \ldots, X_g; Y_1, \ldots, Y_g) \in \mathfrak{L}^{2g} : (X_1, Y_1) \ldots (X_g, Y_g) = I\}$. Clearly,

$$\mathfrak{M} = \frac{R}{\mathfrak{L}} \subseteq \frac{N}{\mathfrak{L}} \subseteq \frac{\mathfrak{L}^{2g}}{\mathfrak{L}} . \tag{9}$$

The quotient spaces $\mathfrak{M}$, $N/\mathfrak{L}$, $\mathfrak{L}^{2g}/\mathfrak{L}$ are taken modulo inner automorphisms and carry the obvious topologies. Certain properties of the space $N/\mathfrak{L}$ have been studied by Gunning [2]; see also Newstead [8].

Next, let T_g denote the usual Teichmüller space. To each point $\tau \in T_g$, there corresponds the complex vector space $Q[\tau]$ of regular quadratic differentials: $\dim Q[\tau] = 3g-3$. If one uses simultaneous uniformization via either quasi-Fuchsian or Schottky groups, it is not particularly difficult to check that locally we can assume that $Q[\tau]$ has a basis $[q_1(t;\tau), \ldots, q_{3g-3}(t;\tau)]$ which varies smoothly in τ. From this, one quickly deduces that the various spaces $Q[\tau]$ can be pasted together to yield a vector bundle TQ over T_g. Locally, of course, TQ can be thought of as $T_g \times C^{3g-3}$. Since T_g is contractible, TQ is actually homeomorphic to $T_g \times C^{3g-3}$. See, for example, Steenrod [14], p. 53.

Let us now choose any point $(\tau, q) \in TQ$. We can then form the Schwarzian DE $\{z, t\} = q(t)$. The solution of this equation is unique only up to an auxiliary LF mapping. The corresponding marked monodromy group $\mathfrak{M}[z]$ is therefore determined only up to an inner automorphism. In this way, we define an obvious monodromy mapping:

$$p : TQ \to \mathfrak{M} . \tag{10}$$

Our main goal is to understand the topological properties of both the mapping $p : TQ \to \mathfrak{M}$ and the set $\mathfrak{M} \subseteq N/\mathfrak{L}$. For motivation, we refer to A-G-F [1], pp. 272-341 and Poincaré [9].

The following remarks may provide a useful perspective. First of all, by counting the constants, we see that TQ and $N/\mathfrak{L}$ both have complex dimension $6g-6$. One also guesses that the mapping p is non-degenerate. It should therefore follow that the fibers of p over $\mathfrak{M}$ are discrete. One might well go a little further and guess that p is actually a homeomorphism. In fact, Poincaré [9], pp. 218-222 seemed to think so. His proof, however, was criticized in A-G-F [1], pp. 310-311, 335-337 and Hilb [3], p. 518. The same criticisms also seem to apply to Schlessinger [12], pp. 312-318. Since then, however, it seems that relatively little progress has been made in getting to the bottom of this unsatisfactory state of affairs.

Caution. This problem should not be confused with the classical Riemann problem (Hilbert problem no. 21); see also Klein [5], pp. 132-137. The linearly polymorphic functions which arise in the solution of the Riemann problem are not, in general, locally schlicht.

§3. *Some known results*

For completeness, we recall several known results concerning marked monodromy groups of the Schwarzian DE (1).

THEOREM A (Poincaré). *Suppose that* z(t) *and* w(t) *are linearly polymorphic functions over the same Teichmüller point* $\tau \in T_g$. *Suppose moreover that* $\mathfrak{M}[z] = \mathfrak{M}[w]$. *Then,* $z(t) \equiv w(t)$.

Proof. See, for example, A-G-F [1], pp. 310-317 and Schiffer-Hawley [11]. ∎

THEOREM B. *Suppose that* $z(t)$ *is a linearly polymorphic function. The following statements are then equivalent*:

(i) $z(U) \neq S$;

(ii) $z : U \to z(U)$ *is a topological covering map*;

(iii) $\mathfrak{M}[z]$ *acts discontinuously on* $z(U)$.

Proof. See Kra [6] and also A-G-F [1], pp. 272-341. ∎

COROLLARY. *Suppose that items* (i)-(iii) *hold. Then,* $\mathfrak{M}[z]$ *can be discontinuous in no domain strictly larger than* $z(U)$.

Proof. See Kra [6]. ∎

THEOREM C. *Let* $z(t)$ *be a linearly polymorphic function. Then, the elements of* $\mathfrak{M}[z]$ *cannot all have a common fixpoint.*

Proof. See A-G-F [1], pp. 297-310. ∎

§4. *Properties of the monodromy mapping*

The linearly polymorphic functions discussed in this section are all understood to be locally schlicht.

THEOREM 1. *The mapping* $p : TQ \to \mathfrak{M}$ *is a local homeomorphism and* $\mathfrak{M}$ *is a smooth complex manifold of dimension* $6g-6$.

For a somewhat weaker version of this theorem, see A-G-F [1], pp. 335-337.

THEOREM 2. *Suppose that* $z(t)$ *and* $w(s)$ *are linearly polymorphic functions such that* $\mathfrak{M}[z] = \mathfrak{M}[w]$. *Then,* $z(U) \cap w(U)$ *is nonvoid.*

Several remarks concerning Theorem 2 are in order. First of all, to appreciate the theorem, one must recall how Poincaré hoped to prove that p is a homeomorphism. Suppose that $z(t)$ and $w(s)$ have $\mathfrak{M}[z] = \mathfrak{M}[w]$. Let $z(t)$ correspond to Teichmüller point τ_t and marked Fuchsian group $\mathcal{G}_t$. Let τ_s and $\mathcal{G}_s$ have a similar meaning for $w(s)$. Choose canonical polygons $\mathcal{F}_t$ and $\mathcal{F}_s$ for $\mathcal{G}_t$ and $\mathcal{G}_s$, respectively. Considered as Riemannian images, $z(\mathcal{F}_t)$ and $w(\mathcal{F}_s)$ are both simply-connected membranes whose sides are identified in pairs. Since $\mathfrak{M}[z] = \mathfrak{M}[w]$, the identification maps are the same in both cases. We also observe that the sum of the vertex angles in both membranes is 2π. One would now like to conclude that the membranes $z(\mathcal{F}_t)$ and $w(\mathcal{F}_s)$ can be made to coincide after making certain admissible deformations. For then, by repeated pasting, the Riemannian images $z(U)$ and $w(U)$ would coincide, and $\tau_t = \tau_s$. Theorem A could then be applied and the mapping p would be 1–1. The main difficulty stems from the fact that $z(\mathcal{F}_t)$ and $w(\mathcal{F}_s)$ can be multi-sheeted.

Moreover, as noted in A-G-F [1], pp. 310-311, it is entirely possible that $z(U) \cap w(U)$ is void, so that the deduction $z(U) = w(U)$ is absurd. Poincaré did not exclude this possibility.

Once the theorem is proved, it is easy to check that by making admissible deformations in $\mathcal{F}_t$ and $\mathcal{F}_s$, one can assume that the vertices of the membranes $z(\mathcal{F}_t)$ and $w(\mathcal{F}_s)$ have the same respective coordinates.

Lastly, it is instructive to check why the exterior half of a marked Fuchsian group does not yield a contradiction to Theorem 2. The unmarked monodromy groups in this case are clearly identical.

In the following two theorems, the marked Fuchsian and marked Schottky groups referred to are assumed to arise from compact RS of genus g.

THEOREM 3. *Let* $z(t)$ *be a linearly polymorphic function such that* $\mathfrak{M}[z]$ *is a marked Fuchsian group* (*element-wise*). *Suppose further that* $z(U) \neq S$. *Then,* $z(t)$ *reduces to a linear fractional mapping. The corresponding assertion for* $z(U) = S$ *is false.*

THEOREM 4. *The analogous theorem holds for the case in which* $\mathfrak{M}[z]$ *is a marked Schottky group.*

The following theorem shows that the mapping p is far from being a homeomorphism.

THEOREM 5. *There exist linearly polymorphic functions* $z(t)$ *and* $w(s)$ *on conformally inequivalent surfaces* F_z *and* F_w *such that* $\mathfrak{M}[z] = \mathfrak{M}[w]$.

Since p fails to be 1–1, one might guess that $p: TQ \to \mathfrak{M}$ is simply a topological covering.

THEOREM 6. *The mapping* $p: TQ \to \mathfrak{M}$ *is a topological covering at least over the part of* $\mathfrak{M}$ *corresponding to marked quasi-Fuchsian groups.*

THEOREM 7. *The mapping* $p: TQ \to \mathfrak{M}$ *is not a covering map, since certain paths in the base space* $\mathfrak{M}$ *cannot be lifted to* TQ.

Our last result gives important information about the size of the set $\mathfrak{M}$. Recall Theorem C.

THEOREM 8. *Let* $(L_1, \ldots, L_g)$ *be a point of* $\mathfrak{L}^g$ *such that:* (i) $L_1(z) = \lambda_1 z$, $|\lambda_1| > 1$; (ii) $L_2, \ldots, L_g$ *are loxodromic or hyperbolic;* (iii) *neither* 0 *nor* ∞ *is a fixpoint of any* L_k *for* $k \geqq 2$. *There then exists a linearly polymorphic function* $z(t)$ *such that* $\mathfrak{M}[z] = [L_1, \ldots, L_g; I, \ldots, I]$.

COROLLARY. *There exist monodromy groups* $\mathfrak{M}[z]$ *arbitrarily close to the identity.*

This corollary should be contrasted with A-G-F [1], pp. 337-341.

§5. *About the proofs*

To illustrate our method, we shall sketch the proofs of Theorems 2, 3, and 7.

Let us first recall Theorem 2 and the remarks which followed. Let f be a diffeomorphism connecting τ_t and τ_s. This clearly lifts to a diffeomorphism $s = \phi(t)$ taking U mod $\mathcal{G}_t$ onto U mod $\mathcal{G}_s$. Assume that $z(U) \cap w(U)$ is void. We may then define

$$E(t) = \frac{[w(s) - z(t)]^2}{w'(s)z'(t)\phi_t(t)}, \tag{11}$$

where $s = \phi(t)$ and $\phi_t = \partial\phi/\partial t$. We easily check that: (a) $E(t) \neq 0, \infty$; (b) $E(Lt) = E(t)L'(t)^2$, $L \in \mathcal{G}_t$. Define a single-valued branch $D(t) = \sqrt{E(t)}$ on U. It is easily seen that $D(Lt) = \chi(L)\,D(t)\,L'(t)$ for a unique homomorphism $\chi : \mathcal{G}_t \to \{-1, 1\}$. It follows that D(t) induces a singularity-free vector field on an appropriate two-sheeted covering of τ_t. This contradicts the Poincaré index theorem, as in Stoker [15], p. 244, since the genus of the covering is $2g-1$. Hence, $z(U) \cap w(U)$ must be nonvoid, as asserted.

We now turn our attention to Theorem 3. In order to prove the first assertion of the theorem, we may suppose that z(t) and $w(s) = s$ satisfy $\mathfrak{M}[z] = \mathfrak{M}[w]$ and that $z(U) \neq S$. Using Theorem 2, Theorem B, and the corollary, we quickly deduce that $z(U) = U$. But, then, z(t) must be schlicht and hence LF.

It remains to construct an appropriate counterexample for the case $z(U) = S$. For simplicity, we consider only the case $g = 2$. Let R denote the regular non-Euclidean octagon with center at the origin, first vertex along the positive real axis, and vertex angle $\pi/4$. By a classical theorem of Poincare (cf. Siegel [13], p. 84), R will be the canonical polygon of some marked surface F of genus $g = 2$. In obvious notation, then, $\partial R = \alpha_1^+ \beta_1^+ \alpha_1^- \beta_1^- \alpha_2^+ \beta_2^+ \alpha_2^- \beta_2^-$.

There will then exist unique (orientation-reversing) LF maps L_1, L_2, L_3, L_4 such that:

$$(12) \qquad \left\{\begin{matrix} L_1 : \alpha_1^- \to \alpha_1^+ & L_2 : \alpha_2^- \to \alpha_2^+ \\ L_3 : \beta_1^- \to \beta_1^+ & L_4 : \beta_2^- \to \beta_2^+ \\ \multicolumn{2}{c}{(L_1, L_3^{-1})(L_2, L_4^{-1}) = I} \end{matrix}\right\} .$$

By simple area considerations, it follows that the circles determined by the geodesics $\alpha_1^+, \alpha_1^-, \alpha_2^+, \alpha_2^-$ are mutually disjoint. Let their centers be c_1, c_2, c_3, c_4, respectively.

Let W_0 denote the planar surface defined by $w^2 = (z-c_1)(z-c_2)$. The surface W_0 will be a two-sheeted ramified covering of the sphere S; we let $\pi : W_0 \to S$ denote the obvious projection.

By means of the geodesics $\alpha_k^{\pm}$, $\beta_k^{\pm}$, we may now define a simply-connected subregion W of W_0 as indicated in Figure 1. Because $W \subseteqq W_0$, it makes sense to write ∂W. We have thus written $\partial W = A_1^+ B_1^+ A_1^- B_1^- A_2^+ B_2^+ A_2^- B_2^-$. Corresponding to (12), we now have:

$$(13) \qquad \left\{\begin{matrix} L_1 : A_1^- \to A_1^+ & L_2 : A_2^- \to A_2^+ \\ L_3 : B_1^- \to B_1^+ & L_4 : B_2^- \to B_2^+ \end{matrix}\right\} .$$

If we now identify the various $A_k^{\pm}$, $B_k^{\pm}$ using the L_k, we obtain a marked compact surface E of genus $g = 2$ (at least topologically). We supply E with a conformal structure by lifting that of S via $\pi : W \to S$. It is not difficult to check that this conformal structure is well-defined. Of course, the only real difficulty occurs at the vertices.

It is now quite clear that F and E are compact RS with obvious Teichmüller markings. We may therefore form the customary Fuchsian data for both:

$$(14) \qquad \begin{matrix} F \to (U, \mathcal{G}_F, R, L_1, L_2, L_3, L_4) \\ \\ E \to (U, \mathcal{G}_E, Q, K_1, K_2, K_3, K_4, \phi) \end{matrix} .$$

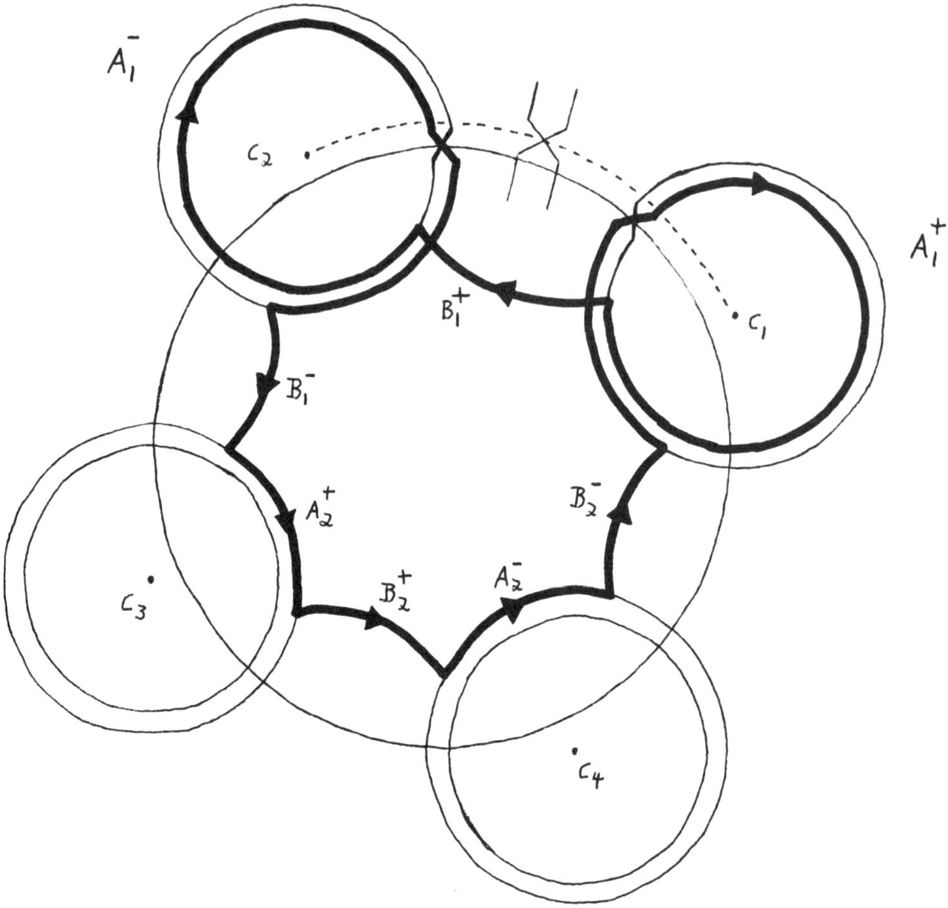

Fig. 1

The sides of the polygons R and Q are identified under the L_i and K_i, respectively, and $\phi : U \to E$ is simply the universal covering map. We may also assume WLOG that $\phi(\partial Q)$ corresponds exactly to ∂W.

Let us now define the function

(15) $$z(t) = \pi[\phi(t)], \quad t \in Q .$$

By means of the conformal structure given to E, we immediately see that z(t) is a locally schlicht meromorphic function on Q. Moreover, the

Riemannian image $z(Q)$ is precisely the surface W. From this, $z(t)$ obviously has a unique (spherically) continuous extension to $Q \cup \partial Q$.

Using the identifications, we readily check that

(16) $$z(K_m t) = L_m z(t), \quad 1 \leqq m \leqq 4 ,$$

along ∂Q as appropriate.

We must now define $z(t)$ for all $t \in U$. To do so, we exploit the polygonal decomposition $\{K(Q) : K \in \mathcal{G}_E\}$ and the canonical isomorphism $\mathcal{G}_E \cong \mathcal{G}_F$, $K \leftrightarrow \tilde{K}$ generated by $\tilde{K}_m = L_m$, $1 \leqq m \leqq 4$. We take as our definition:

(17) $$z(Kt) = \tilde{K}z(t), \quad t \in Q .$$

It is not especially difficult to check that $z(t)$ is a locally schlicht linearly polymorphic function and that $z(U) = S$.

It follows at once that the linearly polymorphic functions $z = z(t)$, $w = w(s) = s$ for $(U, \mathcal{G}_E, Q)$ and $(U, \mathcal{G}_F, R)$ have identical marked monodromy groups: $\mathfrak{M}[z] = \mathfrak{M}[w] = \mathcal{G}_F$. This completes the proof of Theorem 3.

REMARK 1. The geometric meaning of this example is best appreciated if one recalls the discussion following Theorem 2. This explains why the vertices of the membranes R and W have the same coordinates.

REMARK 2. I learned recently that a similar example was discussed by Maskit [7], pp. 6-7.

We now prove Theorem 7. For simplicity, we consider the case $g = 3$. Define a circular domain $D(r)$ for $0 < r < 1$ as depicted in Figure 2. We take $C_1^- = \{|z| = 1\}$, $C_1^+ = \{|z| = 3\}$, $C_2^-(r) = \{|z+2| = r\}$, $C_2^+(r) = \{|z-2| = r\}$, $C_3^- = \{|z-a| = \delta\}$, $C_3^+ = \{|z-b| = \delta\}$. In addition,

$$
(18) \qquad \left\{ \begin{aligned} T_1(z) &= 3z \\ T_2(r)(z) &= 2 - \frac{r^2}{z+2} \\ T_3(z) &= b + \frac{\delta^2}{z-a} \end{aligned} \right\}.
$$

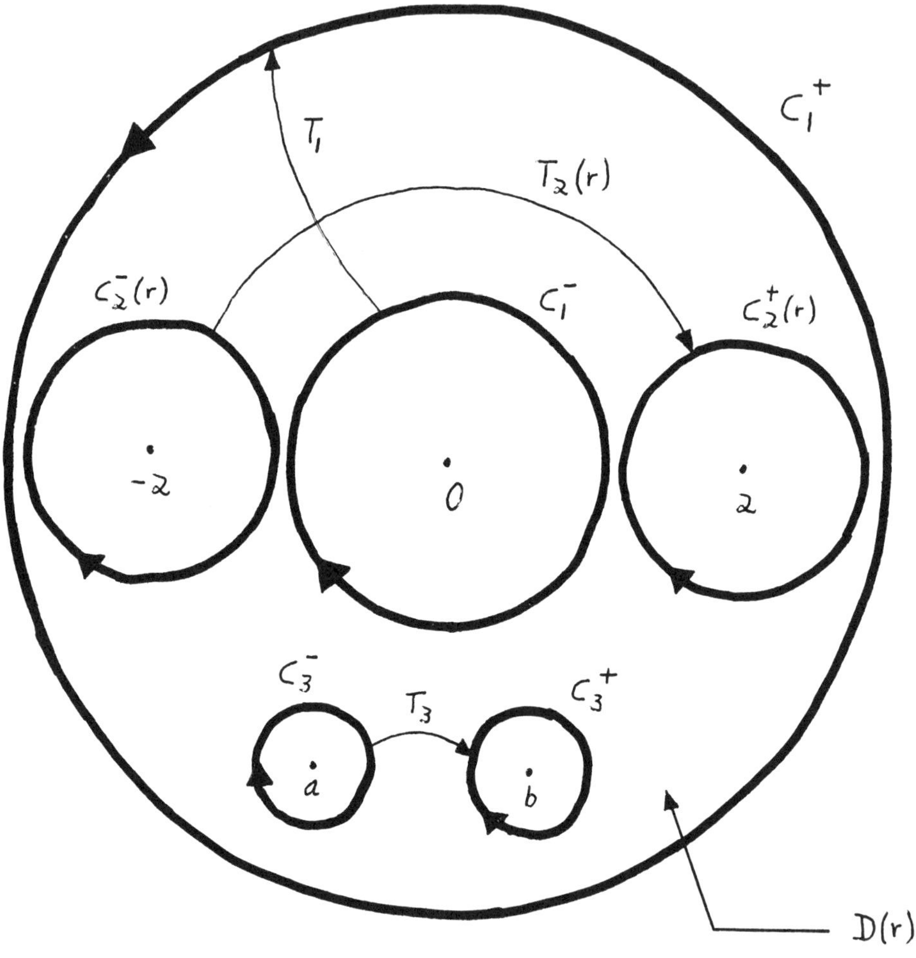

Fig. 2

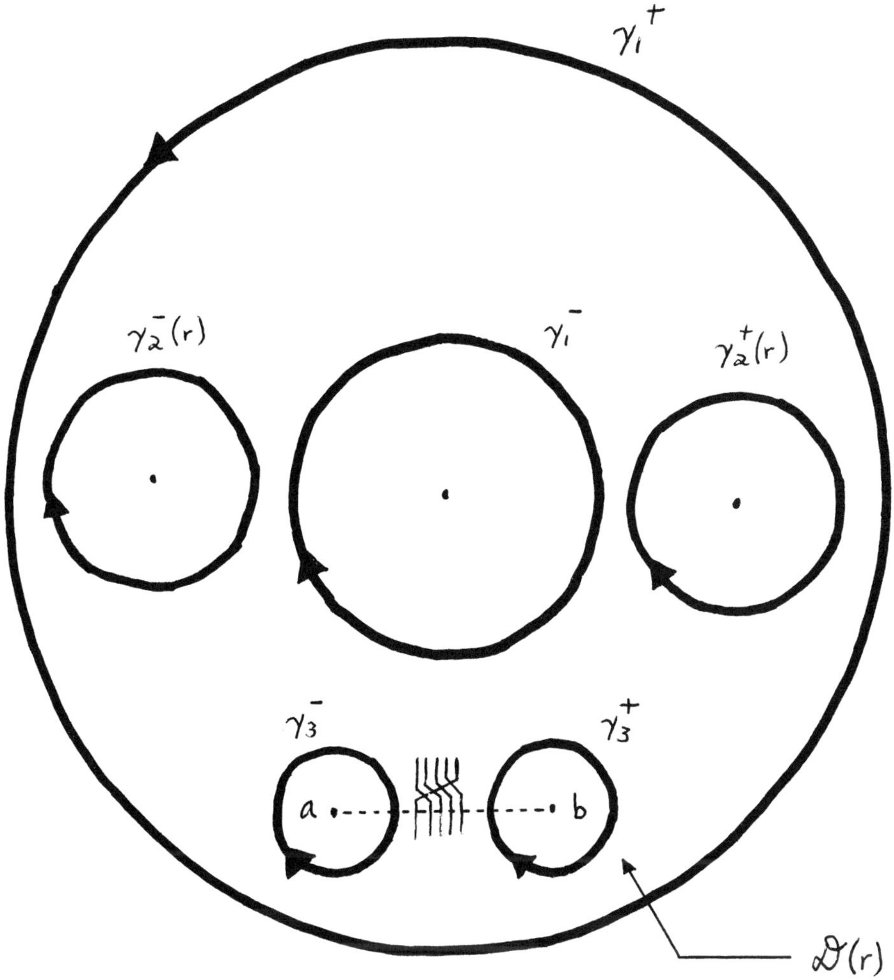

Fig. 3

Using the region D(r), we may clearly define a marked Schottky group $[T_1, T_2(r), T_3]$. As $r \to 1$, we will thus obtain a sequence of linearly polymorphic functions z_r such that $\mathfrak{M}[z_r] = [T_1, T_2(r), T_3; I, I, I]$. The corresponding Teichmüller points $\tau(r)$ may be assumed to vary continuously. The corresponding path $\{z_r\}$ in TQ must tend to $\partial(TQ)$. This is an easy consequence of the fact that, when $r = 1$, the commutator $(T_1, T_2(r))$ is parabolic.

But, now construct an analogous circular subregion $\mathfrak{D}(r)$ on the planar surface $w^5 = (z-a)^2(z-b)^3$. The boundary components $\gamma_1^-, \gamma_1^+, \gamma_2^-(r), \gamma_2^+(r)$ are situated over $C_1^-, C_1^+, C_2^-(r), C_2^+(r)$, respectively, and are placed in *separate* sheets. The components γ_3^- and γ_3^+ are located over C_3^- and C_3^+, and circulate around the branch points $z = a$ and $z = b$ in an obvious way. See Figure 3.

Let us now identify the various components of $\partial\mathfrak{D}(r)$ using $T_1, T_2(r)$, and T_3. In this way, we obtain a compact RS of genus $g = 3$ with Schottky marking. Corresponding to this, there will clearly exist a continuous path $\{w_r\}$ in TQ such that $\mathfrak{M}[w_r] = [T_1, T_2(r), T_3; I, I, I]$. Moreover, since the circles $\gamma_1^+, \gamma_1^-, \gamma_2^+(r), \gamma_2^-(r)$ all lie in separate sheets, they cannot interfere with each other, and thus the same construction continues on past $r = 1$. This is in striking contrast to the situation for $\{z_r\}$. Since $\mathfrak{M}[z_r] = \mathfrak{M}[w_r]$ for $0 < r < 1$, Theorem 7 follows at once.

DEPARTMENT OF MATHEMATICS
HARVARD UNIVERSITY
CAMBRIDGE, MASSACHUSETTS 02138

REFERENCES

[1]. Appell, P., Goursat, E., Fatou, P., *Théorie des Fonctions Algébriques*, Tome 2, Gauthier-Villars, Paris, 1930.

[2]. Gunning, R., Analytic structures on the space of flat vector bundles over a compact Riemann surface, in *SCV (II) Maryland 1970*, Springer Lecture Notes No. 185 (1971), pp. 47-62.

[3]. Hilb, E., Lineare Differentialgleichungen im komplexen Gebiet, Encyklopädie der Math. Wiss. II B 6 (1913), pp. 471-562.

[4]. Klein, F., *Über lineare Differentialgleichungen der zweiten Ordnung*, Göttingen, 1894.

[5]. Klein, F., *Vorlesungen über die Hypergeometrische Funktion*, Springer-Verlag, Berlin, 1933.

[6]. Kra, I., Deformations of Fuchsian groups, Duke Math. J. 36 (1969), pp. 537-546.

[7]. Maskit, B., On a class of Kleinian groups, Ann. Acad. Sci. Fenn., No. 442 (1969), 8 pp.

[8]. Newstead, P., Topological properties of some spaces of stable bundles, Topology 6 (1967), pp. 241-262.

[9]. Poincaré, H., Sur les groupes des équations linéaires, Acta Math 4 (1884), pp. 201-312.

[10]. Poincaré, H., Mémoire sur les fonctions zétafuchsiennes, Acta Math. 5 (1884), pp. 209-278.

[11]. Schiffer, M., and Hawley, N., Half-order differentials on Riemann surfaces, Acta Math. 115 (1966), pp. 199-236.

[12]. Schlessinger, *Handbuch der Theorie der linearen Differentialgleichungen*, Band 2, B. G. Teubner, Leipzig, 1897.

[13]. Siegel, C. L., *Topics in Complex Function Theory*, Vol. 2, Wiley-Interscience, New York, 1971.

[14]. Steenrod, N., *The Topology of Fibre Bundles*, Princeton Univ. Press, 1951.

[15]. Stoker, J. J., *Differential Geometry*, Wiley-Interscience, New York, 1969.

COLLARS ON RIEMANN SURFACES

Linda Keen

Let S be a compact Riemann surface of genus g. The metric on S is induced by factoring the upper half plane U, which has the Poincaré metric on it, by the Fuchsian group G which represents S; $S = U/G$. In what follows all curves on S are assumed to be geodesic in this metric; moreover they are taken to be the unique geodesics in the free homotopy classes determined by the curves. All lengths and areas are given in this metric.

In the study of Teichmüller space, much attention has been focused on the geodesics of S and in particular on sets of disjoint geodesics of S (which contain at most $3g-3$ curves). Changing the length of a geodesic γ of S changes the conformal structure of S and letting the length of γ tend to zero amounts to going to the boundary of Teichmüller space. In order to prove this kind of deformation is well defined, it is necessary to know that only the length of the geodesic in question goes to zero and that the piece of surface which contains it doesn't disappear as well. L. Bers suggested the precise statement of this and in this paper I prove it.

I want first to define a collar on a Riemann surface S.

DEFINITION. Consider the region T of U bounded by the curves $r=1$, $r = e^{\ell}$, $\theta = \theta_0$, $\theta = \pi - \theta_0$. Let γ be a geodesic on S with length ℓ. Then there is a conformal isometric mapping $\phi: T \to S$ such that $\phi(iy) = \gamma$. $\phi(T)$ is called a *collar* on S about γ.

Fact. The area $\mathcal{A}$ of T, and hence of $\phi(T)$ is $2\ell \cot \theta_0$.

The main result is the following:

THEOREM. *For every simple closed geodesic γ on S there is a collar about it of area $\mathfrak{A}$ at least $8/\sqrt{5}$.*

Before proving the theorem let me state two corollaries.

COROLLARY 1. *For every γ as above, consider all closed geodesics δ which intersect γ and which are not freely homotopic to γ^n, $n = \pm 1, \pm 2, \ldots$. Then γ and δ cannot both have length less than $\frac{2\sqrt{2}}{\sqrt[4]{5}}$.*

Proof. δ must have length at least twice the width of the collar; that is, $\delta > 2 \cot \theta$. Therefore $\delta\rho > 2\rho \cot \theta = \mathfrak{A} > 8/\sqrt{5}$ and δ or ρ must be greater than $2\sqrt{2}/\sqrt[4]{5}$.

REMARK. In [4], I prove a similar fact where δ and γ are both non-dividing simple geodesics. The bound I obtain is slightly weaker, it is $\sqrt{2}$. Mumford has a similar unpublished result.

COROLLARY 2. *There are at most 3g–3 simple closed geodesics on S with length less than $\frac{2\sqrt{2}}{\sqrt[4]{5}}$.*

Proof. There are at most 3g–3 disjoint simple closed geodesics on S.

As a preliminary to the proof of the theorem, let me recall some theory of Fuchsian groups and Riemann surfaces. Let γ on S be given. There are many ways to choose curves γ_1^1, γ_2^1 and γ_1^2, γ_2^2 so that $(\gamma, \gamma_1^1, \gamma_1^2)$ and $(\gamma, \gamma_1^2, \gamma_2^2)$ bound adjacent three-holed spheres of S. Pick these curves in some way and let S^1 be the sphere bounded by $(\gamma, \gamma_1^1, \gamma_2^1)$. Orient the curves so that $\gamma \circ \gamma_1^1 \circ \gamma_2^1 \quad 0$. We can, without loss of generality, by relabelling and reorienting if necessary, assume that γ_2^1 is shorter than γ_1^1.

Let G be the Fuchsian group which represents S; $S = U/G$. Let C, C_1 and C_2 be the elements of G which correspond to γ, γ_1^1 and γ_2^1 respectively. Assume G is normalized so that the repelling and attracting fixed points of C are 0 and ∞ respectively and so that the repelling fixed point of C_2 is $+1$. In [2], the following theorem is proved:

THEOREM. *Let* $F = \langle C_1, C_2, C; CC_2C_1 = 1\rangle$. *The axes of* CC_2^{-1} *and* $C_2^{-1}C$ *intersect in a point* $q \in U$. *Let* $q_1 = C(q)$ *and* $q_2 = C_2^{-1}(q)$. *Draw lines* m *and* n *from* q *which are orthogonal to the axes of* C_2 *and* C *respectively. Draw lines* m_1 *and* n_1 *from* q_1 *which are orthogonal to the axes of* C *and* C_1 *respectively and draw lines* m_2 *and* n_2 *from* q_2 *which are orthogonal to the axes of* C_1 *and* C_2 *respectively. The fixed points of* C, C_1 *and* C_2 *separate properly (as in the figure) and the region* P *of* U *bounded by the lines* m, n, m_1, n_1, m_2 *and* n_2 *is a polygon and a fundamental domain for* F. *Moreover, let* H *be the non-empty intersection of the half planes determined by the axes of* C, C_1, *and* C_2; *then* $Q = P \cap H$ *is that piece of the canonical modular polygon for* S, *defined in* [2], *which corresponds to the part of* S *bounded by the geodesics* γ, γ_1^1 *and* γ_2^1. *(See figure.)*

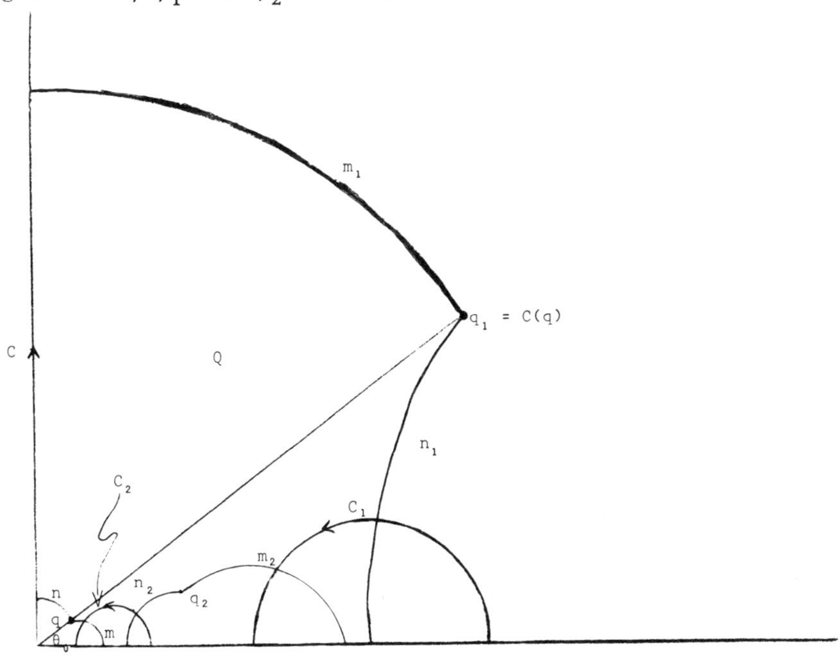

The Collar

To prove the existence of a collar at γ it is sufficient to show that half the collar lies on S_1. The same proof shows that the other half lies on S_2. In Q, n and m_1 have equal length; this length is just $\cot\theta_0$. There is a curve joining q and q_1 such that the distance from any point of the curve to the axis of C is $\cot\theta_0$. It is drawn in the figure. The region of Q bounded by this curve, the axis of C, m, and n is half of a collar at γ. What I must now prove, is that the area A of this region is bounded from below.

Let $k = \frac{1}{2}$ trace (C), $k_1 = \frac{1}{2}$ trace (C_1), and $k_2 = \frac{1}{2}$ trace (C_2). Since the fixed points of the axes of C, C_1 and C_2 separate properly $k_1^2 + k_2^2 + k^2 - 2k_1 k_2 k - 1 > 1$. The constructions of [2] imply that I can assume without loss of generality that k, k_1, k_2 are all less than -1. Since length γ_2^1 is less than length γ_1^1, k_1 is less than k_2.

Write the transformations as:

$$C = \begin{pmatrix} k-k & 0 \\ 0 & k+k \end{pmatrix} \qquad C_2 = \begin{pmatrix} k_2 - J & L \\ M & k_2 + J \end{pmatrix}$$

$$C_1 = \begin{pmatrix} k_1 + k_2 K + kJ & -(k-k)\cdot L \\ (k+K)M & k_1 - k_2 K - kJ \end{pmatrix}$$

where

$$K = \sqrt{k^2-1}, \qquad K_1 = \sqrt{k_1^2 - 1}, \qquad K_2 = \sqrt{k_2^2 - 1},$$

$$J = (k_1 - k_2 k)/K, \qquad L = J + K_2, \qquad M = K_2 - J\ .$$

Compute:

$$C_2^{-1}C = \begin{pmatrix} 2kk_2 - k_1 - k_2 K + Jk & -L(k+K) \\ -M(k-K) & 2kk_2 - k_1 + k_2 K + Jk \end{pmatrix}$$

$$CC_2^{-1} = \begin{pmatrix} 2kk_2 - k - k_2 K + Jk & -L(k-K) \\ -M(k+K) & 2kk_2 - k_1 + k_2 K - Jk \end{pmatrix}$$

The fixed points of $C_2^{-1}C$ are

$$x_1, x_2 = \left\{(Jk-k_2K) \pm [(2kk_2-k_1)^2-1]^{\frac{1}{2}}\right\} / -M(k-K) .$$

The fixed points of CC_2^{-1} are

$$x_1^1, x_2^1 = \left\{(Jk-k_2K) \pm [(2kk_2-k_1)^2-1]^{\frac{1}{2}}\right\} / -M(k+K) .$$

Therefore the equations of the axes of these transformations are:

$$[x+(Jk-k_2K)M^{-1}(k-K)^{-1}]^2 + y^2 = [(2kk_2-k_1)^2-1]M^{-2}(k-K)^{-2}$$

$$[x+(Jk-k_2K)M^{-1}(k+K)^{-1}]^2 + y^2 = [(2kk_2-k_1)^2-1]M^{-2}(k+K)^{-2} .$$

The intersection point q of the axes has coordinates:

$$x = Lk(Jk-kK)^{-1}, \qquad y = \{-LM^{-1}-L^2k^2(Jk-k_2K)^{-2}\}^{\frac{1}{2}} .$$

Consequently,

$$\cot\theta = \{k^2LM[(Jk-k_2K)^2-k^2LM]^{-1}\}^{\frac{1}{2}}$$

and the area A is

$$A = 2\{k^2LM[(Jk-k_2K)^2-k^2LM]^{-1}\}^{\frac{1}{2}} \log|k-K| .$$

I want to find a positive lower bound for $\lim_{k\to -1} A$. However it is easier to compute $\lim_{k\to -1} 1/A^2$.

Let $k=-1-\varepsilon$; then $K=\sqrt{\varepsilon^2+2\varepsilon}$ and $(k-K)^2 = 1+2\sqrt{\varepsilon^2+2\varepsilon}+2(\varepsilon^2+2\varepsilon)+2\varepsilon\sqrt{\varepsilon^2+2\varepsilon}$ so that $2\log|k-K| = 2\sqrt{\varepsilon^2+2\varepsilon}+0(\varepsilon)$ and $(2\log|k-K|)^{-2} = (4(\varepsilon^2+2\varepsilon))^{-1}+0(\varepsilon)$.

Use the identity, which comes from the fact that the determinants of the above matrices are 1,

$$(Jk-k_2K)^2 = (2kk_2-k_1)^2 - LM - 1,$$

to write

$$\tan^2\theta = -1-[(2kk_2-k_1)^2-1]\cdot L^{-1}M^{-1}k^{-2}+k^{-2}$$

$$= \frac{\varepsilon^2+2\varepsilon}{(1+\varepsilon)^2}[-1+([2k_2+k_1+2\varepsilon k_2]^2-1)[(k_1+k_2+k_2\varepsilon)^2-(k_2^2-1)(2\varepsilon+\varepsilon^2)]^{-1}]$$

$$= \frac{\varepsilon^2+2\varepsilon}{(1+\varepsilon)^2}[-1+[(2k_2+k_1)^2-1](k_1+k_2)^{-2}+0(\varepsilon)] \ .$$

Now

$$1/A^2 = \frac{(\varepsilon^2+2\varepsilon)}{4(\varepsilon^2+2\varepsilon)}\frac{1}{(1+\varepsilon)^2}\left[\frac{2k_2}{k_1+k_2}+\frac{k_2^2-1}{(k_1+k_2)^2}\right]+0(\varepsilon)$$

and

$$\lim_{k\to -1} 1/A^2 = \frac{1}{4}\left[\frac{2k_2}{k_1+k_2}+\frac{k_2^2-1}{(k_1+k_2)^2}\right].$$

Recall that $k_1 \leq k_2 < -1$ so that

$$1/A^2 \leq 1/4\cdot 5/4 \quad \text{or} \quad A \geq 4/\sqrt{5}\ .$$

Therefore the area $\mathcal{A}$ of the collar about γ satisfies $\mathcal{A} \geq 2A \geq 8/\sqrt{5}$.

REMARK 1. The bound is sharp since A^2 approaches it when k_2 is kept fixed and k_1 approaches $-\infty$.

REMARK 2. An interesting special case occurs when k_1 and k_2 both approach -1, then $\lim_{k_1, k_2, k\to -1} = 2.$

LEHMAN COLLEGE
GRADUATE CENTER
CITY UNIVERSITY OF NEW YORK

REFERENCES

[1]. Keen, Linda, Intrinsic moduli on Riemann surfaces, Ann. of Math. 84 (1966).

[2]. ———, On Fricke moduli, Advances in the Theory of Riemann Surfaces, Princeton, 1971.

[3]. ———, Correction to "On Fricke Moduli," Proc. Amer. Math. Soc., July, 1973.

[4]. ———, On Fundamental domains and the Teichmüller modular group, to appear.

DEFORMATIONS OF CERTAIN COMPLEX MANIFOLDS

Garo K. Kiremidjian*

In this paper we summarize recent results extending the Kuranishi deformation theory ([8], [9]) to certain non-compact manifolds and compact manifolds with boundary. At the same time our methods enable us to describe sufficiently small complex structures on Riemann surfaces without making use of the theory of quasi-conformal mappings and Schwarzian derivatives. Some of the proofs have already been published ([5]), others will soon appear elsewhere ([6], [7]).

Let M be a complete Kähler manifold such that

a. the holomorphic sectional curvature is bounded;

b. every point of M admits a normal coordinate neighborhood (with respect to the fixed metric on M) whose geodesic is bounded from below.

We denote by $\mathbf{M}$ the underlying C^∞ manifold and by Θ the holomorphic tangent bundle of M. Let $C^{0,1}(M,\Theta)$ be the space of $C^\infty(0,1)$ Θ-valued differential forms. Then we have the following classification result.

THEOREM 1. *If* Θ *is* $W^{0,0}$ *elliptic every complex structure* M_ω *on* $\mathbf{M}$ *represented by* $\omega \in C^{0,1}(M,\Theta)$ *with* $\|\omega\|_k$ *sufficiently small is equivalent to a unique complex structure* M_ϕ *such that* $\bar\partial^*\phi = 0$ *and* $\|\phi\|_k$ *is also sufficiently small.*

---

* This research was supported by grants from the National Science Foundation, GP P019572-002 and 33942X.

Here $\| \ \|_k$ is the Sobolev k-norm with respect to covariant differentiation, k is an integer depending on $\dim_{\mathbf{C}} M$ and $\bar{\partial}^*$ is the adjoint of the operator $\bar{\partial}$ with respect to the L_2-inner product. We refer to [1] for the definition of $W^{p,q}$ ellipticity.

THEOREM 2. *If* Θ *is also* $W^{0,1}$ *elliptic, then all complex structures* M_ω *with* $\|\omega\|_k$ *sufficiently small are trivial, i.e., equivalent to* M.

As an application of Theorem 1 we have the following result.

THEOREM 3. *Let* M *be a Riemann surface satisfying condition b. Then the universal family for complex structures* M_ω *represented by* $\omega \in C^{0,1}(M,\Theta)$ *with finite* $\| \ \|_k$*-norm is the set* $T_k = \{\phi \in C^{0,1}(M,\Theta)/\phi = \lambda^{-2}\bar{\psi}, \|\phi\|_k < \infty, \sup|\phi| < 1\}$ *where* $\lambda^2\, dz d\bar{z}$ *is the complete Poincaré metric on* M *and* ψ *is a holomorphic quadratic differential.*

Next, let M' be a complex Kähler-Einstein manifold of negative scalar curvature. Let M be a relatively compact submanifold of M' such that $n = \dim_{\mathbf{C}} M = \dim_{\mathbf{C}} M'$ and the boundary bM is a C^∞ submanifold of M of real dimension $2n-1$. We further assume that the following condition holds: there exists a constant $c > 0$ such that for all $\phi \in C^{0,q}(\bar{M},\Theta)$, $q = 1,2$, $((2\Box' - \tilde{\Delta})\phi, \phi) \geq c(\phi,\phi)$ where $C^{p,q}(\bar{M},\Theta)$ is the space of all C^∞ Θ-valued (p,q) forms extendible to a neighborhood of $\bar{M}$, $\Box'$ (resp., $\tilde{\Delta}$) is the complex (resp., the real) Laplacian on $C^{p,q}(\bar{M},\Theta)$ and $(\ ,\ )$ is the L_2-inner product.

THEOREM 4. *There exists a universal family of deformations of* M *whose parameter space is, in general, a Banach analytic set.*

For a bordered compact Riemann surface M and its double M' the preceding result yields the following important special case.

THEOREM 5. *Let* T *be the space of Beltrami differentials* $\phi = \lambda^{-2}\bar{\nu}d\bar{z}\,\frac{\partial}{\partial z}$, $\sup|\lambda^{-2}\bar{\nu}| < 1$, *where* νdz^2 *is a holomorphic quadratic differential and* ν *is real on* bM. *Then* T *represents a universal family of deformations of* M *in the category of real analytic families of* M.

§1. *Operators on complex manifolds*

Let $E \xrightarrow{\pi} M$ be a holomorphic vector bundle with fiber $\mathbb{C}^m$ defined on the coordinate covering $\mathfrak{U} = \{U_i\}_{i\epsilon I}$ of M by the holomorphic cocycle $e_{ij} : U_i \cap U_j \to GL(m, \mathbb{C})$. The dual bundle $E^* \to M$ is defined on the same covering by ${}^te_{ij}{}^{-1}$. For example, the holomorphic tangent bundle Θ will be defined in terms of a choice of local coordinates $\left(z_i^{\,1}, \ldots, z_i^{\,n}\right)$ on U_i by the cocycle $J_{ij} = \frac{\partial(z_i)}{\partial(z_j)}$.

A C^∞ form of type (p, q) with values in E is a C^∞ section of the bundle $E \otimes \Theta^{*p} \otimes \bar{\Theta}^{*q}$, where $\Theta^{*p} = \Lambda^p\Theta^*$ and $\bar{\Theta}^*$ is the complex conjugate of Θ^*. Thus, locally on U_i, such a form is given by the column vector $\phi_i = {}^t\left(\phi_i^{\,1}, \ldots, \phi_i^{\,m}\right)$ such that $\phi_i^{\,j}$ are scalar C^∞ forms of type (p, q) and, on $U_i \cap U_j$, $\phi_i = e_{ij}\phi_j$. The vector space of these forms will be denoted by $C^{p,q}(M, E)$.

Let g be a hermitian metric on M and h a hermitian metric on the fibers of E. If $E = \Theta$, then $g = h$. For each $x \in M$, g and h induce an inner product $\langle\ ,\ \rangle$ on the vector space $E_x \otimes \Theta_x^{\,*p} \otimes \bar{\Theta}_x^{\,*q}$ where E_x (resp., $\Theta_x^{\,*}$) is the fiber of E (resp., Θ^*) over x.

We now introduce the following operators on $\oplus\, C^{p,q}(M, E)$.

$$\bar{\partial} : C^{p,q}(M, E) \to C^{p,q+1}(M, E)\ , \tag{1.1}$$

the exterior differentiation with respect to the complex conjugate of local holomorphic coordinates.

$$\# : C^{p,q}(M, E) \to C^{q,p}(M, E^*)\ , \tag{1.2}$$

defined locally, on U_i, by $(\#\phi)_i = \bar{h}_i\bar{\phi}_i$ where h_i is the positive definite hermitian matrix, representing h on U_i.

$$(1.3) \qquad * : C^{p,q}(M,E) \to C^{n-q,n-p}(M,E), \quad n = \dim_{\mathbb{C}} M ,$$

given by the condition ${}^t\psi \wedge \# * \phi = \langle\psi,\phi\rangle\, dM$, dM being the volume element in the metric g.

We note that $\#$ and $*$ commute.

$$(1.4) \qquad \bar\partial^* : C^{p,q}(M,E) \to C^{p,q-1}(M,E), \quad \bar\partial^* = -*\#\bar\partial\#* .$$

Let θ be the connection of the metric h. The local expression of θ on U_i is $h_i^{-1}\partial h_i$.

$$(1.5) \qquad \tilde\partial : C^{p,q}(M,E) \to C^{p+1,q}(M,E), \quad \tilde\partial\phi = \partial\phi + \theta \wedge \phi .$$

Thus the covariant differentiation associated with θ is given by $D\phi = \tilde\partial\phi + \bar\partial\phi$. If N is a relatively compact open submanifold of M with smooth bounding bN, then, by Stokes' theorem, we have that

$$(1.6) \qquad \int_N \langle\bar\partial\phi,\psi\rangle\, dM = \int_N \langle\phi,\bar\partial^*\psi\rangle\, dM + \int_{bN} {}^t\phi \wedge \# * \psi$$

$$(1.7) \qquad \int_N \langle D\phi,\psi\rangle\, dM = \int_N \langle\phi,D^*\psi\rangle\, dM + \int_{bN} {}^t\phi \wedge \# * \psi$$

for all $\phi \in C^{p,q}(M,E)$, $\psi \in C^{p,q+1}(M,E)$. Here $D^* = -*D* = -*\bar\partial* + \bar\partial^*$.

Finally, the L_2-inner product on $C^{p,q}(M,E)$ is given by $(\phi,\psi) = \int_M \langle\phi,\psi\rangle\, dM$, the L_2-norm is $\|\phi\| = (\phi,\phi)^{\frac{1}{2}}$ and the Sobolev k-norm is defined by

$$(1.8) \qquad \|\phi\|_k^{\,2} = \sum_{\alpha \le k} \|D^\alpha\phi\|^2 ,$$

where D^α stands for covariant differentiation (with respect to θ) of order α.

§2. *Complex structures on complex manifolds*

In this section we collect some well known facts about almost complex and complex structures. For a complete discussion we refer to [8] and [9].

We assume now that M is a complex manifold of dimension n and let $\mathbf{M}$ be the underlying C^∞ manifold of dimension 2n. If $C T\mathbf{M}$ is the complexification of the real tangent bundle $T\mathbf{M}$, then $CT\mathbf{M} = \Theta \oplus \overline{\Theta}$ where Θ is the holomorphic tangent bundle of M.

DEFINITION 2.1. An almost complex structure on $\mathbf{X}$ is given by a C^∞ vector sub-bundle Θ' of $CT\mathbf{M}$ (over $\mathbf{C}$) such that $CT\mathbf{M} = \Theta' \oplus \overline{\Theta}'$.

Let $\rho''(M)$ be the projection of $CT\mathbf{M}$ onto $\overline{\Theta}'$.

DEFINITION 2.2. The almost complex structure Θ' is of finite distance from the given complex structure M on $\mathbf{M}$ if $\rho''(M)$ induces an isomorphism from $\overline{\Theta}'$ onto $\overline{\Theta}$.

The set of almost complex structures of finite distance from M is in bijective correspondence with the set of all $\phi \in C^{0,1}(M,\Theta)$ such that, at each point $p \in M$, $\omega \circ \omega : \overline{\Theta}_p \to \overline{\Theta}_p$ does not have an eigenvalue 1. An almost complex structure determined by such ϕ will be denoted by M_ϕ.

THEOREM 2.3 (Newlander-Nirenberg). M_ϕ *is induced by a complex structure if and only if* $\bar{\partial}\omega - [\omega,\omega] = 0$.

Here $\bar{\partial}$ is the exterior differentiation operator with respect to the complex conjugates of local holomorphic coordinates and $[\phi,\phi] \in C^{0,2}(M,\Theta)$ has a local expression

$$[\phi,\phi] = \sum_{a,\beta,\gamma,r} \left(\phi_\beta{}^r \frac{\partial \phi_\gamma{}^a}{\partial z^r} - \phi_\gamma{}^r \frac{\partial \phi_\beta{}^a}{\partial z^r} \right) d\bar{z}^\beta \wedge d\bar{z}^\gamma \frac{\partial}{\partial z^a} ,$$

$$\phi = \sum_{a,\beta} \phi^a_\beta \, d\bar{z}^\beta \frac{\partial}{\partial z^a} .$$

Let f be a diffeomorphism of $\mathbf{M}$ and let Θ_1' and Θ_2' be almost complex structures on $\mathbf{M}$.

DEFINITION 2.4. f is an isomorphism between Θ_1' and Θ_2' if its differential sends Θ_1' to Θ_2'.

If $\phi \in C^1(M,\Theta)$ such that M_ϕ can be defined and if f is sufficiently close to the identity map of M in C^1-topology, then there exists a unique almost complex structure M_ω induced by an element $\omega \in C^{0,1}(M,\Theta)$ such that $f: M_\omega \to M_\phi$ is an isomorphism. We denote ω by $\phi \circ f$ and in terms of local coordinates $z = (z^1, \ldots, z^n)$

$$(2.1)\quad \sum_\beta \omega_\alpha{}^\beta(z)\left(\frac{\partial f^\gamma}{\partial z^\beta} + \sum_\mu \frac{\partial \bar{f}^\mu}{\partial z^\beta}\phi_\mu{}^\gamma(f(z))\right) = \frac{\partial f^\gamma}{\partial \bar{z}^\alpha} + \sum_\mu \frac{\partial \bar{f}^\mu}{\partial z^\alpha}\phi_\mu{}^\gamma(f(z)) ,$$

$$f(z) = (f^1(z), \ldots, f^n(z)) .$$

PROPOSITION 2.5. Let V be a normed vector space and let $\xi : V \to C^{0,0}(M,\Theta)$ be an R-linear map. Let f^s ($s \in V$, $|s|$ is small) be a C^∞ family of diffeomorphisms of M such that f^0 is the identity map. Assume that, for each chart $z = (z^1, \ldots, z^n)$ of M,

$$f^{\alpha,s}(z) \equiv z^\alpha + \xi^\alpha(z,s) \pmod{|s|^2}$$

where $\xi(s) = \sum_\alpha \xi^\alpha(z,s)\frac{\partial}{\partial z^\alpha}$. Let $\phi \in C^{0,1}(M,\Theta)$ such that $\phi \circ f^s$ can be defined. Then

$$(2.2)\quad \phi \circ f^s = \phi + \bar{\partial}_\phi(\xi(s) + \phi \circ \overline{\xi(s)}) + (\bar{\partial}\phi - [\phi,\phi]) \circ \overline{\xi(s)} + 0(|s|^2) ,$$

where

$$\bar{\partial}_\phi \xi = \bar{\partial}\xi - [\phi,\xi],\ [\phi,\xi] = \sum_{\alpha,\beta,\gamma}\left(\frac{\partial \xi^\beta}{\partial z^\gamma}\phi_\alpha{}^\gamma - \frac{\partial \phi_\alpha{}^\beta}{\partial z^\gamma}\xi^\gamma\right) d\bar{z}^\alpha \frac{\partial}{\partial z^\beta} ,$$

$$\phi \circ \bar{\xi} = \sum_{\alpha,\beta} \phi_\alpha{}^\beta \bar{\xi}^\alpha \frac{\partial}{\partial z^\beta} \quad \text{and} \quad \psi \circ \bar{\xi} = \sum_{\alpha,\beta,\gamma} \psi_{\alpha\beta}{}^\gamma \bar{\xi}^\alpha d\bar{z}^\beta \frac{\partial}{\partial z^\gamma} ,$$

$$\psi = \sum_{\alpha,\beta,\gamma} \psi_{\alpha\beta}{}^\gamma d\bar{z}^\alpha \wedge d\bar{z}^\beta \frac{\partial}{\partial z^\gamma} \in C^{0,2}(M,\Theta) .$$

Let $C_k^{p,q}(M,\Theta) = \{\phi \in C^{p,q}(M,\Theta)/\|\phi\|_k < \infty\}$ and let $\mathcal{H}_k^{p,q}$ be the completion of $C_k^{p,q}(M,\Theta)$ in the norm $\| \ \|_k$.

DEFINITION 2.6. A complex analytic family of deformations of M is a triple (S, ψ, s_0) where S is a (Banach) analytic set, $\psi : S \to C_k^{0,1}(M,\Theta)$ is an analytic map (i.e., the composition of ψ with the inclusion $C_k^{0,1}(M,\Theta) \subset \mathcal{H}_k^{p,q}$ is analytic) such that for each $s \in S$, $\psi(s)$ represents a complex structure $M_{\psi(s)}$ on M and $\psi(s_0) = 0$.

REMARK 2.7. If $M \subset M'$ is a compact manifold with boundary, then we replace $C_k^{0,1}(M,\Theta)$ by $C^{0,1}(\bar{M},\Theta)$.

DEFINITION 2.8. A complex analytic family (T, ϕ, t_0) is a universal family of deformations of M if for any other family (S, ψ, s_0) there exists a neighborhood S' of s_0 in S, an analytic map $\tau : S' \to T$ and a C^∞ family of diffeomorphisms f^s of M such that $\tau(s_0) = t_0$, $f^{s_0} = \mathrm{id}$ and $\psi(s) \circ f^s = \phi(\tau(s))$ for each $s \in S'$.

§3. *Complex structures on non-compact manifolds*

In this section we assume that M is a non-compact manifold with a complete Kähler metric h satisfying conditions a and b given in the introduction of this paper. We now describe the tools used in proving Theorem 1.

LEMMA 3.1. *Let $\xi \in C^{0,0}(M,\Theta)$ and let ξ_x (resp., $\|\xi\|_x$) be the tangent vector at $x \in M$ determined by ξ (resp., the length of ξ_x with respect to h). We denote by $e(\xi)(x)$ the end point of the geodesic of length $\|\xi+\bar{\xi}\|_x$ drawn from x in the direction $\xi_x+\bar{\xi}_x$. Let $\varepsilon > 0$ be such that $\|\xi\|_x + \|D\xi\|_x \leq e$ for all $x \in M$. Then the differentiable mapping $e(\xi) : M \to M$, $x \mapsto e(\xi)(x)$, is a diffeomorphism if ε is sufficiently small.*

LEMMA 3.2. *Let ζ be a tensor field on M. Then, for all $x \in M$, $\|\zeta\|_x \leq c_n \sum_{\alpha \leq 2n} \|D^\alpha \zeta\|$ where $n = \dim_{\mathbb{C}} M$, $\|\zeta\|_x$ is the length of ζ at x with* *respect to* h, $\|D^{\alpha}\zeta\|$ *is the* L_2*-norm of the* α-th *covariant differential of* ζ *(also with respect to* h) *and* c_n *is a constant depending only on* n.

LEMMA 3.3. *Let* Θ *be* $W^{p,q}$ *elliptic (see* [1]). *For* $\phi \in C_k^{p,q}(M,\Theta)$ *we denote by* $G\phi \in C^{p,q}(M,\Theta)$ *the unique solution of* $\square'\psi = (\bar{\partial}^*\bar{\partial} + \bar{\partial}\bar{\partial}^*)\psi = \phi$ *with finite* L_2*-norm. Then* $G\phi \in C_{k+2}^{p,q}(M,\Theta)$ *and there exists a constant* $c_k > 0$ *(depending only on* k) *such that* $\|G\phi\|_{k+2} \leq c_k\|\phi\|_k$*, i.e.,* $G: C_k^{p,q}(M,\Theta) \to C_{k+2}^{p,q}(M,\Theta)$ *is continuous.*

The proof of Lemma 3.1 is based on the Rauch comparison theorem and some additional properties of Jacobi fields. Lemma 3.2 and Lemma 3.3 can be obtained by modifying the classical proofs of Sobolev's and Friedrichs' inequalities. We refer to [6] for further details.

Proof of Theorem 1. Let V_1 and W_1 be neighborhoods of 0 in $C_k^{0,1}(M,\Theta)$ and $C_{k+1}^{0,0}(M,\Theta)$, respectively. If V_1 and W_1 are sufficiently small, then, for each $(\omega,\xi) \in V_1 \times W_1$, $\omega \circ e(\xi)$ can be defined. The construction of $e(\xi)$ and (2.1) imply that

$$\omega \circ e(\xi) = \bar{\partial}\xi + \omega + F(\omega,\xi) \tag{3.1}$$

where $F(\omega,\xi)$ is of at least order 2 in ω and ξ, i.e., it involves expressions which are at least quadratic in $\omega_\alpha{}^\beta(e(\xi))$, ξ^α and the first order derivatives of ξ^α. Here $\omega_\alpha{}^\beta$ and ξ^α are the components of ω and ξ.

Let $j_1(\omega,\xi) = \bar{\partial}^*(\omega \circ e(\xi))$. If k is sufficiently large it follows from Lemma 3.2 that $j_1(\omega,\xi) \in C_{k-1}^{0,0}(M,\Theta)$ and, moreover, j_1 can be extended to a map $j: V \times W \to \mathcal{H}_{k-1}^{0,0}$ which is a C^∞ map in the sense of the theory of Banach spaces. Here V and W are now neighborhoods of 0 in $\mathcal{H}_k^{0,1}$ and $\mathcal{H}_{k+1}^{0,0}$. Lemma 3.3 implies that the derivative of j with respect to ξ at (0,0) is an isomorphism. Then, by the implicit function theorem for Banach spaces, we have that there is a C^∞ map of V into W, $\omega \mapsto \xi(\omega)$ such that $j(\omega,\xi(\omega)) = 0$ for all $\omega \in V$ (if V is sufficiently small). Let $\phi = \omega \circ e(\xi(\omega))$. Then $\bar{\partial}^*\phi = 0$ and if ω is integrable, so is ϕ., Q.E.D.

Proof of Theorem 2. We have that each complex structure M_ω with $\|\omega\|_k$ sufficiently small is equivalent to M_ϕ with $\overline{\partial}^*\phi = 0$. We apply the operator $G\overline{\partial}^*$ to both sides of $\overline{\partial}\phi - [\phi,\phi] = 0$. Then $\phi = G\overline{\partial}^*[\phi,\phi]$. Lemma 3.2 and Lemma 3.3 imply that $\|\phi\|_k \leq c'_k\|\phi\|_k^{\,2}$ for some constant c'_k depending only on k. Hence, if $\|\omega\|_k$ is sufficiently small, $\|\phi\|_k < \frac{1}{c'_k}$ and the preceding inequality can hold if and only if $\phi = 0$.

Outline of the proof of Theorem 3. We first observe that $\phi \in T_k$ if and only if $\phi \in C_k^{\,0,1}(M,\Theta)$ and $\overline{\partial}^*\phi = 0$ where $\overline{\partial}^*$ is taken with respect to the Poincare metric $\lambda^2 dzd\overline{z}$. The proof of Theorem 1 implies that if (S,ψ,s_0) is a complex analytic family of deformations of M and if S′ is a sufficiently small neighborhood of s_0 in S, then there is a C^∞ map $\tau : S' \to T_k$ such that $\tau(s_0) = 0$ and $\tau(s) = \psi(s) \circ f^s$ for a C^∞ family of diffeomorphisms f^s with $f^{s_0} = \mathrm{id}$. By using the fact that ψ depends complex analytically on s and Proposition 2.5 one can show that τ is also complex analytic (see [5]). Q.E.D.

§4. *The case of compact manifolds with boundary*

Let M be a relatively compact submanifold of M′ such that the conditions satisfied at the beginning of this paper are satisfied.

First of all, we point out that if $\xi \in C^{0,0}(\overline{M},\Theta)$, then $\#\xi$ can also be regarded as a scalar (1,0)-form. Let Δ be the real Laplacian on the space of such forms. It is not difficult to see that

$$\square'\xi = \frac{1}{2}\,\#^{-1}\left(\Delta\#\xi - \frac{R}{n}\,\#\xi\right) \tag{4.1}$$

where the operators $\square'$, Δ and $\#$ are defined with respect to the Kähler-Einstein metric h on M′ and R is the scalar curvature of h.

Let $P = \{\xi \in C^{0,0}(\overline{M},\Theta)/*(\#\xi+\#\overline{\xi}) = *d(\#\xi+\#\overline{\xi}) = 0 \text{ on } bM\}$. Since $R < 0$, the theory developed in [4] and [10] enables us to conclude that for each $\xi \in C^{0,0}(\overline{M},\Theta)$ there exists a unique $\eta \in P$ such that

$$\square'\eta = \xi, \quad \|\eta\|_{k+2} \leq c_k\|\xi\|_k \tag{4.2}$$

for some constant c_k depending only on k.

The same theory also shows that for each $\phi \in C^{0,q}(\overline{M},\Theta)$, $q = 1,2$, there exists a unique $\psi \in C^{0,q}(\overline{M},\Theta)$ such that

$$(4.3) \quad \square'\psi = \phi, \quad *\psi = *D\psi = 0 \text{ on } bM, \quad \|\psi\|_{k+2} \leq \text{const. } \|\phi\|_k ,$$

since $\square' = \frac{1}{2}(\tilde{\Delta}+\tilde{A})$ where $\tilde{\Delta} = DD^* + D^*D$ and $(\tilde{A}\phi,\phi) \geq \text{const. } \|\phi\|^2$ for all $\phi \in C^{0,q}(\overline{M},\Theta)$, $q = 1,2$.

In the sequel we will denote η and ψ by $G\xi$ and $G\phi$, respectively. Moreover, ϕ can be uniquely represented as $\phi = \phi_1 + \phi_2$ where $\phi_1 = G\square'\phi$ and $\square'\phi_2 = 0$. We set $H\phi = \phi_2$.

We are now ready to outline the proof of Theorem 4.

For $\xi \in C^{0,0}(\overline{M},\Theta)$ the differentiable map $e(\xi): M \to M'$ (see Lemma 3.1) is a diffeomorphism of M if $\xi + \overline{\xi}$ is tangent to bM, $e(\xi)$ is constructed in terms of a metric σ on M' with respect to which bM is a totally geodesic submanifold and ξ is sufficiently small in C^1-topology. It is well known that such σ always exists. On the other hand, it can easily be checked that $\xi + \overline{\xi}$ is tangent to bM if and only if $*(\#\xi+\#\overline{\xi}) = 0$ on bM, with respect to any Riemannian metric on M'. Thus, for $\xi \in P$, $e(\xi)$ is a diffeomorphism if $\|\xi\|_k$ is sufficiently small, $k > \dim_{\mathbf{C}} M + 2$.

We can now repeat the proof of Theorem 1 by replacing $C_{k+1}{}^{0,0}(M,\Theta)$ by P and $\mathcal{H}_{k+1}{}^{0,0}$ by $\hat{P}_{k+1}$, the completion of P in $\| \ \|_{k+1}$-norm. Hence there exists a neighborhood V of the origin in $\mathcal{H}_k{}^{0,1}$, the completion of $C_k{}^{0,1}(\overline{M},\Theta)$ in $\| \ \|_k$-norm, and a C^∞ map $\xi: V \to \hat{P}_{k+1}$ such that $\overline{\partial}^*(\phi \circ e(\xi(\phi)) = 0$ for $\phi \in V$. This follows from the implicit function theorem which can be applied because of (4.2). Moreover, $\xi(\phi) \in C^{0,0}(\overline{M},\Theta)$ whenever $\phi \in C^{0,1}(\overline{M},\Theta)$ since $\overline{\partial}^*(\phi \circ e(\xi(\phi))) = \overline{\partial}^*\overline{\partial}\xi(\phi) + \overline{\partial}^*\phi + \overline{\partial}^*F(\phi,\xi(\phi)) = 0$, which is a second-order elliptic differential equation with respect to ξ.

The preceding result suggests that it is natural to construct a universal family by considering the set $\Phi = \{\phi \in C^{0,1}(\overline{M},\Theta)/\overline{\partial}\phi - [\phi,\phi] = 0,\ \overline{\partial}^*\phi = 0\}$. If $\phi \in \Phi$, then $\phi - G\overline{\partial}^*[\phi,\phi] = H\phi$ and thus $\Phi \subset \Psi = \{\phi \in C^{0,1}(\overline{M},\Theta/\phi - G\overline{\partial}^*[\phi,\phi] \in H^{0,1}\}$ where $H^{0,1} = \{\phi \in C^{0,1}(\overline{M},\Theta)/\square'\phi = 0\}$.

The map $J : C^{0,1}(\overline{M},\Theta) \to C^{0,1}(\overline{M},\Theta)$ given by $J\phi = \phi - G\overline{\partial}^*[\phi,\phi]$ is uniformly continuous in $\| \|_k$-norm topology and therefore it can be considered as a continuous map $J : \mathcal{H}_k^{0,1} \to \mathcal{H}_k^{0,1}$. Moreover, it is obvious that J is also complex analytic. Since the differential of J at the origin is the identity map, by the inverse mapping theorem for Banach spaces there exists a neighborhood N of 0 in $\mathcal{H}_k^{0,1}$ and a complex analytic map $t \mapsto \phi(t)$, $t \in N$, such that $J(\phi(t)) = t$. Let $N' = N \cap H^{0,1}$. Then it is not difficult to see that $\phi(t) \in \Phi$ if and only if $t \in T = T_1 \cap T_2$, where $T_1 = \{t \in N'/w(t) = 0\}$, $T_2 = \{t \in N'/\overline{\partial}^*\phi(t) = 0\}$ and $w : N' \to \mathcal{H}_{k-1}^{0,2}$ is the complex analytic map defined by $w(t) = \overline{\partial}t - G\overline{\partial}^*\overline{\partial}[\phi(t),\phi(t)] - H[\phi(t),\phi(t)] + H\overline{\partial}G\overline{\partial}^*[\phi(t),\phi(t)]$. Thus $(T,\phi,0)$ represents a complex analytic family of deformations of M.

Let (S,ψ,s_0) be any other family. If S' is a sufficiently small neighborhood of s_0 in S, then there exists a C^∞ map $\xi : S' \to P$ such that $\overline{\partial}^*(\psi(s) \circ e(\xi(s))) = 0$. Hence $\psi(s) \circ e(\xi(s)) \in \Phi$, so there is a C^∞ map $\tau : S' \to T$ such that $\psi(s) \circ e(\xi(s)) = \phi(\tau(s))$. Moreover, $\xi(s_0) = \tau(s_0) = 0$. Again, by using Proposition 2.5, the analyticity of $\psi(s)$ and $\| \|_k$-norm estimates, it can be shown that τ is also complex analytic (see [5] and [7]. Q.E.D.

Outline of the proof of Theorem 5. Let M be a bordered compact Riemann surface and let M' be the double of M, i.e., if, for some Fuchsian group Γ, $M = U/\Gamma$, then $M' = \Omega/\Gamma$, where U is the upper half plane of the complex plane $\mathbb{C}$ and Ω is the complement of the limit set of Γ. The symmetry of M', changing its orientation, is an anticonformal involution and can be lifted to the involution $z \mapsto \overline{z}$ of $\mathbb{C}$. It is shown in [2] that there is a bijective correspondence between $C^{0,1}(\overline{M},\Theta)$ and $C_r^{0,1}(M',\Omega) = \{\phi = \mu(z)\,d\overline{z}\,\frac{\partial}{\partial z} \in C^{0,1}(M',\Theta)/\overline{\mu(z)} = \mu(\overline{z})\}$. It is also known that $T = \{\phi \in C_r^{0,1}(M',\Theta)/\overline{\partial}^*\phi = 0,\ \sup_{M'} |\mu| < 1\}$ is an open subset of a finite dimensional real vector space.

We observe that M' is a Kähler-Einstein manifold of negative (scalar) curvature. Let (S,ψ,s_0) be a real analytic family of deformations of M, i.e., S is a real analytic manifold, $\psi(s_0) = 0$ and the composition of $\psi : S \to C_r^{0,1}(M', \Theta)$ with the inclusion $C_r^{0,1}(M', \Theta) \subset \dot{\mathcal{H}}_{r,k}^{0,1}$ (the completion of $C_r^{0,1}(M, \Theta')$ in the $\| \|_k$-norm topology) is an analytic map of real Banach spaces. The proof of Theorem 4 implies that there is a neighborhood S' of s_0 in S and a C^∞ map $\xi : S' \to P$ such that $\overline{\partial}^*(\psi(s) \circ e(\xi(s))) = 0$. Moreover, by using Jacobi fields (for complete details see [6] and [7]), it can be shown that $\psi(s) \circ e(\xi(s)) \in C_r^{0,1}(M', \Theta)$. Thus we also obtain a C^∞ map $\tau : S' \to T$ defined by $\tau(s) = \psi(s) \circ e(\xi(s))$.

Finally, we will show that τ is real analytic. We may assume that S' is an open subset of R^m. Let S^* be an open subset of C^m such that $S^* \cap R^m = S'$ and $\psi^* : S^* \to C^{0,1}(M', \Theta)$ is a complex analytic extension of ψ. The proof of Theorem 4 allows us to conclude that the C^∞ map $\xi : S' \to P$ can be extended to a C^∞ map $\xi^* : S^* \to P$ such that $\overline{\partial}^*(\psi(s^*) \circ e(\xi(s^*))) = 0$. Therefore, τ can be extended to a C^∞ map $\tau^* : S^* \to T^* = \{\omega \in C^{0,1}(\overline{M}, \Theta) / \overline{\partial}^* \omega = 0\}$. Theorem 4 also shows that τ^* is complex analytic if S' is a sufficiently small neighborhood of s_0. Hence $\tau : S' \to T$ is real analytic. Q.E.D.

STANFORD UNIVERSITY

REFERENCES

[1]. Andreotti, A., and Vesentini, E., On deformations of discontinuous groups, Acta Math., 112 (1963), pp. 249-298.

[2]. Earle, C. J., Teichmüller spaces of groups of the second kind, Acta Math., 112 (1964), pp. 91-97.

[3]. ________, Reduced Teichmüller spaces, Trans. Amer. Math. Soc., 126 (1967), pp. 54-63.

[4]. Fichera, G., Linear Elliptic Differential Systems and Eigenvalue Problems, Springer-Verlag, Berlin-Heidelberg-New York (1965).

[5]. Kiremidjian, G., Complex structures on Riemann surfaces, Trans. Amer. Math. Soc., 169 (1972), 317-336.

[6]. ———, Deformations of complex structures on certain non-compact manifolds, Ann. of Math. (to appear).

[7]. ———, On deformations of complex compact manifolds with boundary (submitted to Pacific J. Math.).

[8]. Kuranishi, M., New proof for the existence of locally complete families of complex structures, Proc. Conf. Complex Analysis (Minneapolis, 1964), Springer, Berlin (1965), pp. 142-154.

[9]. ———, Deformations of complex analytic structures on compact manifolds, Seminar on Global Analysis, Montreal (1969).

[10]. Morrey, Jr., C. B., Multiple Integrals in the Calculus of Variations, Springer-Verlag, New York (1966).

ON THE $A_q(\Gamma) \subset B_q(\Gamma)$ CONJECTURE FOR INFINITELY GENERATED GROUPS

Joseph Lehner

§1. Let Γ be a Fuchsian group acting on the unit disk $U : |z| < 1$ and let $q > 0$. The holomorphic function f defined in U is called an automorphic form of weight q (or degree $-2q$) if

$$f(Az)\,A'(z)^q = f(z), \quad z \in U, \quad A \in \Gamma \,. \tag{1}$$

If q is not an integer, we assume the existence of a "multiplier system" $\{v(A) : A \in \Gamma, |v| = 1\}$; then (1) is replaced by

$$f(Az)\,A'(z)^q = v(A)\,f(z) \,,$$

where a definite branch of A'^q has been chosen.

Bers (see for example [1]) has introduced the Banach spaces $A_q(\Gamma)$, $B_q(\Gamma)$ of *integrable* and *bounded* automorphic forms. The form $f \in A_q(\Gamma)$ if

$$\|f\|_1 = \iint_R (1-|z|^2)^{q-2}|f(z)|\,dxdy < \infty \,, \tag{2}$$

where R is a fundamental region for Γ in U; $f \in B_q(\Gamma)$ if

$$\|f\|_\infty = \sup_{z \in U} (1-|z|^2)^q|f(z)| < \infty \,. \tag{3}$$

It has been conjectured that

$$A_q(\Gamma) \subset B_q(\Gamma) \tag{4}$$

for all Fuchsian groups Γ and this has been verified for finitely generated Γ ([3], [5], [6], [8]). To my knowledge it has not been known up to now whether (4) is true – or false – for a single infinitely generated group.

In this note it is proved that the conjecture (4) is indeed true for an infinite class of infinitely generated groups.

THEOREM 1. *Let* Γ *satisfy the following conditions*:

$$\Gamma \text{ consists of } 1 \text{ and hyperbolic elements,} \tag{5}$$

There exists a constant $t_0 > 2$ *such that*

$$|\text{trace } A| \geq t_0 \quad \text{for all} \quad A \in \Gamma,\ A \neq 1. \tag{6}$$

Then (4) *is true, and the inclusion map is continuous.*

We remark that the trace condition (6) is satisfied for the hyperbolic elements of every finitely generated group. (A proof is apparently contained in the Nielsen-Fenchel manuscript.)

THEOREM 2. *Let* G *be a finitely generated Fuchsian group without parabolic elements. Let* H *be a subgroup of* G. *Then* $A_q(H) \subset B_q(H)$ *with a continuous inclusion map.*

In the last section we show that the infinitely generated groups satisfying the conditions of Theorem 1 or Theorem 2 are infinite in number.

§2. Let R be the Ford fundamental region for Γ. For $|\zeta| < 1$ define

$$\Delta(\zeta) = \left\{ w : |w-\zeta| < \frac{1-|\zeta|}{2} \right\} .$$

Then $\Delta(\zeta) \subset U$. Let $m_1, m_2, \ldots$ be positive constants depending only on Γ and q. Let $n(\Delta(\zeta), z)$ be the number of points in $\Delta(\zeta)$ equivalent under Γ to z. We recall that $(1-|z|^2)^{-2}dxdy$, $z = x+iy$, as well as $(1-|z|^2)^q|f(z)|$, are invariant under Γ. This shows, in particular, that the definition (2) is independent of the choice of fundamental region.

LEMMA 1. *Let* Γ *satisfy the conditions* (5) *and* (6). *Then for* $|\zeta| < 1$ *we have*

$$n(\Delta(\zeta), z) \leq m_1, \quad z \in U .$$

Theorem 1 follows from Lemma 1. In fact, the elementary identity

$$f(\zeta) = \frac{4}{\pi(1-|\zeta|)^2} \iint_{\Delta(\zeta)} f(z)dxdy, \quad |\zeta| < 1$$

leads to

$$(1-|\zeta|^2)^q |f(\zeta)| \leq m_2 \iint_{\Delta(\zeta)} (1-|z|^2)^{q-2} |f(z)| dxdy \tag{7}$$

$$= m_2 \iint_R n(\Delta(\zeta), z)(1-|z|^2)^{q-2} |f(z)| dxdy$$

$$\leq m_2 m_1 \|f\|_1 ,$$

where

$$m_2 = m_2(q) = \begin{cases} 4^q \pi^{-1}, & q \geq 2 \\ 4^3 \pi^{-1}(4/3)^{q-2}, & q < 2 . \end{cases}$$

Hence $\|f\|_\infty \leq m_3 \|f\|_1$ and Theorem 1 follows.

§3. We now prove Lemma 1. Let d denote hyperbolic distance; σ, hyperbolic area. Every element A of Γ, $A \neq 1$, is hyperbolic and has a multiplier $k(A)$, which we may assume to be greater than 1, and the trace condition (6) is equivalent to

$$k(A) = k(A^{-1}) \geq k_0 > 1 . \tag{8}$$

Write $A = (a\bar{c} \,|\, c\bar{a})$, $a\bar{a} - c\bar{c} = 1$. Let $I_A : |cz+\bar{a}| = 1$, $I'_A : |-cz+a| = 1$, be the isometric circles of A, A^{-1}, respectively. A circle through the fixed points of A cuts the isometric circles in two points equivalent by A. The hyperbolic length of the intercepted arc equals the translation length of A; as we shall see this is $2 \log k$.

Map U on $H = \{w : \mathrm{Im}\, w > 0\}$ by a linear fractional transformation T so that the fixed points of A are carried to $0, \infty$. Then A becomes $A_1 = TAT^{-1} : w \to kw$, where $k = k(A_1) = k(A)$. Since T carries the hyperbolic metric on U into that on H, we can write d for the latter. Now I_A, I'_A go into concentric circles about $w = 0$ whose Euclidean radii are in the ratio $1 : k$; let these be $|w| = r$, $|w| = kr$.

We have

$$d(ir, ikr) = 2 \int_r^{kr} \frac{dt}{t} = 2 \log k ,$$

the translation length. This is the minimum distance between any two equivalent points lying on the concentric circles. In fact, the locus of points at a hyperbolic distance $2 \log k$ from $|w| = r$ consists of two circles, one of which, C, passes through r, $-r$, ikr. Let L be the hyperbolic line through w_0, kw_0, $|w_0| = r$, and cutting C in w_1. Then $d(w_0, kw_0) \geq d(w_0, w_1) = 2 \log k$, as asserted, and so for z on I_A we have

(9) $$d(z, Az) = d(w_0, kw_0) \geq 2 \log k \geq 2 \log k_0 = m_4 > 0 .$$

Obviously the same inequality holds if z is on I'_A. By the triangle inequality, (9) holds for z lying outside both I_A and I'_A.

Now if $z \in R$, z does lie outside both I_A and I'_A for every A in Γ, $A \neq 1$. Thus each z in R is the center of a disk Δ_0 of fixed radius $m_4/2$ that contains no image of z except z.

Let $\Delta_n = A_n(\Delta_0)$, where $\Gamma - \{1\} = \{A_n, n = 1, 2, \ldots\}$. The Δ_n do not intersect and all Δ_n have the same hyperbolic radius $m_4/2$. We regard $\Delta(\zeta)$ as a hyperbolic disk; it is easily checked that its radius is less than $\left(\frac{1}{2}\right) \log 5 = m_5$. Hence all Δ_n that meet $\Delta(\zeta)$ lie in a hyperbolic disk Δ' of radius less than $m_5 + m_4/2 = m_6$. We have $\sigma(\Delta') = m_7$, $\sigma(\Delta_0) = m_8$. Ther.

$$n(\Delta(\zeta), z)\sigma(\Delta_0) \leq \sigma(\Delta') ,$$

i.e., $n(\Delta(\zeta), z) \leq m_7/m_8$. This concludes the proof of Lemma 1 and therefore of Theorem 1.

§4. *Proof of Theorem 2*

We write more explicitly $n_G(\Delta, z)$ in place of $n(\Delta, z)$, and note first that $n_G(\Delta(\zeta), z)$ is bounded for all $|\zeta| < 1$, $|z| < 1$. In fact we need consider only $z \in R$. Every $z \in R$ can be covered by a region D_z, which is a disk or a portion of a disk lying wholly in R and for which

$$m_9 < \sigma(D_z), \quad \text{diam } D_z < m_{10}, \quad z \in R .$$

This follows from the finite generation of G and the absence of cusps in R. Hence by the argument of Lemma 1, $n_G(\Delta(\zeta), z) < m_{11}$, and obviously $n_G(\Delta, z) \leq n_H(\Delta, z)$. We then have by (7), for $f \in A_q(H)$,

$$(1-|\zeta|^2)|f(\zeta)| < m_2 \int_{R_H} n_H(\Delta(\zeta), z)(1-|z|^2)^{q-2}|f(z)|dxdy$$
$$\leq m_{12}\|f\|_1 ,$$

where $\|f\|_1$ is the $A_q(H)$ norm of f. Q.E.D.

§5. In this section we give examples of infinitely generated groups that satisfy the hypotheses of Theorem 1 or Theorem 2. Let $\Gamma(1) = SL(2, Z)$ be the classical modular group. For $A = (ab \mid cd) \in \Gamma$ we have $|a+d| \geq 3$ when A is hyperbolic. Thus the hyperbolic elements of any $H \subset \Gamma(1)$ satisfy the trace condition (6). Let $T_j = (r_j, -1-r_j^2 | 1, -r_j)$, $j = 1, 2, \ldots,$ where r_j is a rational integer and $r_{j+1}-r_j \geq 3$, $r_1 \geq 2$. Then $T_j \in \Gamma(1)$ and $I(T_j)$ is $|z-r_j| = 1$, $I(T_j^{-1})$ is $|z+r_j| = 1$. Let H be the subgroup of $\Gamma(1)$ generated by $\{T_j, \ j = 1, 2, \ldots\}$. Since $I(T_j)$, $I(T_j^{-1})$ are mutually disjoint, H is the free product of the infinite cyclic groups $\{T_j\}$ and is therefore infinitely generated ([4], p. 56). Furthermore, H consists entirely of hyperbolic elements. In fact, every $W \in H$ can be written uniquely in the form $S_k S_{k-1} \cdots S_1$, where each S_i is a T_j or T_j^{-1} and S_{i+1} is never equal to S_i^{-1}. Now $I(W)$ is inside $I(S_1)$ – see [4], p. 57 – and $I(W^{-1})$ is inside $I(S_k^{-1})$. If $S_k^{-1} \neq S_1$, $I(W)$ and $I(W^{-1})$ do not intersect and so W is hypervolic. If $S_k^{-1} = S_1$, we consider $W' = S_k^{-1} W S_1^{-1} = S_k^{-1} W S_k = S_{k-1} \cdots S_2$, where we may assume $S_{k-1}^{-1} \neq S_2$. Then W' is hyperbolic and so is W. Hence H satisfies (5) and (6).

The group H is of the second kind. An example of a group of the first kind is G', the commutator subgroup of G, where $G = \Gamma(2)$, the principal congruence subgroup of level 2. G is free, so of course contains no elliptic elements; moreover, it is not hard to see that G′ has no parabolic elements. Since G′ is normal in G and of infinite index, it is infinitely generated.

In his lecture at this Conference, W. Magnus ([7]) discussed a family of infinitely generated subgroups of $\Gamma(1)$ considered originally by B. H. Neumann, which have no elliptic or parabolic elements. There are uncountably many such groups.

Next, let $G = \{x, y : x^2 = y^3 = (xy)^7 = 1\}$; this can be realized as a Fuchsian group without parabolic elements. G has a normal subgroup N that is a surface group of genus 3, and N has infinite index (and therefore infinitely generated) subgroups obtained, for example, as the set of words for which the exponent sum in a given generator vanishes. These subgroups are examples for the group H of Theorem 2.

UNIVERSITY OF PITTSBURGH
PITTSBURGH, PENNSYLVANIA 15260

REFERENCES

[1]. Bers, L., Completeness theorems for Poincaré series in one variable, Proc. International Symposium on Linear Spaces, Jerusalem, 1961.

[2]. Bers, L., Automorphic forms and Poincaré series for infinitely generated Fuchsian groups, Amer. J. Math. 87 (1965), pp. 196-214.

[3]. Drasin, D., and Earle, C., On the boundedness of automorphic forms, Proc. Amer. Math. Soc. 19 (1968), pp. 1039-1042.

[4]. Ford, L. R., Automorphic functions, Chelsea, New York, 1951.

[5]. Knopp, M. I., Bounded and integrable automorphic forms, Indiana University Math. Jour. 22 (1973), pp. 769-778.

[6]. Lehner, J., On the $A_q(\Gamma) \subset B_q(\Gamma)$ conjecture, Modular Functions of One Variable I, Springer Lecture Notes 320.

[7]. Magnus, W., Rational representations of Fuchsian groups, (to appear in Sitzungsber Akad. Wiss. Göttingen).

[8]. Metzger, T. A., and Rao, K. V., On integrable and bounded automorphic forms, Proc. Amer. Math. Soc. 28 (1971), pp. 562-566; On integrable and bounded automorphic forms II, Proc. Amer. Math. Soc. 32 (1972), pp. 201-204.

THE FUNDAMENTAL GROUPS OF CERTAIN SUBGROUP SPACES

A. M. Macbeath

§1. *Introduction*

Let Γ be a NEC group – that is to say, a discrete subgroup with compact quotient space of $G = PGL(2, \mathbb{R})$. Let $S_0(\Gamma, G)$ denote the family consisting of all NEC groups isomorphic to Γ. We endow $S_0(\Gamma, G)$ with the topology defined (for general locally compact G) by Chabauty [3]. It is proved in [5] that $S_0(\Gamma, G)$ is a manifold of finite dimension; also that it is a $K(\pi, 1)$ and that its fundamental group is an extension of an infinite cyclic group $\mathbb{Z}$ by a certain subgroup $\mathrm{Aut}^+\Gamma$ of index 2 in the group $\mathrm{Aut}\,\Gamma$ of all automorphisms of Γ. We shall prove here that $\mathbb{Z}$ is central in $\pi_1(S_0(\Gamma, G))$.

NOTATION. For any group H and any element $h \in H$, $I(h)$ will denote the inner automorphism $I(h)(k) = hkh^{-1}$. We regard I as a homomorphic mapping of H into Aut H, so that, for instance, if K is a subgroup I(K) will denote the set of all maps $\{I(k) : k \in K\}$. For the group G, it is known that I is an isomorphic map of G onto Aut G [7].

An NEC group contained in $G_0 = PSL(2, \mathbb{R})$, the identity component of G, is a Fuchsian group, and, if also it contains no elements of finite order, it is isomorphic to the fundamental group of a compact orientable surface and will be called a *surface group*.

§2. *The rectangle of spaces*

In the present state of knowledge our only way of obtaining information about the space $S_0(\Gamma, G)$ is the study of its relationship with other better known spaces. The parent space is the Weil representation space $R_0(\Gamma, G)$

and the two offspring are the Teichmüller space $T(\Gamma, G)$ and the subgroup space $S_0(\Gamma, G)$. The modulus space, which will hardly concern us here, is the grandchild. (This "family tree" represents the mathematical situation and has nothing to do with the historical order in which these spaces were discovered.)

The space $R_0(\Gamma, G)$ defined by Weil [8] in a much more general setting, is the subspace of the topological product G^Γ consisting of those mappings $r : \Gamma \to G$ which (i) are homomorphisms (ii) are injective and (iii) have discrete image group $r(\Gamma)$ with compact quotient space $G/r(\Gamma)$.

A map $r : \Gamma \to G$ can be composed on the right by an automorphism $\alpha \in \text{Aut}\,\Gamma$ or on the left by an automorphism $\beta \in \text{Aut}\,G$

$$\Gamma \xrightarrow{\alpha} \Gamma \xrightarrow{r} G \xrightarrow{\beta} G .$$

We thus obtain two group actions

$$r \mapsto r \circ \alpha, \qquad \alpha \in \text{Aut}\,\Gamma,$$
$$r \mapsto \beta \circ r, \qquad \beta \in \text{Aut}\,G.$$

These two actions commute and the subspace $R_0(\Gamma, G) \subset G^\Gamma$ is stable under each action. The Teichmüller space $T(\Gamma, G)$ may be defined as the quotient space $R_0(\Gamma, G)/\text{Aut}\,G$, and the subgroup space $S_0(\Gamma, G)$ is homeomorphic to $R_0(\Gamma, G)/\text{Aut}\,\Gamma$. (For the relationship with other definitions and details of results sketched here, see [5].)

Since the actions of Aut G, Aut Γ commute, the action of Aut Γ is compatible with the partitioning of $R_0(\Gamma, G)$ into Aut G-orbits. Thus there is an induced action of Aut Γ on $T(\Gamma, G)$. The same procedure with Aut G, Aut Γ interchanged yields an action of Aut G on $S_0(\Gamma, G)$.

We may also consider the combined action of $\text{Aut}\,G \times \text{Aut}\,\Gamma$ on $R_0(\Gamma, G)$ given by

$$(\beta, \alpha) r = \beta \circ r \circ \alpha^{-1} .$$

The quotient space $R_0(\Gamma, G)/\text{Aut}\,G \times \text{Aut}\,\Gamma$ is called the *small modulus space* $M(\Gamma, G)$. It can, of course, be obtained from $R_0(\Gamma, G)$ in two steps, first factoring out Aut Γ to obtain $S_0(\Gamma, G)$ and then factoring out Aut G. Alternatively, one could first factor out Aut G to obtain Teichmüller space and then factor out Aut Γ. The situation can be represented in the diagram

$$(1) \qquad \begin{array}{ccc} R_0(\Gamma, G) & \xrightarrow{\text{Aut } \Gamma} & S_0(\Gamma, G) \\ {\scriptstyle \text{Aut } G}\downarrow & & \downarrow{\scriptstyle \text{Aut } G} \\ T(\Gamma, G) & \xrightarrow{\text{Aut } \Gamma} & M(\Gamma, G) \end{array}$$

Each arrow represents the quotient map modulo the group written near it.

§3. *The relationship between* R_0 *and* T

Let γ_1, γ_2 be two non-commuting elements of Γ, each of infinite order. If we consider γ_1, γ_2 as acting on the real projective line $\mathbf{R} \cup \{\infty\}$, each γ_i will have an attracting fixed point $q(\gamma_i)$ and a repelling fixed point $p(\gamma_i)$. Let $Q(\gamma_1, \gamma_2, \Gamma)$ denote the subspace of $R_0(\Gamma, G)$ consisting of those r such that

$$p(r(\gamma_1)) = 0, \quad q(r(\gamma_1)) = \infty, \quad p(r(\gamma_2)) = 1 .$$

It is shown in [5] that every $r \in R_0(\Gamma, G)$ is uniquely expressible in the form

$$r = I(g) \circ r_0, \qquad g \in G, \qquad r_0 \in Q(\gamma_1, \gamma_2, \Gamma) .$$

Moreover, the expression of r in this form depends continuously on r. Thus

3.1. $R_0(\Gamma, G)$ *is a product space* $G \times Q$ *and the quotient map of* R_0 *onto* T *can be regarded as the projection of the product onto the factor* Q. *Thus* $Q(\gamma_1, \gamma_2, \Gamma)$ *is homeomorphic to* $T(\Gamma, G)$.

It is also known (see, for instance, [2] for the classical case of surface groups) that

3.2. $T(\Gamma, G)$ *is homeomorphic to a Euclidean space.*

Now G is a union of two open and closed components, the subgroup $G_0 = PSL(2, \mathbb{R})$ and one other G_0-coset. It is known that G_0 is a product $\mathbb{R}^2 \times S^1$. Thus $R_0(\Gamma, G)$ is a union of two open and closed subspaces $R_0^+(\Gamma, G)$ and $R_0^-(\Gamma, G)$, each homeomorphic to a product of S^1 and a Euclidean space.

If $\alpha \in \mathrm{Aut}\,\Gamma$, the homeomorphism $r \mapsto r \circ \alpha$ of $R_0(\Gamma, G)$ either interchanges the two components R_0^+ and R_0^- or else it maps R_0^+ on itself. Let $\mathrm{Aut}^+\,\Gamma$ denote the subgroup consisting of those which map R_0^+ on itself. It can be shown that $\mathrm{Aut}^+\,\Gamma$ is a proper subgroup of $\mathrm{Aut}\,\Gamma$, so it has index 2 and we can obtain $S_0(\Gamma, G)$ as the quotient space

$$S_0(\Gamma, G) = R_0^+(\Gamma, G)/\mathrm{Aut}^+\,\Gamma \ .$$

It is also known that the group $\mathrm{Aut}\,\Gamma$ acts discontinuously on $R_0(\Gamma, G)$ without fixed points. Thus the map of $R_0^+(\Gamma, G)$ onto $S_0(\Gamma, G)$ is a regular covering map and since $R_0^+(\Gamma, G)$ is a connected manifold, the same is true of $S_0(\Gamma, G)$. Since R_0^+ is the product of a contractible space with S^1, we have

$$\pi_1(R_0^+) \cong \pi_1(S^1) \cong \mathbf{Z} \ ,$$

and, as mentioned in the introduction, $\pi_1(S_0)$ is, by standard covering space theory, an extension

$$1 \to \mathbf{Z} \to \pi_1(S_0(\Gamma, G)) \to \mathrm{Aut}^+\,\Gamma \to 1 \ .$$

Since the fundamental group $\mathbf{Z}$ of R_0^+ is abelian, each homeomorphism $h : R_0^+ \to R_0^+$ defines a unique automorphism $h^* : \pi_1(R_0^+) \to \pi_1(R_0^+)$, independent of the choice of basepoint. Our result will be proved if we can show that, for each of the right translations ρ_α, $\alpha \in \mathrm{Aut}^+\,\Gamma$, defined by $\rho_\alpha(r) = r \circ \alpha$, we have

$$\rho_\alpha^* = \mathrm{id} \ .$$

This will be proved first for a surface group, then deduced for general Γ.

§4. *The rectangle of connected spaces*

We can also consider the effect of the transformations of Aut G on the components R_0^+, R_0^-, obtaining a subgroup $\mathrm{Aut}^+ G$ which maps R_0^+ onto itself. A glance at the product representations $R_0 = I(G) \circ Q$, $R_0^+ = I(G_0) \circ Q$ shows at once that $\mathrm{Aut}^+ G = I(G_0)$, so that $\mathrm{Aut}^+ G$ is also a proper subgroup of Aut G and $T(\Gamma, G) = R_0^+(\Gamma, G)/\mathrm{Aut}^+ G$. In order to deal only with connected spaces, we consider the modified diagram

$$(2) \qquad \begin{array}{ccc} R_0^+(\Gamma, G) & \xrightarrow{\mathrm{Aut}^+ \Gamma} & S_0(\Gamma, G) \\ \mathrm{Aut}^+ G \downarrow & & \downarrow \mathrm{Aut}^+ G \\ T(\Gamma, G) & \xrightarrow{\mathrm{Aut}^+ \Gamma} & M^+(\Gamma, G) \end{array}$$

Here $M^+(\Gamma, G)$ is the *large modulus space*. (If Γ is a surface group $M^+(\Gamma, G)$ agrees with the classical concept of modulus space, since its points represent the directly conformal equivalence-classes of Riemann surfaces of the appropriate genus. The small modulus space is then derived from the large one by identifying each Riemann surface with its mirror image.)

The equivalence class $I(G_0) \circ q$, $q \in Q$, will be called a *fibre*. The formula

$$I(h) \circ I(g) \circ q = I(hg) \circ q$$

shows that G_0 acts by left translation along each fibre. As before, the quotient map on the left of the rectangle (2) is, essentially, the projection of the product space onto the factor Q.

§5. *The stabilizer of a fibre*

Let $q \in Q$. Consider the subgroup S_q of $\mathrm{Aut}^+ \Gamma$ which maps the fibre $I(G_0) \circ q$ onto itself. If $\alpha \in \mathrm{Aut}^+ \Gamma$, then $\alpha \in S_q$ if and only if there exists $g = g(\alpha) \in G_0$ such that

$$(3) \qquad q \circ \alpha = I(g(\alpha)) \circ q .$$

The element $g(\alpha)$, if it exists, is uniquely determined by α and equation (3), because the centralizer of an NEC group in G consists of the unit element only. From (3) follows $q(\Gamma) = I(g)(q(\Gamma)) = gq(\Gamma)g^{-1}$, so that $g \in N_G(q(\Gamma)) \cap G_0$. Conversely, if $g \in N_G(q(\Gamma)) \cap G_0$, then $I(g)$ is an automorphism of $q(\Gamma)$ and the automorphism $\alpha \in \mathrm{Aut}^+ \Gamma$ defined by $\alpha = q^{-1} \circ I(g) \circ q$ belongs to S_q. We deduce

5.1. *The stabilizer* S_q *of the fibre of* q *in* $\mathrm{Aut}^+ \Gamma$ *is isomorphic to the intersection with* G_0 *of the normalizer of the image group* $q(\Gamma)$. *The isomorphism is given by the formula*

$$g \mapsto q^{-1} \circ I(g) \circ q \ .$$

Thus S_q *is isomorphic to a Fuchsian group.*

If $\alpha \in S_q$, let us now study the action of ρ_α on the fibre of q. Operating with $I(h)$ on both sides of (3) we find

$$I(h) \circ q \circ \alpha = I(hg(\alpha)) \circ q \ . \tag{4}$$

Let ρ_g denote right translation $h \mapsto hg$ in G_0 and let $j_q : G_0 \to R_0^+(\Gamma, G)$ denote the embedding of G_0 into R_0^+ given by $j_q(g) = I(g) \circ q$. We can re-write (4) in the form

$$\rho_\alpha \circ j_q = j_q \circ \rho_{g(\alpha)} \ . \tag{5}$$

For each mapping f let f^* denote as before the induced homomorphism of fundamental groups. Since j_q is an embedding of G_0 as a fibre in the product of G_0 with the simply connected space Q, j_q^* is an isomorphism, and since G_0 is a connected Lie group $\rho_g^* = \mathrm{id}$. Thus from (5) we deduce

$$\rho_\alpha^* \circ j_q^* = j_q^*$$

so that

$$\rho_\alpha^* = \mathrm{id} \ .$$

We have proved

5.2. *If* $\alpha \in \mathrm{Aut}^+ \Gamma$ *and* ρ_α *maps any fibre* $I(G_0) \circ q$ *onto itself, then* $\rho_\alpha^* = \mathrm{id}$.

It is now easy to prove our main result for surface groups. Let Γ be a surface group and let H be the subgroup of $\mathrm{Aut}^+ \Gamma$ consisting of all those α such that $\rho_\alpha^* = \mathrm{id}$. By 5.2, H contains the stabilizers of all fibres, and hence the subgroup generated by the stabilizers. By Maclachlan's theorem [6], the stabilizers of the fibres generate the whole of $\mathrm{Aut}^+ \Gamma$ when Γ is a surface group, and this case of our theorem is proved. Thus

5.3. *If* Γ_1 *is a surface group and* $\alpha \in \mathrm{Aut}^+ \Gamma_1$ *then* $\rho_\alpha^* = \mathrm{id}$.

§6. *Proof that* Z *is central*

Suppose now that Γ is an NEC group and Γ_1 is a subgroup of Γ of finite index. Each $r \in R_0(\Gamma, G)$ has a restriction $r_1 = r|_{\Gamma_1} \in R_0(\Gamma_1, G)$. If $\beta \in \mathrm{Aut}\, G$ then $\beta \circ r_1 = (\beta \circ r)|_{\Gamma_1}$ and, if γ_1, γ_2 are two non-commuting translations in Γ_1 then $r \in Q(\gamma_1, \gamma_2, \Gamma) = Q$, say, if and only if $r_1 \in Q(\gamma_1, \gamma_2, \Gamma) = Q_1$, say. It is known that Q is embedded in Q_1 in topologically the same manner as a linear subspace is embedded in a Euclidean space [5].

Suppose further that Γ_1 is a surface group and that $\alpha \in \mathrm{Aut}^+ \Gamma$ maps Γ_1 into itself. Then ρ_α maps $R_0(\Gamma, G)$, regarded as a subspace of $R_0(\Gamma_1, G)$, into itself. If $f(e^{2\pi i t})$, $0 \leq t \leq 1$, is a closed path in $R_0^+(\Gamma, G) \subset R_0^+(\Gamma_1, G)$, then $\rho_\alpha \circ f$ is also in $R_0^+(\Gamma, G)$. Let f be a path representing a generator of the fundamental group of $R_0^+(\Gamma_1, G) = Q_1 \times G_0$ lying in $R_0^+(\Gamma, G) = Q \times G_0$. By 5.3 $\rho_\alpha \circ f$ is homotopic to f in $Q_1 \times G_0$ and hence, because of the nature of the embedding of Q in Q_1 as a linear subspace, $\rho_\alpha \circ f$ is homotopic to f in $Q \times G_0$. Thus we have proved

6.1. *If* α *is an element of* $\mathrm{Aut}^+\,\Gamma$ *which maps a surface group* Γ_1 *on itself, where* Γ_1 *is a subgroup of* Γ *of finite index, then* $\rho_\alpha^* = \mathrm{id}$.

However, it is known by a theorem of Fox and others ([4], see also [9], pp. 85-87) that there is at least one subgroup N of Γ which has finite index in Γ and is a surface group. Moreover, Γ is finitely generated, so there is a subgroup N_1 of N, still of finite index in Γ, which is a characteristic subgroup, i.e., every $\alpha \in \mathrm{Aut}\,\Gamma$ maps N_1 onto itself (see Baumslag [1]). The theorem that $\rho_\alpha^* = \mathrm{id}$ for all α, and therefore that $\mathbf{Z}$ is central, now follows from 6.1.

REFERENCES

[1]. Baumslag, G., Automorphism groups of residually finite groups. J. London Math. Soc. 38 (1963), pp. 117-118.

[2]. Bers, Lipman, Quasi-conformal mappings and Teichmüller's theorem. Princeton conference on Analytic Functions, 1960.

[3]. Chabauty, C., Limites d'ensembles et géometrie des nombres. Bull Soc. Math. France. 78 (1950), pp. 143-151.

[4]. Fox, R. H., On Fenchel's conjecture about F-groups. Mat. Tidsskrift (1952), pp. 61-65.

[5]. Macbeath, A. M., and Singerman, D., Spaces of subgroups and Teichmüller spaces. (To appear.)

[6]. Maclachlan, C., Modulus space is simply connected. Proc. Amer. Math. Soc. 29 (1971), pp. 185-186.

[7]. Schreier, O., und van der Waerden, B. L., Die Automorphismen der projektiven Gruppen. Hamburg Abh. VI (1928), pp. 303-322.

[8]. Weil, A., Discrete subgroups of Lie groups. Annals of Math. 72 (1960), pp. 369-384.

[9]. Zieschang, H., Coldewey, H.-D., and Vogt, E., Flächen und ebene diskontinuierliche Gruppen. Springer Lecture Notes No. 122.

MODULAR GROUPS AND FIBER SPACES OVER TEICHMÜLLER SPACES

C. Maclachlan

§1. *Introduction*

In [3], Bers constructed a fiber space $F(\Gamma)$ over the Teichmüller space $T(\Gamma)$ of Γ, a Fuchsian group. He also showed that the Teichmüller modular group, $\text{Mod}\,\Gamma$, which acts on $T(\Gamma)$, can be extended to a group $\text{mod}\,\Gamma$ acting on $F(\Gamma)$. In the case where Γ is finitely generated and of the first kind, this gives a normal complex space $\dfrac{F(\Gamma)}{\text{mod}\,\Gamma} = Y(\Gamma)$ which is a fiber space over the space of moduli $X(\Gamma) = \dfrac{T(\Gamma)}{\text{Mod}\,\Gamma}$, with a fiber over a point of $X(\Gamma)$ representing the class of a group Γ' being isomorphic to $\dfrac{U}{N_G(\Gamma')}$, where U denotes the upper half-plane, G the group $PSL(2, R)$ and $N_G(\Gamma')$, the normalizer of Γ' in G.

In this article, topologically-equivalent constructions of the space $F(\Gamma)$ are given, in terms of fiber bundles. First, using the Weil space $R_0(\Gamma, G)$ (see [13]), of faithful representations of Γ in G, which preserve the type of the elements of Γ, the orientation and the discreteness of the group. Secondly, using the space $M(\Gamma)$ of complex structures introduced by Earle and Fells [7], [8]. Both the spaces constructed have "modular groups," which are isomorphic to $\text{mod}\,\Gamma$, acting on them in a natural way. These actions are shown to be compatible with the action of $\text{mod}\,\Gamma$ on $F(\Gamma)$ and so alternative topological descriptions of $Y(\Gamma)$ are obtained.

When Γ is torsion-free, Bers proves in [3], that there is a canonical biholomorphic map between $F(\Gamma)$ and $T(\dot{\Gamma})$ where $\dot{\Gamma}$ is a uniformizing group of U/Γ with an additional puncture. This induces an isomorphism between $\text{mod}\,\Gamma$ and a subgroup of $\text{Mod}\,\dot{\Gamma}$ of finite index.

Using the extension of the Nielsen isomorphism between Mod Γ and the mapping class group associated with surface $^U/_\Gamma$ [12], this isomorphism is studied independently of the canonical map, and an extension of the isomorphism theorem is obtained in the case where Γ has elements of finite order.

Finally, using the isomorphism theorem and the compatibility of the actions on the spaces concerned, some results on the nature of mod Γ are obtained. First, the fact that mod Γ is residually finite is used to show that Mod Γ, for Γ a free group, is residually finite. Second, every finite subgroup of mod Γ is cyclic. Third, when Γ is torsion-free, that mod Γ can be generated by elements of finite order, except for some cases when the genus of $^U/_\Gamma$ is 2, which implies that $Y(\Gamma)$ is simply-connected.

§2. *Fiber spaces over Teichmüller space*

Let G denote the group $PSL(2, \mathbf{R})$. Γ, throughout, will denote a finitely-generated Fuchsian group of the first kind.

First let us describe the Bers fiber space $F(\Gamma)$ over $T(\Gamma)$ as given in [3].

Let Q denote the group of quasi-conformal self-homeomorphisms of U. Each element of Q extends to $\mathbf{R}$. Let

$$Q_{norm} = \{w \in Q \mid w(0) = 0,\ \ w(1) = 1,\ \ w(\infty) = \infty\}$$

and $Q_0 = \{w \in Q \mid w|\mathbf{R} = Id\}$. Also $Q(\Gamma) = \{w \in Q \mid w\Gamma w^{-1} \subseteq G\}$.

$L_\infty(U)$ denotes the Banach space of bounded measurable functions and $L_\infty(U,\Gamma)$, the subspace of Γ-compatible functions, i.e., $L_\infty(U,\Gamma) = \{\mu \in L_\infty(U) \mid (\mu_0\gamma)\overline{\gamma}' = \gamma'\mu,\ \forall\gamma \in \Gamma\}$. For $w \in Q$, the Beltrami coefficient $\mu = \dfrac{w_{\overline{z}}}{w_z} \in L_\infty(U)_1$, the unit ball, and each $\mu \in L_\infty(U)_1$ is the Beltrami coefficient of a unique member of Q_{norm}, which is written as w_μ. Also $w_\mu \in Q_{norm}(\Gamma)$ if and only if $\mu \in L_\infty(U,\Gamma)_1$.

$w, w' \in Q$ are said to be equivalent if $w|R = w'|R$, and the equivalence class of w is denoted by $[w]$. Then $T(\Gamma)$ is the space $\{[w_\mu] | \mu \in L_\infty(U,\Gamma)_1\}$ with topology inherited from $L_\infty(U)$.

Let w^μ denote the unique normalized quasi-conformal self-homeomorphism of $\hat{C}(= C \cup \{\infty\})$ such that $w^\mu|U$ has Beltrami coefficient μ and $w^\mu|L$ is conformal. Then $w_\mu = k_\mu \circ w^\mu|U$ where k_μ is that conformal mapping of $w^\mu(U)$ to U required to normalize w_μ.

Then $F(\Gamma)$ is the space of pairs $([w_\mu], z)$ where $z \in w^\mu(U)$ and the projection $p: F(\Gamma) \to T(\Gamma)$ is the projection onto the first co-ordinate.

Construction A

From results in [13], an equivalent topological description of $F(\Gamma)$ is obtained as follows. Elements of G are classed as hyperbolic, parabolic or elliptic according as they have two real, one real, or two complex fixed points in $\hat{C}$ respectively. An isomorphism $\phi: \Gamma \to G$ whose image is discrete is type-preserving if it preserves the classes of hyperbolic, parabolic and elliptic elements. Such an isomorphism can always be induced by a homeomorphism $w: U \to U$ [20] and ϕ is orientation-preserving if w is orientation-preserving. $R_0(\Gamma, G)$ is the space of all type-and orientation-preserving isomorphisms of Γ into G such that the image is discrete, topologized as a subspace of $G^{|\Gamma|}$.

G acts by conjugation on $R_0(\Gamma, G)$ thus: if $r \in R_0(\Gamma, G)$, $g \in G$, then $g \cdot r(\gamma) = g^{-1} r(\gamma) g$ for all $\gamma \in \Gamma$. The quotient is Teichmüller space $T(\Gamma, G)$ whose elements we write as Gr, $r \in R_0(\Gamma, G)$. The mapping $j_1: T(\Gamma) \to T(\Gamma, G)$ given by $j_1[w_\mu] = Gr$ where $r(\gamma) = w_\mu \gamma w_\mu^{-1}$, is a homeomorphism.

The natural map $p_1: R_0(\Gamma, G) \to T(\Gamma, G)$ defines a principal G-bundle ξ with fiber G; in fact, it is trivial since it is isomorphic to the product principal G-bundle, $G \times T(\Gamma, G)$. [13]

Let $a \in U$ and $H = \mathrm{Stab}_G(a)$ so that H is a closed subgroup of G homeomorphic to S^1. G is homeomorphic to $G/H \times H$.

Restricting the structure group in the bundle ξ to H, we obtain a fiber space $R_0(\Gamma, G) \bmod H$ over $T(\Gamma, G)$ with fiber the coset space G/H.

This is isomorphic over $T(\Gamma, G)$ to the associated homogeneous bundle $\xi[G/H]$ with fiber model G/H, which is a trivial fiber bundle.

THEOREM 1. *The fiber space* $p_1 : R_0(\Gamma, G) \bmod H \to T(\Gamma, G)$ *is isomorphic as a fiber space to* $p : F(\Gamma) \to T(\Gamma)$.

Proof. A homeomorphism α_1 is constructed such that the following diagram commutes.

$$\begin{array}{ccc} F(\Gamma) & \xrightarrow{\alpha_1} & R_0(\Gamma, G) \bmod H \\ \downarrow p & & \downarrow p_1 \\ T(\Gamma) & \xrightarrow{j_1} & T(\Gamma, G) \end{array} \tag{1}$$

Elements of $R_0(\Gamma, G) \bmod H$ will be denoted by Hr, $r \in R_0(\Gamma, G)$. Let $\alpha_1([w_\mu], z) = Hr$ where r is induced by $t^{-1}w_\mu$ where t is any element of G such that $t(a) = k_\mu(z)$. Since k_μ depends only on $[w_\mu]$, α_1 is well-defined.

Let w be any quasi-conformal homeomorphism of U which induces r [20]. Let $w_\mu = t_0 w$ where $t \in G$. Define

$$\alpha'_1 : R_0(\Gamma, G) \bmod H \to F(\Gamma)$$

by $\alpha'_1(Hr) = ([w_\mu], k_\mu^{-1}t(a))$. If w_1, w_2 both induce r, let $w_{\mu_i} = t_i \circ w_i$ $(i = 1, 2)$. Clearly $[w_{\mu_1}] = [w_{\mu_2}]$ and so $k_{\mu_1} = k_{\mu_2}$ since these depend only on the class. Also $w_2^{-1}w_1$ induces the identity on Γ and so must fix R pointwise. Thus $t_2 = t_1$. Thus $k_{\mu_1}^{-1}t_1(a) = k_{\mu_2}^{-1}t_2(a)$. Now if $r = h \cdot s$ for some $h \in H$, and r is induced by w, then s is induced by $h^{-1}w$. Thus $\alpha'_1(Hs) = ([w_\mu], k_\mu^{-1}th(a))$ since $w_\mu = th \cdot h^{-1}w$. Thus α'_1 is well-defined and readily checked to be the inverse of α_1.

Also α_1 is continuous since t can be chosen continuously in G/H such that $t(a) = k_\mu(z)$.

Using the normalization given in [13], it is shown that α'_1 is continuous. Denote the repelling and attracting fixed points of a hyperbolic element γ by $p(\gamma), q(\gamma)$ respectively. For a pair of non-commuting hyperbolic elements γ_1, γ_2 of Γ, set

$$T(\gamma_1, \gamma_2, \Gamma) = \{r \in R_0(\Gamma, G) \mid p(r(\gamma_1)) = 0 \ \ p(r(\gamma_2)) = 1 \ \ q(r(\gamma_1)) = \infty\} .$$

For $r \in R_0(\Gamma, G)$, there exists a unique $\alpha(r) \in G$ such that $\alpha(r) \cdot r \in T(\gamma_1, \gamma_2, \Gamma)$ and $r \mapsto \alpha(r)$ is continuous [13].

Assume without loss of generality, that the elements γ_1, γ_2 are such that $p(\gamma_1) = 0$, $p(\gamma_2) = 1$, $q(\gamma_1) = \infty$. In that case, if w induces $r \in T(\gamma_1, \gamma_2, \Gamma)$, then w is normalized.

α'_1 is induced by the mapping $r \mapsto ([w_\mu], k_\mu^{-1} t(a))$. Now $r = \alpha(r)^{-1} \cdot s$ where $s \in T(\gamma_1, \gamma_2, \Gamma)$. Thus $r \mapsto ([w_\mu], k_\mu^{-1} \alpha(r)(a))$. The mapping $r \mapsto [w_\mu]$ is continuous, k_μ depends continuously on $[w_\mu]$ and $r \mapsto \alpha(r)$ is continuous. Thus since the mapping $R_0(\Gamma, G) \to R_0(\Gamma, G) \bmod H$ is open, α'_1 is continuous.

α_1 is then a homeomorphism such that (1) commutes.

Construction B

To obtain the topological description of $F(\Gamma)$ from the fiber bundle presentation of Teichmüller theory as in [7], [8], we first summarize those parts of that theory which are required.

Let $M(U) = C^\infty(U, \Delta)$, where Δ is the unit disc with the C^∞-topology, and $M(\Gamma)$ the subspace of Γ-compatible functions. Let $D(\Gamma)$ denote the normalizer of Γ in $D(U)$, the topological group of all diffeomorphisms of U with C^∞-topology, and $D_0(\Gamma)$, the centralizer of Γ in $D(\Gamma)$.

$D(\Gamma)$, and hence $D_0(\Gamma)$, act on the right of $M(\Gamma)$ as follows. If $\mu \in M(\Gamma)$, then $w_\mu \in D(U)$ and for $f \in D(\Gamma)$, $\mu \cdot f = \mu_1$, where μ_1 is the Beltrami coefficient of $w_\mu \circ f$. The quotient space $T_0(\Gamma)$ of $M(\Gamma)$ by $D_0(\Gamma)$, whose elements are written as $\mu D_0(\Gamma)$ is homeomorphic to $T(\Gamma)$ under the mapping $j_0 : T_0(\Gamma) \to T(\Gamma)$ given by $j_0(\mu D_0(\Gamma)) = [w_\mu]$.

Now $M(\Gamma) \to T_0(\Gamma)$ is a principal $D_0(\Gamma)$-bundle Φ. Further $D_0(\Gamma)$ is contractible and Φ is trivial being homeomorphic to the product principal $D_0(\Gamma)$ bundle [7], [8].

Form the associated bundle $\Phi[U]$ with fiber model U. Precisely, let $D_0(\Gamma)$ act on $M(\Gamma) \times U$ by $(\mu, t)f = (\mu \cdot f^{-1}, f(t))$ to form the quotient space $\Phi[U]$ whose elements we denote by $[(\mu, t)]$, with projection p_0 such that the diagram

$$\begin{array}{ccc} M(\Gamma) \times U & \longrightarrow & \Phi[U] \\ \downarrow & & \downarrow p_0 \\ M(\Gamma) & \longrightarrow & T_0(\Gamma) \end{array}$$

commutes.

THEOREM 2. *The fiber bundle* $p_0 : \Phi[U] \to T_0(\Gamma)$ *is isomorphic as a fiber space to* $p : F(\Gamma) \to T(\Gamma)$.

Proof. α_0 is constructed so that the following diagram commutes

$$\begin{array}{ccc} \Phi[U] & \xrightarrow{\alpha_0} & F(\Gamma) \\ p_0 \downarrow & & \downarrow p \\ T_0(\Gamma) & \xrightarrow{j_0} & T(\Gamma) \end{array} \tag{2}$$

$\alpha_0[(\mu, t)] = ([w_\mu], w^\mu(t))$. Since for $f \in D_0(\Gamma)$, f is the identity in $\mathbf{R}$, α_0 is well-defined.

Since Φ is trivial, there is a continuous mapping $\sigma : T_0(\Gamma) \to M(\Gamma)$. Define $\alpha'_0 : F(\Gamma) \to \Phi[U]$ by $\alpha'_0([w_\mu], z) = [(\mu_1, (w^{\mu_1})^{-1}(z))]$ where $\mu_1 = \sigma j_0^{-1}[w_\mu]$. α'_0 is continuous and the inverse of α_0. Since $D_0(\Gamma)$ is contractible, $\Phi[U]$ is trivial and it follows that α_0 is continuous.

REMARK. These spaces which have been constructed are homeomorphic to $T(\Gamma, G) \times {}^{G}\!/_{H}$ and $T_0(\Gamma) \times U$, so that it is not surprising that they are homeomorphic to $F(\Gamma)$ which is homeomorphic to $T(\Gamma) \times U$. However, the homeomorphisms are so constructed that they are compatible with the various natural actions of "modular groups" on these spaces, as is shown in the next section.

§3. *Modular groups*

With notation as in Section 2, let $N(\Gamma)$ denote the normalizer of Γ in Q.

DEFINITION (Bers). $\operatorname{mod}\Gamma = \{[v] : v \in N(\Gamma)\} \cong \dfrac{N(\Gamma)}{N(\Gamma) \cap Q_0}$. $\operatorname{mod}\Gamma$ acts on $F(\Gamma)$ as follows

$$[v]([w_\mu], z) = ([w_\nu], \hat{z})$$

where $w_\nu = \alpha w_\mu v^{-1}$ where $\alpha \in G$ and $\hat{z} = w^\nu v(w^\mu)^{-1}(z)$. Each element of $\operatorname{mod}\Gamma$ defines a homeomorphism (in fact, a holomorphic automorphism) of $F(\Gamma)$ and the group acts properly discontinuously on $F(\Gamma)$ [3].

$\operatorname{mod}\Gamma$ can be considered to act on $T(\Gamma)$. The normal subgroup Γ of $\operatorname{mod}\Gamma$ acts trivially on $T(\Gamma)$.

DEFINITION. $$\operatorname{Mod}\Gamma = \frac{\operatorname{mod}\Gamma}{\Gamma}.$$

$\operatorname{Mod}\Gamma$ is the Teichmüller modular group and so the quotient space of $T(\Gamma)$ by $\operatorname{Mod}\Gamma$ is the space of moduli $X(\Gamma)$. The quotient space $Y(\Gamma)$ of $F(\Gamma)$ by $\operatorname{mod}\Gamma$ is a fiber space over $X(\Gamma)$,

(3)
$$\begin{array}{ccc} F(\Gamma) & \xrightarrow{\ \operatorname{mod}\Gamma\ } & Y(\Gamma) \\ {\scriptstyle p}\downarrow & & \downarrow \\ T(\Gamma) & \xrightarrow{\ \operatorname{Mod}\Gamma\ } & X(\Gamma) \end{array}$$

The fiber over a point of $X(\Gamma)$ representing the class of a group Γ' is homeomorphic to $\dfrac{U}{N_G(\Gamma')}$ [3].

Construction A

On $R_0(\Gamma, G)$, the group $\operatorname{Aut}\Gamma$ of type- and orientation-preserving automorphisms acts in a natural way. Thus, for $r \in R_0(\Gamma, G)$, $\alpha \in \operatorname{Aut}\Gamma$,

$\phi \cdot r(\gamma) = r_0\phi(\gamma), \gamma \in \Gamma$. The actions of Aut Γ and G on $R_0(\Gamma, G)$ commute so that Aut Γ acts naturally on $R_0(\Gamma, G)$ mod H.

mod Γ is isomorphic to Aut Γ via the isomorphism $\theta_1([v]) = \alpha$ where $\alpha(\gamma) = v^{-1}\gamma v$. Since Γ is of the first kind it is an injection and it is a surjection by the Fenchel-Nielsen Theorem ([20], [16]).

THEOREM 3. *Under* θ_1 *and* α_1, *as defined in Theorem* 1, *the actions of* mod Γ *on* $F(\Gamma)$ *and* Aut Γ *on* $R_0(\Gamma, G)$ mod H *are compatible.*

Proof. It must be shown that $\alpha_1([v]([w_\mu], z)) = \theta_1[v]\alpha_1([w_\mu], z)$. Now $\theta_1[v]\alpha_1([w_\mu], z) = H(r_0\alpha)$ where $r_0\alpha(\gamma) = t^{-1}w_\mu v^{-1}\gamma v w_\mu^{-1}t$ where $t(a) = k_\mu(z)$. Also $\alpha_1([v]([w_\mu], z)) = Hs$ where $s(\gamma) = t_1^{-1}w_\nu \gamma w_\nu^{-1}t_1$ where $t_1(a) = k_\nu(\hat{z}) = w_\nu v(w^\mu)^{-1}(z) = aw_\mu(w^\mu)^{-1}z = ak_\mu(z)$. Replacing t above by $a^{-1}t_1$ then gives compatibility.

Construction B

$\frac{D(\Gamma)}{D_0(\Gamma)}$ acts on $\Phi[U]$ by $[(\mu, t)]\overline{f} = [(\mu \cdot f^{-1}, f(t)]$ for $\overline{f} \in \frac{D(\Gamma)}{D_0(\Gamma)}$.

Again the groups $\frac{D(\Gamma)}{D_0(\Gamma)}$ and $\frac{N(\Gamma)}{N(\Gamma) \cap Q_0}$ are isomorphic via $\theta_0(\overline{f}) = \tilde{f}$ where $\overline{f}, \tilde{f}$ denote the images of $f \in D(\Gamma)$ in these groups.

THEOREM 4. *Under* θ_0 *and* α_0, *as described in Theorem* 2, *the actions of* mod Γ *on* $F(\Gamma)$ *and* $\frac{D(\Gamma)}{D_0(\Gamma)}$ *on* $\Phi[U]$ *are compatible.*

Proof. Straightforward.

THEOREM 5. $\frac{R_0(\Gamma, G) \text{ mod } H}{\text{Aut } \Gamma}$ *and* $\frac{\Phi[U]}{D(\Gamma)/D_0(\Gamma)}$ *are homeomorphic to* $Y(\Gamma)$ *as fiber spaces over* $X(\Gamma)$.

§4. *Isomorphisms between modular groups*

In this section, the modular groups, regarded as Aut Γ and $\frac{\text{Aut } \Gamma}{\text{Inn } \Gamma}$, are represented as mapping class groups of corresponding surfaces and the isomorphism theorem between modular groups of [3] is generalized to the case where Γ has torsion.

Any finitely generated Fuchsian group of the first kind has a presentation of the form.

$$\text{(4)}\qquad \begin{aligned} &\text{Generators: } e_1, e_2, \ldots, e_r, p_1, \ldots, p_s, a_1, b_1, \ldots, a_g, b_g \\ &\text{Relations: } e_i^{m_i} = 1 \ (i = 1, 2, \ldots, r) \ \prod_{i=1}^{r} e_i \prod_{j=1}^{s} p_j \prod_{k=1}^{g} [a_k, b_k] = 1 . \end{aligned}$$

Γ is said to have signature $(g; m_i, \ldots, m_r; s)$, the m_i are the periods of Γ and s the number of parabolic classes, i.e., the number of conjugacy classes of maximal cyclic subgroups generated by parabolic elements. g is the genus of the quotient surface U/Γ.

U/Γ can be given the structure of a surface X of genus g with s punctures and r distinguished points. These points are partitioned into sets Σ_i, $i = 1, 2, \ldots, k$ where all points in one set Σ_i have the same ramification in the branched covering $U \to U/\Gamma$. Since the homeomorphism $f: X \to X$ permutes the punctures, let us consider these as an additional set Σ_{k+1} of distinguished points on the compact surface. Let $\mathcal{H}(g, \{\Sigma_i; i = 1, 2, \ldots, k+1\})$ be the group of all orientation-preserving homeomorphisms f of X such that $f(\Sigma_i) = \Sigma_i$, $i = 1, 2, \ldots, k$. Let $\mathcal{H}_0(g, \{\Sigma_i\}) = \{f \in \mathcal{H}(g, \{\Sigma_i\}) \mid f$ is homotopic to identity$\}$ and $M(g, \{\Sigma_i\})$ be the quotient.

Nielsen's theorem generalizes to give an isomorphism

$$\text{(5)}\qquad \lambda : M(g, \{\Sigma_i\}) \to \frac{\text{Aut } \Gamma}{\text{Inn } \Gamma}$$

defined as follows: $f \in \mathcal{H}(g, \{\Sigma_i\})$ can be lifted to a homeomorphism w of U which induces an automorphism of Γ by conjugation. Then $\lambda\{f\} = \bar{\phi}$ where $\{f\}$ denotes the homotopy class of f and $\bar{\phi}$ the inner automorphism class of ϕ [12].

These mapping class groups or 0'th homeotopy groups are first considered using results of Birman [4]. Using homotopy exact sequences, the following sequence is shown to be exact

$$
\begin{aligned}
(6) \qquad & 1 \to \Pi_1(X_g \setminus \{x_1, \ldots, x_n\}) \to M(g, \{x_i;\ i = 0, 1, \ldots, n\}) \\
& \to M(g, \{x_i;\ i = 1, 2, \ldots, n\}) \to 1
\end{aligned}
$$

where X_g denotes a compact surface of genus g and Π_1 the fundamental group, and the partition of points in the mapping class groups is into singletons.

For any partition $P = \{\Sigma_i;\ i = 1, 2, \ldots, k\}$ on n points, let $S(P)$ denote the subgroup of S_n, the symmetric group, which preserves P. The sequence

$$
(7) \qquad 1 \to M(g; \{x_i;\ i = 1, 2, \ldots, n\}) \to M(g; \{\Sigma_i;\ i = 1, 2, \ldots, k\}) \to S(P) \to 1
$$

is exact [4], [22]. Likewise, if a singleton set $\{x_0\}$ is adjoined to $\{\Sigma_i\}$, the sequence

$$
(8) \qquad 1 \to M(g; \{x_i;\ i = 0, 1, \ldots, n\}) \to M(g; \{x_0, \Sigma_i\}) \to S(P) \to 1
$$

is exact. From (6), (7), (8) we obtain the commutative diagram

$$
(9) \qquad
\begin{array}{ccccccccc}
 & & & & 1 & & 1 & & \\
 & & & & \downarrow & & \downarrow & & \\
1 & \to & \Pi_1(X_g \setminus \{x_1, \ldots, x_n\}) & \to & M(g; \{x_i; i = 0, 1, \ldots, n\}) & \to & M(g; \{x_i, i = 1, 2, \ldots, n\}) & \to & 1 \\
 & & & & \downarrow & & \downarrow & & \\
1 & \longrightarrow & K & \longrightarrow & M(g; \{x_0, \Sigma_i\}) & \longrightarrow & M(g, \{\Sigma_i\}) & \longrightarrow & 1 \\
 & & & & \downarrow & & \downarrow & & \\
 & & & & S(P) & \longrightarrow & S(P) & & \\
 & & & & \downarrow & & \downarrow & & \\
 & & & & 1 & & 1 & &
\end{array}
$$

Thus K, the kernel of $M(g; \{x_0, \Sigma_i\}) \to M(g; \{\Sigma_i\})$ is isomorphic to $\Pi_1(X_g \setminus \{x_1, \ldots, x_n\})$.

Let Γ have signature $(g; m_1, m_2, \ldots, m_r; s)$ and Γ_0 have signature $(g; - ; r+s)$. Then there is a natural epimorphism $p: \Gamma_0 \to \Gamma$. Let $K_0 =$ Kernel of p. Let the partition corresponding to U/Γ on $r+s$ points be denoted by $\{\Sigma_i : i = 1, 2, \ldots, k\}$. Note that Σ_k has s points.

THEOREM 6. $\mathrm{Aut}\,\Gamma \cong \dfrac{M(g; \{x_0, \Sigma_i\})}{K_0}$.

Proof. Delete all fixed points of elliptic elements in Γ from U to obtain U_0. The covering $U_0 \to \dfrac{U_0}{\Gamma}$ is smooth and has covering group Γ. Fix $x_0 \in \dfrac{U_0}{\Gamma}$. If Π denotes the fundamental group of $\dfrac{U_0}{\Gamma}$ based at x_0, then U_0 corresponds to a normal subgroup Π' of Π. Precisely, U_0 is homeomorphic to the space of path classes with initial point x_0 where two paths ℓ_1, ℓ_2 are in the same class if $\ell_1 \ell_2^{-1}$ is defined and determines an element of Π'. The projection $U_0 \to \dfrac{U_0}{\Gamma}$ maps the class $[\ell_1]$ of ℓ_1 onto its end point. Now if $q: \Pi \to \Gamma$ with kernel Π' then $q([\sigma])[\ell] = [\sigma \cdot \ell]$ where σ is a loop on $\dfrac{U_0}{\Gamma}$ based at x_0 and $[\sigma]$ denotes the corresponding element of the fundamental group.

Consider the following diagram

$$\begin{array}{ccccccccc} 1 & \longrightarrow & \Gamma_0 & \xrightarrow{d} & M(g;\{x_0, \Sigma_i\}) & \xrightarrow{\beta} & M(g;\{\Sigma_i\}) & \longrightarrow & 1 \\ & & \downarrow{\scriptstyle p} & & \downarrow{\scriptstyle \mu} & & \downarrow{\scriptstyle \lambda} & & \\ 1 & \longrightarrow & \Gamma & \longrightarrow & \mathrm{Aut}\,\Gamma & \longrightarrow & \dfrac{\mathrm{Aut}\,\Gamma}{\mathrm{Inn}\,\Gamma} & \longrightarrow & 1 \end{array} \tag{10}$$

The top row is exact by (9), the map β being the obvious projection and d being defined via the homotopy exact sequence. λ is the isomorphism defined at (5). It remains to define μ so that the diagram commutes.

Let $[f] \in M(g:\{x_0, \Sigma_i\})$. Thus f fixes x_0 and maps Π' into itself. Representing U_0 as the space of path classes, define $w: U_0 \to U_0$ by $w[\ell] = [f(\ell)]$ and extend w to U. Then w is a homeomorphism of U

which induces f. Define $\mu[f] = \alpha$ where $\alpha(\gamma) = w^{-1}\gamma w$. Then $\alpha \in \mathrm{Aut}\,\Gamma$ (see [12]). If [f] is the identity, then f is isotopic to the identity by an isotopy fixing the base point [9]. Thus, in that case, $w^{-1}\gamma w[\ell] = w^{-1}\gamma[f(\ell)] = w^{-1}[\sigma \cdot f(\ell)]$ (where $q[\sigma] = \gamma$) $= [f^{-1}(\sigma \cdot f(\ell))] = [\sigma \cdot \ell]$ since f is isotopic to the identity as above. Thus $w^{-1}\gamma w[\ell] = \gamma[\ell]$, and μ is well-defined.

One checks immediately that the right hand square of (10) commutes. Chasing through the definition of d in [4], Birman shows that, if σ is a simple closed loop on $\frac{U_0}{\Gamma}$ then $d[\sigma] = \xi_\sigma$, the back and forth twist about σ. Checking on generators of Γ_0 gives that the left hand square commutes also.

Thus μ is a surjection since λ and p are, and since λ is an isomorphism, the kernel of μ is isomorphic to the kernel of p.

THEOREM 7. $M(g;\{x_0, \Sigma_i, i = 1,2,\dots,k\})$ *is a subgroup of index* $s+1$ *in* $M(g;\{\Sigma_i, i = 1,2,\dots,k-1, \Sigma'_k\}) \cong \frac{\mathrm{Aut}\,\dot\Gamma}{\mathrm{Inn}\,\dot\Gamma}$ *where* $\dot\Gamma$ *has signature* $(g; m_1,\dots,m_r; s+1)$, Σ'_k *has* $s+1$ *members.*

Proof. Call the two mapping class groups M, M' and the partitions P, P'. Clearly $M \subseteq M'$ and both contain $M(g;\{x_i, i = 0,1,\dots,r+s\})$ as a subgroup of indices $|S(P)|$ and $|S(P')|$ respectively, by (7).

COROLLARY 8. *If* Γ *has signature* $(g; -; s)$ *and* $\dot\Gamma$ *is a uniformizing group of* $U/\Gamma \setminus \{x_0\}$, *then* $\mathrm{Aut}\,\Gamma$ *is isomorphic to a subgroup of index* $s+1$ *in* $\frac{\mathrm{Aut}\,\dot\Gamma}{\mathrm{Inn}\,\dot\Gamma}$.

COROLLARY 9. *The projection* $p:\Gamma_0 \to \Gamma$ *induces an isomorphism from* $\mathrm{Aut}\,\Gamma$ *to* $\frac{\mathrm{Aut}_1\,\Gamma_0}{K_0}$ *where* $\mathrm{Aut}_1\,\Gamma_0$ *is a subgroup of finite index in* $\mathrm{Aut}\,\Gamma_0$.

Proof. $\mathrm{Aut}\,\Gamma \cong M(g;\{x_0, \Sigma\})$ where Σ has $r+s$ members. Let $\mathrm{Aut}_1\,\Gamma_0$ be the subgroup corresponding, under this isomorphism, to

$M(g; \{x_0, \Sigma_i, i = 1, 2, \ldots, k\})$. This subgroup has index $\frac{|S_{r+s}|}{|S(P)|}$ in $\mathrm{Aut}\,\Gamma_0$. The isomorphism of Theorem 6 is then simply that one induced by p.

§5. *Applications of the isomorphism theorem*

I. Let Γ have signature $(g; - ; s)$ where $s \geq 1$. It has been proved in the case $g = 0$, that $\frac{\mathrm{Aut}\,\Gamma}{\mathrm{Inn}\,\Gamma}$ is residually finite [19] and in fact, this result is now known for all finitely-generated torsion-free Fuchsian groups (see remarks and references in [5]). Here we offer an alternative proof for the free groups defined above.

THEOREM 10. *If* Γ *has signature* $(g; - ; s)$ *where* $s \geq 1$, *then* $\frac{\mathrm{Aut}\,\Gamma}{\mathrm{Inn}\,\Gamma}$ *is residually finite.*

Proof. Let Γ_1 have signature $(g; - ; s-1)$. By Corollary 8, $\mathrm{Aut}\,\Gamma_1$ is isomorphic to a subgroup of index s in $\frac{\mathrm{Aut}\,\Gamma}{\mathrm{Inn}\,\Gamma}$. Now Γ_1 is residually finite, and as a subgroup of the full automorphism group of Γ_1, $\mathrm{Aut}\,\Gamma_1$ is residually finite [2]. Since $\frac{\mathrm{Aut}\,\Gamma}{\mathrm{Inn}\,\Gamma}$ is isomorphic to the mapping class group of U/Γ it is finitely generated [4]; but any finitely-generated group which contains a residually finite subgroup of finite index, is itself, residually finite.

II. This concerns the short exact sequence

$$1 \to Z \to \pi \to \mathrm{Aut}^+\,\Gamma \to 1 \tag{11}$$

discussed in [13]. (Since we have been considering only orientation-preserving homeomorphisms throughout, our $\mathrm{Aut}\,\Gamma$ is automatically the $\mathrm{Aut}^+\,\Gamma$ of [13].) $\mathrm{Aut}\,\Gamma$ acts properly discontinuously without fixed points on $R_0(\Gamma, G)$, the quotient being $S_0(\Gamma, G)$ which is a $K(\pi, 1)$-space, with fundamental group π. Since $R_0(\Gamma, G)$ is homeomorphic to $G \times T(\Gamma, G)$, it has fundamental group Z, and the sequence (11) is obtained.

THEOREM 11. *If* Γ *is torsion-free, all finite subgroups of* Aut Γ *are cyclic.*

Proof. Let Γ have signature $(g; - ; s)$. Let Γ_1 have signature $(g; p, p, \ldots, p; 1)$ where the period p is repeated s times. By Theorem 6 and the Nielsen isomorphism, Aut Γ is isomorphic to $\dfrac{\mathrm{Aut}\,\Gamma_1}{\mathrm{Inn}\,\Gamma_1}$.

Let A be a finite subgroup of $\dfrac{\mathrm{Aut}\,\Gamma_1}{\mathrm{Inn}\,\Gamma_1}$ which is not cyclic. Then A contains a soluble subgroup B which is not cyclic, for if A is simple, take the subgroup generated by two distinct elements of order 2, which must exist by the Feit-Thompson theorem. Using this, one can always obtain such a subgroup B. Using Smith's fixed point theorem (as in [11], [14]), B has a fixed point Gr in $T(\Gamma_1, G)$. Thus there exists a Fuchsian group Γ_2 such that $r(\Gamma_1)$ is normal in Γ_2 and $\dfrac{\Gamma_2}{r(\Gamma_1)} \cong B$.

Let Γ_2 have signature $(\gamma; m_1, \ldots, m_t; k)$ where the parabolic classes are generated by $p_1, p_2, \ldots, p_k$ as at (4). Let p_i have exponent μ_i mod $r(\Gamma_1)$. Thus the number of parabolic classes in $r(\Gamma_1) = o(B) \sum_{i=1}^{k} \frac{1}{\mu_i} = 1$. Clearly $\mu_i | o(B)$. Thus the only possibility is $k = 1$, $\mu_1 = o(B)$ whence $\dfrac{\Gamma_2}{r(\Gamma_1)}$ is cyclic. This contradiction gives the result.

It is shown in [15] that Z is central in π. Then the above result can be deduced for general Γ; i.e., with periods.[1] The fundamental group π is torsion-free [6]. Also for trivial finite group B actions on Z, i.e., central extensions, the group $H^2(B, Z) \cong {}^{B}/_{B'}$ (see e.g., [12]). Calculating the actual extensions one finds that the only torsion-free ones occur when B is cyclic.

---

1 I am grateful to Hugh Morton of Liverpool for this proof.

THEOREM 12. *The extension* (11) *does not split.*

Proof. Since π is torsion-free, it suffices to show that Aut Γ contains elements of finite order. If Γ has torsion, then the inner automorphism by that element gives the result. Thus suppose Γ is torsion-free. Using the representation of Aut Γ as $\frac{\text{Aut }\Gamma_1}{\text{Inn }\Gamma_1}$ as in Theorem 11, one can readily verify that Aut Γ has elements of finite order. Since element of finite order correspond to groups such as Γ_2 in Theorem 11, this is also a consequence of the results in [18]. Furthermore, it will emerge as a consequence of the next result which shows that, in fact, Aut Γ is generated by elements of finite order.

III. If Γ is torsion-free, it is shown that $Y(\Gamma)$ is usually simply-connected. It is known that $X(\Gamma)$ is usually simply-connected, [17], [21], and the method of proof here is the same.

The effective and properly-discontinuous action of $\frac{\text{Aut }\Gamma}{\text{Inn }\Gamma}$ on $T(\Gamma, G)$ in almost all cases can be used to show that Aut Γ acts effectively and properly discontinuously on $R_0(\Gamma, G)$ mod H [3]. Since $R_0(\Gamma, G)$ mod H is simply-connected, the fundamental group of $Y(\Gamma) = \frac{R_0(\Gamma, G) \text{ mod } H}{\text{Aut }\Gamma}$ is isomorphic to the quotient of Aut Γ by the subgroup generated by those elements which have fixed points [1]. But an element $\alpha \in$ Aut Γ has a fixed point if and only if it has finite order. For if $\alpha \in$ Aut Γ has a fixed point, then $\overline{\alpha} \in \frac{\text{Aut }\Gamma}{\text{Inn }\Gamma}$ has a fixed point in $T(\Gamma, G)$ and so $\overline{\alpha}$ has finite order n. Thus α^n is an inner automorphism by some element $\gamma_1 \in \Gamma$, which necessarily has a fixed point in $R_0(\Gamma, G)$ mod H. Thus there exists $r \in R_0(\Gamma, G)$, $h \in H$ such that $hr(\gamma)h^{-1} = \gamma_1 r(\gamma)\gamma_1^{-1}$, for all $\gamma \in \Gamma$. But $r(\Gamma)$ has trivial center and so $\gamma_1 \in H$, and so fixes $a \in U$. But Γ is torsion-free, thus $\gamma_1 = \text{Id}$ and so $\alpha^n = \text{Id}$. Conversely, if α has finite order n, then $\overline{\alpha}$ fixes, say, $Gr \in T(\Gamma, G)$. Thus there exists $g \in G$ such that $r_0\alpha(\gamma) = gr(\gamma)g^{-1}$, $\gamma \in \Gamma$. Then $r_0\alpha^n(\gamma) = g^n r(\gamma) g^{-n} = r(\gamma)$. Thus g is elliptic and has a fixed point b. Let $g' \in G$ be such that $g'(a) = b$. Define $r': \Gamma \to G$ by $r'(\gamma) = g'^{-1} r(\gamma) g'$. Then $\alpha(Hr') = Hr'$.

It remains to show that Aut Γ is generated by elements of finite order. Now Aut Γ is isomorphic to $M(g; \{x_0, \Sigma\})$ which is a subgroup of index $s+1$ in the mapping class group of the $s+1$ punctured surface of genus g, $M(g; \Sigma')$. Denote $M(g, \Sigma')$ by $M(g, s+1)$ and $M(g; \{x_0, \Sigma\})$ by $M_1(g, s+1)$. Let $F(g, s+1)$ and $F_1(g, s+1)$ denote the respective subgroups which are generated by elements of finite order. The method of proof in [21] can be adapted almost verbatim as we only require the additional restriction that, throughout, one point be kept fixed.

Using the generators of [4] for $M(g, s+1)$ one immediately obtains generators for $M_1(g, s+1)$. For $g \geq 1$, Patterson proves $\frac{M(g,s+1)}{F(g,s+1)}$ is cyclic generated by the common class of twists and since at each stage the relevant homeomorphisms have, or can be made to have a fixed point, the result is also true for $\frac{M_1(g,s+1)}{F_1(g,s+1)}$. For $g = 0$, $M_1(0, s+1)$ can be generated by $\sigma_1, \ldots, \sigma_{s-1}, \sigma_s^2, \sigma_s \sigma_{s-1}^2 \sigma_s^{-1}, \ldots, \sigma_s \sigma_{s-1} \cdots \sigma_3 \sigma_2^2 \sigma_3^{-1} \cdots \sigma_s^{-1}$ [10]. But $\sigma_1 \ldots \sigma_{s-1}$ and $\sigma_1^2 \sigma_2 \ldots \sigma_{s-1}$ have finite order. Thus $\sigma_1 \in F_1(0, s+1)$. Using the relations between the generators one deduces that $\sigma_2, \ldots, \sigma_{s-1}$, $\sigma_s^2 \in F_1(0, s+1)$ and so that $M_1(0, s+1) = F_1(0, s+1)$.

Precisely as in [21], the cyclic group $\frac{M_1(g,s+1)}{F_1(g,s+1)}$ is finite and its order divides 12 if $g = 1$, 10 if $g = 2$ and 2 if $g \geq 3$. If $g = 1$, then provided $s \neq 1$, the α and β of [21] can be considered as elements of $M_1(1, s+1)$ and so $M_1(1, s+1) = F_1(1, s+1)$. When $s = 1$, α^2 and β give the same result. If $g = 2$, α^2 and β^3 can be considered as elements of $M_1(2, s+1)$ if and only if $s+1 \not\equiv 0, 4 \pmod 5$, and in these cases $M_1(2, s+1) = F_1(2, s+1)$. In the cases $s+1 \equiv 0, 4 \pmod 5$ the same argument as in [21] gives $\frac{M_1(2,s+1)}{F_1(2,s+1)} \cong Z_5$. If $g \geq 3$, the elements β, γ both have two fixed points and, as in [21], $M_1(g, s+1) = F_1(g, s+1)$.

THEOREM 13. *If* Γ *has signature* $(g; -; s)$ *then* $Y(\Gamma)$ *has trivial fundamental group in all cases except for* $g = 2$, $s \equiv 3, 4 \pmod 5$ *when the fundamental group is* Z_5.

DEPARTMENT OF MATHEMATICS
UNIVERSITY OF ABERDEEN
KING'S COLLEGE
ABERDEEN SCOTLAND

REFERENCES

[1]. Armstrong, M. A., "The fundamental group of the orbit space of a discontinuous group." Proc. Camb. Phil. Soc. 64 (1968), pp. 299-301.

[2]. Baumslag, G., "Automorphism groups of residually finite groups." Jour. London Math. Soc. 38 (1963), pp. 117-118.

[3]. Bers, L., "Fiber spaces over Teichmüller spaces." Acta Math. 130 (1973), pp. 89-126.

[4]. Birman, J. S., "Mapping class groups and their relationship to Braid Groups." Comm. on Pure and Applied Math. 22 (1969), pp. 213-238.

[5]. Birman, J. S., "Mapping class groups of surfaces." These Proceedings.

[6]. Conner, P. E., and Montgomery, D., "Transformation groups on a $K(\pi,1)$." Michigan Math. Journal 6 (1959), pp. 405-417.

[7]. Earle, C. J., and Eells, J., "A fibre bundle description of Teichmüller theory." Jour. of Diff. Geom. 3 (1969), pp. 19-43.

[8]. Earle, C. J., "On holomorphic families of pointed Riemann surfaces." Bull. Amer. Math. Soc. 79 (1973), pp. 163-166.

[9]. Epstein, D. B. A., "Curves on 2-manifolds and isotopies." Acta Math. 115 (1966), pp. 83-107.

[10]. Gillette, R., and Van Buskirk, J., "The word problem and consequences for the braid groups and mapping-class groups of the 2-sphere." Trans. Amer. Math. Soc. 131 (1968), pp. 277-296.

[11]. Kravetz, S., "On the geometry of Teichmüller spaces and the structure of their modular groups." Ann. Acad. Sci. Fenn. Series A. No. 278 (1959).

[12]. Learner, A., "Cohomology of groups." Queen Mary College Notes, London 1965.

[13]. Macbeath, A. M., and Singerman, D., "Spaces of subgroups and Teichmüller space" to appear.

[14]. Macbeath, A. M., "On a theorem of J. Nielsen." Quart. Jour. Maths. 13 (1962), pp. 235-236.

[15]. Macbeath, A. M., "The fundamental groups of certain subgroup spaces." These proceedings.

[16]. Maclachlan, C., and Harvey, W. J., "On mapping-class groups and Teichmüller spaces" to appear.

[17]. Maclachlan, C., "Modulus space is simply-connected." Proc. Amer. Math. Soc. 29 (1971), pp. 85-86.

[18]. Maclachlan, C., "Maximal normal Fuchsian groups." Illinois Jour. of Maths. 15 (1971), pp. 104-113.

[19]. Maclachlan, C., "On a conjecture of Magnus on the Hurwitz Monodromy group," Math Z. 132 (1973), pp. 45-50.

[20]. Maskit, B., "On boundaries of Teichmüller spaces and on Kleinian groups II". Ann. Math. 91 (1970), pp. 607-639.

[21]. Patterson, D. B., "The fundamental group of the moduli space" to appear.

[22]. Scott, G. P., "Braid groups and groups of homeomorphisms of surfaces." Proc. Camb. Phil. Soc. 68 (1970), pp. 605-617.

UNIVERSAL PROPERTIES OF FUCHSIAN GROUPS IN THE POINCARÉ METRIC

Albert Marden*

§1. *Introduction*

We are going to deal here with arbitrary Fuchsian groups Γ acting in the unit disk Δ. That is with groups that may not be finitely generated and may have elliptic elements. We can form the Poincaré fundamental polygon with center at $z = 0$,

$$P(\Gamma) = \{z \in \Delta : d(z, 0) \leq d(z, T(0)), \quad \text{for all} \quad T \in \Gamma\}$$

for those Γ for which $z = 0$ is not a fixed point (and this will be assumed to be true of Γ when we form it). Here $d(.\,,.)$ is the hyperbolic distance in Δ determined by the Poincaré metric $|dz|/(1-|z|^2)$. This will be the only metric used in this paper.

We will also use the notation $D(p; r)$ for the (non-Euclidean) disk of radius r with center at p, and for a given Fuchsian group Γ,

$$\mathcal{I}(p; r) = \{T \in \Gamma : T(D(p; r)) \cap D(p; r) \neq \phi\} ,$$

$$\mathcal{G}(p; r) = \text{subgroup of } \Gamma \text{ generated by } \mathcal{I}(p; r) .$$

The purpose of this paper is to prove the following results.

THEOREM 1. (a) *There exists* $r > 0$ *with the following property. Given any point* $p \in \Delta$ *and any Fuchsian group* Γ, *either the subgroup* $\mathcal{G}(p; r)$ *is cyclic, or for some* $E, F \in \mathcal{G}(p; r)$,

* Supported in part by the National Science Foundation.

(1) $$\mathcal{G}(p;r) = \langle E, F;\ E^2 = F^2 = \text{id.}\rangle\ .$$

(b) *Given any* $R < \infty$ *there exists* $r = r(R) > 0$ *with the following property. Suppose* Γ *is any Fuchsian group such that for some* $p \in D(0;R)$, $\mathcal{G}(p;r)$ *contains an element of infinite order. Then either* $\mathcal{G}(0;R) = \mathcal{G}(p;r)$ *or* $\mathcal{G}(0;R)$ *is of the form* (1) *and* $\mathcal{G}(p;r)$ *is a cyclic subgroup.*

COROLLARY 1. *There exists* $r > 0$ *with the following property. Given any* $p \in \Delta$ *and any group* Γ, *if* $D(p;r)$ *contains more than one fixed point of* Γ *then the set of fixed points* $\mathcal{F}$ *of* Γ *in* $D(p;r)$ *arises as follows. There are elliptic transformations* $E, F \in \Gamma$ *of order* 2, *whose fixed points lie in* $\mathcal{F}$, *such that* $\mathcal{F}$ *is contained in the set of fixed points of the subgroup* $\langle E, F;\ E^2 = F^2 = \text{id.}\rangle$.

We remark that the possibility suggested in Corollary 1 that $D(p;r)$ contains two or more fixed points actually occurs. For an example consider the group generated by E_n and F_n,

$$E_n(z) = [-z(1+n^2) + 2n]/[(1+n^2) - 2nz]\ ,$$

$$F_n(z) = [-z(1+n^2) - 2n]/[(1+n^2) + 2nz]\ ,$$

which are of order 2 with fixed points $1/n$ and $-1/n$ respectively.

THEOREM 2. *There exists* $r > 0$ *with the following property. Given any Fuchsian group* Γ *there exists a Möbius transformation* T *such that the Poincaré polygon* $P(T\Gamma T^{-1})$ *contains the disk* $\{z : d(0,z) < r\}$.

These results have their roots in a well-known theorem in general Lie group theory due to Kazhdan and Margulis and to a considerable strenthening of that result by H-C. Wang. We refer to Wang's paper [9] for a clear discussion of these matters. The theorems above include these general results in the context of Fuchsian groups (discrete subgroups of $SL(2, R)/\pm 1$) but they are also more than that. For in the particular situation here one is able to obtain much more detailed information than has been done in general. Even if the general Lie group results had been assumed, I do not believe this paper could have been substantially shortened.

Theorem 2 is not new. The whole question originated with work of W. Harvey. He first noted that for groups Γ without elliptic elements, Theorem 2 and its interpretation in terms of disks on the quotient surface Δ/Γ is true as a direct consequence of the Kazhdan-Margulis result. Harvey then went on and made very important use of it in forming a compactification of Riemann (moduli) space. In addition he announced a geometric proof for certain cases. More recently, Sturm and Shinnar [8] found a very elementary proof of Theorem 2. They derived it from the observation that any non-Euclidean triangle of area $> \varepsilon$ contains a disk of radius $\delta(\varepsilon)$ depending only on ε. In contrast, we will deduce it from Corollary 1.

Our proofs of the theorems will be based on a fundamental lemma of Chuckrow for Schottky groups which has to be considerably elaborated on in Sections 2-5. In addition we will make use of a recent important and deep result of Purzitsky concerning the classification of 2-generator Fuchsian groups. If we only considered groups without elliptic elements, what would remain of the theorems could actually be proven very easily and quickly (direct consequences of Lemmas 2 and 12). Indeed, practically all of our effort is spent in trying to locate in a "canonical" manner a torsion-free two-generator subgroup of any given non-elementary two-generator Fuchsian group.

§2. *Two generator groups*

We will always assume that our Fuchsian groups act on the unit disk Δ. The notation $b(G)$ will be used to designate the set of points on the surface Δ/G over which Δ is branched. If G has N generators then the Bers area theorem [1] gives the inequality

$$2g + \Sigma(1 - \ell(p)^{-1}) \leq N + 1 \tag{1}$$

where g is the genus of Δ/G and the sum is over those points p such that (a) $p \in b(G)$ and Δ is branched of order $\ell(p)$ over p, $2 \leq \ell(p) < \infty$, and (b) p is an ideal boundary component of Δ/G and $\ell(p) = \infty$. This inequality is not best possible but when $b(G) \neq \phi$ in general nothing better is known.

We will be especially concerned here with two generator non-cyclic Fuchsian groups which we shall denote as $G = \langle U, V \rangle$ where U and V are generators.

For two generator groups G if the genus of Δ/G is positive then it must be one. Furthermore from (1), $b(G) = \phi$ if Δ/G is not compact otherwise card $b(G) = 1$ except perhaps for the case that card $b(G) = 2$ and Δ is branched of order 2 over each point of $b(G)$. However this latter possibility can be excluded by elementary homology considerations. That is, if Δ/G is a torus then $b(G)$ contains exactly one point.

On the other hand if Δ/G has genus zero the situation is much harder to analyze, particularly if Δ/G is compact, i.e., is the sphere. In this case we will appeal to the important results of Purzitsky [7]. He has shown that card $b(G) \leq 4$ with equality occurring only in the case that Δ/G is a sphere and Δ is branched of order 2 over three of the points of $b(G)$ and of odd order over the remaining one. In the remaining cases, card $b(G) = 3-n$ where n is the number of ideal boundary points of Δ/G. Inequality (1) of course only gives that card $b(G) \leq 6-2n$.

LEMMA 1. *Suppose* $G = \langle S, T \rangle$ *has no elliptic elements. Then there exists a sequence of Schottky groups* $G_k = \langle S_k, T_k \rangle$ *such that* $\lim S_k = S$, $\lim T_k = T$.

Proof. If G contains no parabolic elements it is already a Schottky group. So assume that it does. In the course of the proof we are going to have to refer extensively to other papers, in particular to the notion of "pairing the punctures by pairing cylinders" in [6] and to the technique of obtaining convergent sequences of groups by shrinking the pairing cylinders to punctures which is developed in [3]. But by doing so we can dispense with a long complicated discussion and present a simplified outline as follows.

The surface Δ/G is either a once punctured annulus, a once punctured torus, a twice punctured disk, or a triply punctured sphere. Let Ω be the regular set of G with respect to its action in the extended plane. Then the punctures on the one or two components of Ω/G can be paired by pairing cylinders to obtain a closed surface R_k of genus 2. The index k is an indication of the size of the pairing cylinders so that $R_k \to \Omega/G$ as $k \to \infty$. By Chuckrow's results [2] the generators S, T of G determine mutually disjoint simple loops α, β in R_k so that S and T generate the cover transformations in the "Schottky covering surface" correspond- to α, β formed from Ω by adjoining lifts of the pairing cylinders. But now actually construct the Schottky covering surface of R_k determined by α and β and realize this in the plane. The group of cover transformations is a Schottky group G_k. After a suitable normalization of G_k by conjugating if necessary, we can express $G_k = \langle S_k, T_k \rangle$ where $S_k \to S$ and $T_k \to T$.

LEMMA 2. *Suppose* $G_n = \langle S_n, T_n \rangle$, $n = 1, 2, \ldots$, *is a sequence of Fuchsian groups without elliptic transformations. Then if* $S = \lim S_n$ *and* $T = \lim T_n$ *exist as Möbius transformations, neither* S *nor* T *is the identity.*

Proof. If no G_n contains parabolic transformations then G_n is a Schottky group if we allow it to act on the extended plane. In this case Lemma 2 was proved by Chuckrow [2].

Next suppose each G_n contains parabolic transformations. By Lemma 1, G_n can be approximated by a Schottky group $G_{nk} = \langle S_{nk}, T_{nk} \rangle$ where S_{nk}, T_{nk} are arbitrarily close to S_n, T_n respectively. But now Chuckrow's result can be applied to a suitable diagonal sequence $\{G_{nk_n}\}$.

§3. *Two generator groups* G *with elliptic elements such that* Δ/G *is not the sphere*

In the remainder of this paper we will use the following terminology. An element $T \in G$ is *determined by the loop* $\gamma \subset \Delta/G - b(G)$ if for some

$O^* \epsilon \Delta$ over the initial point O of γ, the lift of γ from O^* terminates at $T(O^*)$. In each case geometric assumptions are made in order to simplify the group theory. Note that by Purzitsky's results [7], the condition card $b(G) \leq 2$ is equivalent to the condition that Δ/G is not the sphere.

LEMMA 3. *Suppose* $G = \langle S, T\rangle$ *contains elliptic transformations yet* S *and* T *are not elliptic but are determined by lifting from a point* $O^* \epsilon \Delta$ *two simple loops* α, β *from a point* $O \epsilon R' = \Delta/G - b(G)$. *Then if* Δ/G *is not a torus and* card $b(G) \leq 2$, *the subgroup* $H = \langle S^3, T^2\rangle$ *does not contain elliptic transformations.*

Proof. Since each loop α, β is retractible in R' to a boundary component, it follows from the hypothesis $R = \Delta/G$ is a doubly connected plane region and Δ is branched over one point $b \epsilon R$. The topological situation is as indicated in Figure 1.

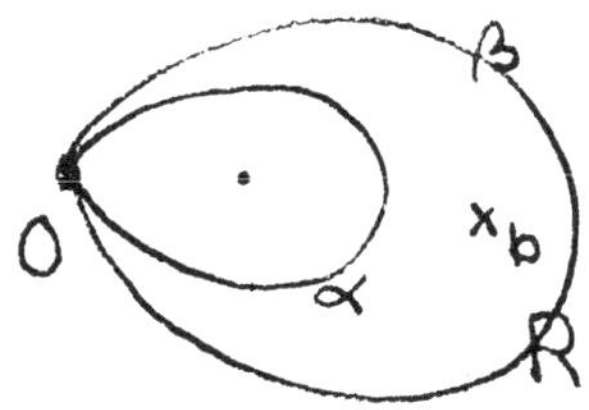

Fig. 1

Now $R^* = \Delta/H$ is a possibly branched (over b) covering surface of R determined by the requirement that for some $O^* \epsilon R^*$, the lifts α^* of α^3 and β^* of β^2 are closed curves. If R^* is triply connected we are finished because Δ is then unbranched over R^*. Otherwise R^* is doubly connected and Δ is branched over one point $b^* \epsilon R^*$ (Figure 2).

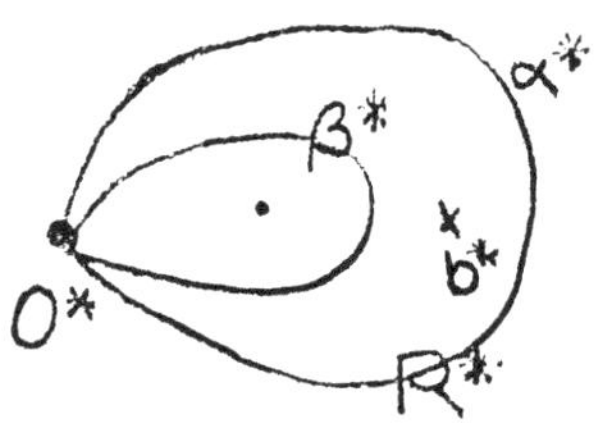

Fig. 2

Each of α^* and β^* are simple loops from O^* which are mutually disjoint except at O^* since α doesn't "cross" β in R. There is a relatively compact region D^* in R^* bounded by $\alpha^* \cup \beta^*$ and containing

b^*. No lift of α^3 or β^2 can cross ∂D^* and furthermore no lift can lie in D^*. For such a lift would either bound a disk in $D^* - b^*$ or be retractible to b^* which would imply that S^3 or T^2 is the identity or elliptic. Thus the projection of D^* to R is exactly that region D containing b with $\partial D = \alpha \cup \beta$. But except perhaps for b, D^* covers each point of D the same number of times [5, Lemma 2]. This is impossible since α is covered three times but β only twice.

LEMMA 4. *Suppose* $G = \langle S, T \rangle$ *is a Fuchsian group such that* S *and* T *are each determined by lifting a simple loop on* $R = \Delta/G$. *Assume also that* G *is not generated by two elliptic transformations of order* 2, card $b(G) \leq 2$, *and that* Δ/G *is not a torus. Then at least one of the four subgroups*

$$\langle (ST)^3, (TS)^2 \rangle, \langle T^3, ST^2S^{-1} \rangle ,$$

and the groups with T *replaced by* T^{-1} *above, contain no elliptic elements.*

Proof. We may assume of course that G contains elliptic transformations so that Δ is branched over a set b(G) of one or two points of R. Consider as usual the triply connected region $R' = R - b(G)$ and fix $O \in R'$. Let α, β be simple loops from O, mutually disjoint except at O, such that (i) one of α, β is retractible to each point of b(G), and (ii) $\alpha\beta$ is homotopic to a simple loop. Now S and T are induced by lifts of α, β, $\alpha\beta$ or their inverses since up to homotopy these are the only simple loops. We can assume that a lift of α induces S. Then T is induced by β, $\alpha\beta$ on their inverses. Replacing T by T^{-1} if necessary we may assume that T is induced by β or $\alpha\beta$ (this change gives rise to the two possibilities of Lemma 4). There is a

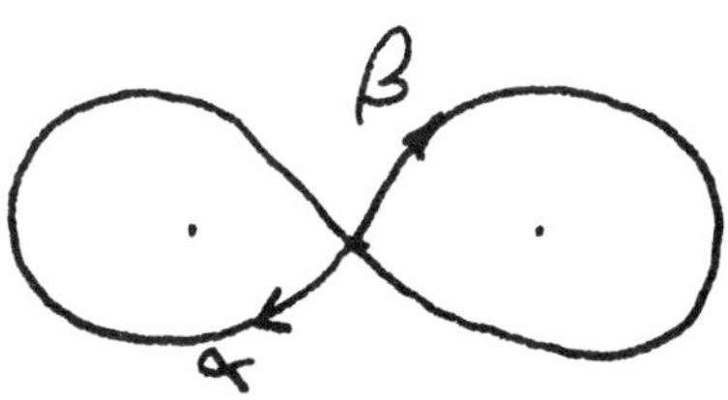

Fig. 3

point $O^* \epsilon \Delta$ over O and a loop $\gamma \subset R'$ from O such that the lift of α from O^* induces S and $\gamma\beta\gamma^{-1}$ or $\gamma\alpha\beta\gamma^{-1}$ from O^* induces T.

Let $T_o \epsilon G$ be determined by lifting β from O^*. For some k we claim that

$$T = S^k T_o S^{-k} \quad \text{or} \quad T = S^k(T_o S)S^{-k}$$

depending on whether T is induced by β or by $\alpha\beta$.

The proof is based on the fact that S and T generate G and so do S and T_o. In terms of S and T_o, G has a standard presentation; for example if there are two branch points, $G = \langle S, T_o; S^p = T_o{}^q = 1\rangle$. For some reduced word W, in terms of S and T_o, either $T = WT_oW^{-1}$ or $T = WT_oSW^{-1}$. By replacing T by some $S^{-k}TS^k$ we may assume that the first letter of W is not $S^{\pm 1}$. Now write T_o as a reduced word in T and S and then in WT_oW^{-1} and S or WT_oSW^{-1} and S. Because no further cancellation can occur we see that in neither case is it possible to have $W \neq 1$.

If T is induced by β set

$$U = ST, \quad V = TS$$

and if T is induced by $\alpha\beta$ set

$$U = T, \quad V = STS^{-1} .$$

In either case U and V are induced by the simple loop $\alpha\beta$ and hence are not elliptic transformations [5, Lemma 3]. The assumption in Lemma 4 made on G insures that the subgroup

$$H = \langle U, V\rangle$$

is indeed non-cyclic [5, Lemma 4]. The argument made above implies that H is conjugate in G to the subgroup $H_o = \langle U_o, V_o\rangle$ where U_o and V_o are the transformations determined by lifting the essentially simple loops $\alpha\beta$ and $\beta\alpha$ from O^*. In fact $H = S^k H_o S^{-k}$.

Consider the sequence of branched covering surfaces

$$\Delta \to \Delta/H_o \to \Delta/G .$$

There is a point $O_1 \in \Delta/H_o$ over $O \in \Delta/G = R$ such that the lifts of $\alpha\beta$ and $\beta\alpha$ to Δ/H_o from O_1 are simple loops which are mutually disjoint except at O_1. (This is true because the two loops $\alpha\beta$, $\beta\alpha$ do not "cross".) In other words either H_o contains no elliptic transformations or it fulfills exactly the conditions of Lemma 3. In either case the assertion of Lemma 4 is true.

LEMMA 5. *Suppose for* $G = \langle S, T\rangle$ *the surface* $R = \Delta/G$ *is a torus with* S *and* T *determined by simple loops* α, β *on* R. *Then if* $H = \langle S, T^2\rangle$, *the surface* Δ/H *is not a torus and the generators* S *and* T^2 *are determined by simple loops on it.*

Proof. Our hypothesis implies that Δ is branched over one point b of R. If Δ/H is a torus too then Δ is branched over one point $b^* \in \Delta/H$ which lies over b. There is a lift α^* of α from R to Δ/H which is a simple loop cutting Δ/H into a doubly connected plane region containing b^*. But since H is a proper subgroup of the non-abelian group G, there are other lifts of α: Take $O \in \Delta/H$ over the initial point of α but such that $O \notin \alpha^*$. Since Δ/H is compact, for some n the lift $\alpha_1{}^*$ of α^n from O will be a simple loop in $R^* = \Delta/H - \alpha^* - b^*$. Now $\alpha_1{}^*$ is not homotopic to 1 or retractible in R^* to b^* since S is not elliptic. But $\alpha_1{}^*$ can't be freely homotopic to α^* in R^* either. We are thus left with a contradiction.

To finish the proof we need only observe that α^2 lifts to a simple loop in Δ/H because $T \notin H$.

§4. *Groups* G *for which* Δ/G *is the sphere*

In this section we will analyze Purzitsky's groups with card $b(G) = 4$ described in Section 2 and triangle groups G of signature (p, q, r). These are the groups for which card $b(G) = 3$ and Δ is branched of order p, q, r respectively over the three points of $b(G)$. These groups exist if and only if

$$p^{-1} + q^{-1} + r^{-1} < 1 . \tag{1}$$

LEMMA 6. *Suppose* $G = \langle S, T \rangle$ *is a triangle group as above with generators* S, T *determined by simple loops on* Δ/G. *Assume* $S^2 \neq \text{id}$. *Then one* G_0 *of the three subgroups*

$$\langle S, TST^{-1} \rangle ,$$
$$\langle S, (TST^{-1})\, S(TS^{-1}T^{-1}) \rangle ,$$
$$\langle S, (TST^{-1}STS^{-1}T^{-1})\, S(TST^{-1}S^{-1}TS^{-1}T^{-1}) \rangle ,$$

is such that Δ/G_0 *is not compact. Furthermore the displayed generators of* G_0 *are determined by simple loops on* Δ/G_0.

Proof. Both S and T are elliptic since they are determined by simple loops necessarily retractible in $\Delta/G - b(G)$ to a point of $b(G)$. In view of (1), it is no restriction to assume that the order p of S is not two.

Set $T_1 = TST^{-1}$ and consider the subgroup

$$G_1 = \langle S, T_1 \rangle$$

which is generated by two elliptic transformations of order $p > 2$. Certainly Δ/G_1 cannot be a torus. If Δ/G_1 is not the sphere we are finished. Otherwise G_1 is a triangle group. In this case fix disjoint simple loops α, β on $\Delta/G_1 - b(G_1)$ which determine S, T_1 respectively.

Now form the subgroup

$$G_2 = \langle S, T_1 S T_1^{-1} \rangle$$

which is also generated by two elliptic elements of order p. If Δ/G_2 is not a sphere we are finished. Otherwise G_2 is also a triangle group. In this case we will analyze the covering surfaces

$$\Delta \to \Delta/G_2 \to \Delta/G_1$$

where Δ/G_2 is finite sheeted over Δ/G_1.

The set $b(G_2)$ of three points in Δ/G_2 lies over the set $b(G_1) \subset \Delta/G_1$. Of course the branch points of the covering $\Delta/G_2 \to \Delta/G_1$ also lie over $b(G_1)$.

Denote the points of $b(G_1)$ as $\{x, y, z\}$ where α is retractible to x and β to y. Denote the points of $b(G_2)$ as $\{u, v, w\}$ so labeled that S is determined by a simple loop surrounding u and $T_1 S T_1^{-1}$ by one surrounding v. In fact both of these loops can be taken as lifts of $\alpha \subset \Delta/G_1$. With this labeling, both u and v lie over x.

Case 1. w lies over z. Since Δ is branched over Δ/G_2 only at the three points of $b(G_2)$ the following assertion is true. Let $O^* \in \Delta/G_2$ be any point over the origin $O \in \Delta/G_1$ of β. Then the lift of β^p from O^* is a simple loop on Δ/G_2. With the exception of the two lifts of α which surround u and v, all simple loops on Δ/G_2 over α are lifts of α^p.

Now for some $N < \infty$ every point of $\Delta/G_1 - b(G_1)$ is covered exactly N times in $\Delta/G_2 - b(G_2)$. This holds for lifts of α and β to Δ/G_2. Therefore for some integers $m, n \geq 0$,

$$2 + mp = np = N .$$

This implies $p = 1$ or 2 which is impossible.

Case 2. w lies over x. Suppose Δ is branched of order $k \geq 2$ over w. Since Δ is branched of order p over x and w lies over x, this implies that $p = p_o k$ for some integer $p_o \geq 1$. A lift of α^{p_o} is a simple loop around w. Applying the same argument as in Case 1 to lifts of α and β to Δ/G_2 we find that for some integers $m, n \geq 0$,

$$2 + p_o + mp = np = N . \tag{2}$$

This implies that $p_o = 1$ or 2. If $p_o = 1$, (2) shows that p divides 3 and hence $p = 3$. But then G_2 would be a triangle group with signature

$(3,3,3)$ which is impossible by (1). If $p_o = 2$ then (2) implies that $p=4$. This would mean that G_2 has signature $(4,4,2)$ which is also impossible.

Case 3. w lies over y. Again let k denote the order of branching of Δ over w and set $p = p_o k$. In this case β^{p_o} lifts to a simple loop around w. For some integers $m, n \geq 0$,

$$2 + mp = p_o + np = N \ . \tag{3}$$

Therefore $p_o = 1$ or 2. The possibility $p_o = 1$ is excluded by (3). But the possibility $p_o = 2$ requires another construction.

If $p_o = 2$ then G_2 has signature $(p, p, p/2)$ and by (1), $p \geq 6$. Take another subgroup

$$G_3 = \langle S, T_2 S T_2^{-1} \rangle, \quad T_2 = T_1 S T_1^{-1}$$

and analyze as above the coverings

$$\Delta \to \Delta/G_3 \to \Delta/G_2$$

under the assumption that G_3 is also a triangle group (if it is not, we are finished). Let α', β', γ' be small simple loops on Δ/G_2 surrounding u, v, w respectively where α' determines S and β' determines T_2. Set $b(G_3) = \{a, b, c\}$ where a and b lie over u. Reapply Case 1 and 2 above to rule out the possibility that c lies over w or u.

The remaining possibility is that c lies over v. Compare the lifts of α' and of γ' to Δ/G_3. If Δ/G_3 is N' sheeted over Δ/G_2, there are integers $m, n \geq 0$ such that

$$2 + mp = n(p/2) = N' \ .$$

This implies that p is at most 4 which is not the case.

LEMMA 7. *Suppose the generators of* $G = \langle S, T \rangle$ *are determined by simple loops on* $\Delta/G - b(G)$ *and* G *has the presentation for some odd* $q \geq 3$

$$G = \langle E_1, E_2, E_3;\ E_1^2 = E_2^2 = E_3^2 = (E_1 E_2 E_3)^q = 1 \rangle .$$

Then for the subgroup $G_o = \langle S, T^2 \rangle$, Δ/G_o *is not compact and the generators* S, T^2 *of* G_o *are hyperbolic transformations determined by simple loops on* Δ/G_o.

Proof. The surface Δ/G is a sphere, card $b(G) = 4$, and S and T are determined by simple loops α, β on Δ/G each of which surrounds two of the four points in $b(G)$. In particular both S and T are hyperbolic.

If Δ/G_o is also a sphere then it is a finite sheeted covering of Δ/G and card $b(G_o) = 3$ or 4. Look at the set of simple loops $\{\alpha_i\}$ on Δ/G_o, necessarily mutually disjoint, where each α_i is a lift of some power of α. An α_i cannot be retractible in $\Delta/G_o - b(G_o)$ to a point of $b(G_o)$ for this would mean that S is elliptic. This leaves us with only one possibility. Namely that card $b(G_o) = 4$ and there is only one loop in the set $\{\alpha_i\}$ this necessarily being a lift of α. Hence Δ/G_o is a 1-sheeted cover of Δ/G. But this is impossible since by construction there is a lift of β^2 which is a simple loop on Δ/G_o.

The same type of argument shows that Δ/G_o cannot be a torus either.

§5. *The basic lemma*

We are now in a position to prove the crucial lemma.

LEMMA 8. *Suppose* $G_n = \langle S_n, T_n \rangle$, $n = 1, 2, \dots,$ *is a sequence of Fuchsian groups such that* S_n *and* T_n *are determined by simple loops on* $\Delta/G_n - b(G_n)$. *Assume also that no* G_n *is generated by two elliptic transformations of order* 2. *Then if* $S = \lim S_n$ *and* $T = \lim T_n$ *exist as Möbius transformations, neither* S *nor* T *is the identity.*

Proof. Assume to the contrary that $S = \mathrm{id}$. By passing to a subsequence it suffices to consider the following cases individually. If Δ/G_n is a torus for all n, Lemma 5 assures us that G_n can be replaced by $G'_n = \langle S_n, T_n^2 \rangle$ where both S_n and T_n^2 are hyperbolic and determined by simple loops on Δ/G'_n which is not a torus. If Δ/G_n is a sphere and card $b(G_n) = 3$ for all n, Lemma 6 provides for a subgroup $G'_n = <S_n, U_n>$ to replace G_n but for which Δ/G'_n is not compact. Finally in the case that card $b(G_n) = 4$ for all n, Lemma 7 allows us to replace G_n by $G'_n = <S_n, T_n^2>$ for which Δ/G'_n is not compact.

Consequently we can put ourselves in the situation where Lemma 4 is applicable to each G_n. Hence there is a subgroup $H_n = <U_n, V_n>$ of G_n, of one of the forms described by Lemma 4, which contains no elliptic elements. Consider the sequence of subgroups of the H_n,

$$H'_n = <U_n, U_n^2 V_n^{-3}> .$$

Because H_n is a free group with the two generators U_n, V_n, H'_n is not cyclic. By examining each of the four possibilities for H_n we see that $S = \text{id.}$ implies $U_n^2 V_n^{-3} \to \text{id.}$ This is impossible by Lemma 2.

REMARK. The example presented in Section 1 shows that Lemma 8 is false in general for a sequence of groups G_n which are generated by two elliptic elements U_n, V_n of order 2. Such groups are of the form $S_n = U_n$, $T_n = U_n V_n$ where both U_n and V_n approach an elliptic transformation S of order 2. Of course these dihedral groups are the only two generator groups with a finite number of limit points.

§6. *Proof of Corollary* 1

We can assume $p = 0$. Assume the corollary is false and that Γ_n is a group that violates it for $D_n = \{z : d(0, z) < 1/n\}$. In what follows, when choosing an elliptic transformation $S \in \Gamma_n$ with fixed point q we will always do it in the following way. Let $M_q = \{T \in \Gamma_n : T(q) = q\}$ be the finite cyclic subgroup which preserves q. Then S is to be chosen not only to be a generator of M_q, but a generator that is determined by the lift of a *simple* loop in Δ/M_q. The order μ of S is the smallest $k > 0$ such that $S^k = \text{id.}$

The group Γ_n has at least two fixed points p_n, q_n in D_n. Let S_n, T_n of order μ_n, ν_n respectively denote the corresponding elliptic transformations. In the following argument we will freely replace $\{\Gamma_n\}$ by a subsequence when it is appropriate without changing notation. There are three cases.

Case 1. Lim $\mu_n = \infty$, $\lim \nu_n = \nu$. Since $\lim p_n = \lim q_n = 0$, in this case $\lim S_n = \text{id.}$ and $\lim T_n = T$ which is elliptic of order ν if $\nu < \infty$ otherwise $T = \text{id.}$ too. However this case cannot occur because of Lemma 8 applied to $G_n = \langle S_n, T_n \rangle$.

Case 2. Lim $\mu_n = \mu < \infty$, $\lim \nu_n = \nu < \infty$, and $\mu \neq 2$. Then $S = \lim S_n$ and $T = \lim T_n$ are elliptic of orders μ, ν respectively and each has $z = 0$ as a fixed point. Apply Lemma 6 if necessary to find a subgroup $G_n = \langle S_n, U_n \rangle$ of $\langle S_n, T_n \rangle$ where Δ/G_n is not compact and U_n is of order μ_n (or ν_n if we can take $U_n = T_n$). Inspecting the possible forms that U_n can take we see that $U = \lim U_n$ exists and is elliptic of order μ (or ν) with fixed point $z = 0$.

The proof of Lemma 4 shows both that $U_n S_n$ is not elliptic and that it is induced by a simple loop on Δ/G_n. Because $US = \lim U_n S_n$ has $z = 0$ as a fixed point but cannot be elliptic it follows that $US = \text{id.}$ Hence we obtain a contradiction on applying Lemma 8 to the subgroup G_n of Γ_n

$$G_n = \langle S_n, U_n \rangle = \langle S_n, U_n S_n \rangle .$$

Case 3. Lim $\mu_n = \lim \nu_n = 2$ and hence $\mu_n = \nu_n = 2$ for all large n. Let L_n be the non-Euclidean line through the fixed points p_n, q_n of S_n, T_n. There may be other fixed points of Γ_n on the segment L_n' of L_n which is bounded by p_n and q_n. Since $L_n' \subset D_n$, we may assume any other fixed points are also of order 2 or one of the previous cases will be applicable. Thus we can assume that in fact there are no fixed points of Γ_n on the interior of L_n'. But this also implies that if any $U_n \in \Gamma_n$ has a fixed point on L_n then $U_n \in \langle S_n, T_n \rangle$. For otherwise after conjugating

U_n by something in $<S_n, T_n>$ we would obtain an elliptic element with fixed point in the interior of L_n' (all the fixed points of $<S_n, T_n>$ lie on L_n).

Since we are assuming Corollary 1 fails for Γ_n, there is a fixed point $x_n \in D_n$, $x_n \notin L_n$, of $U_n \in \Gamma_n$. We can assume that $U_n^2 = \text{id.}$ for otherwise one of the previous cases would be applicable. Let H_n be the subgroup generated by S_n, T_n and U_n. The fixed points p_n, q_n, x_n of these three transformations project to three distinct points $\overline{p_n}, \overline{q_n}, \overline{x_n}$ in $b(H_n) \subset \Delta/H_n$. If necessary, T_n can be replaced by a conjugate in H_n whose fixed point is closer to p_n then q_n is, so that after this is done the segment L_n' between p_n and (the possibly new) q_n projects to a simple arc running between $\overline{p_n}$ and $\overline{q_n}$. This change may require that D_n be enlarged slightly but the crucial property that $\cap D_n = 0$ will not be lost. The purpose of this is that now we can assert that $S_n T_n$ is a hyperbolic transformation which is determined by a simple loop on Δ/H_n. Similarly we can also assume that $S_n U_n$ is hyperbolic and determined by a simple loop.

Now each transformation $S = \lim S_n$, $T = \lim T_n$, $U = \lim U_n$ is elliptic of order 2 with fixed point at $z = 0$. By the same argument as in Case 2, $\lim S_n T_n = \lim S_n U_n = \text{id.}$ The final contradiction is reached on applying Lemma 8 to the sequence of groups

$$G_n = <S_n T_n, S_n U_n> .$$

§7. *Proof of Theorem* 1, *part* (a)

We may assume that $p = 0$. The proof will be accomplished by drawing a contradiction from the assumption that the theorem is false for the sequence of groups $\{\Gamma_n\}$ as follows. There is a nested sequence of disks $\{D_n\}$, centers at $z = 0$, $\cap D_n = 0$, such that the subgroup G_n generated by

$$\mathcal{G}_n = \{T \in \Gamma_n : T(0) \in D_n\}$$

is neither cyclic nor generated by two elliptic transformations of order 2. We can also assume that for all n, $D_n \subset D$ where D is the disk chosen in accordance which Corollary 1 (which we are using as a lemma).

On the surface $R = \Delta/G_n - b(G_n)$ we are going to try to find two simple loops α, β which determine transformations in G_n. To do this draw non-Euclidean lines γ_i^* from $z = 0$ to each point $T_i(0)$ for $T_i \in \mathcal{J}_n$ making slight detours if necessary to avoid the fixed points (for simplicity we will also assume that $z = 0$ is not a fixed point of Γ_n). Each γ_i^* projects to a loop γ_i on R from the projection 0_* of 0. The loop γ_1 has a first point of intersection with itself. This gives us a simple loop α where $\alpha = \gamma_1$ if γ_1 is itself a simple loop. There is a lift α^* of α which is a segment along γ_1^*. Denote by $U_n \in G_n$ the transformation determined by this lift.

Remove the segment α^* from γ_1^* and join the newly exposed ends by using U_n (or U_n^{-1} as the case may be) to get a broken line segment from 0 to (say) $U_n^{-1}T_1(0)$ whose length is less than the radius of D_n. This implies that $U_n^{-1}T_1 \in \mathcal{J}_n$.

Repeat the process with the line from 0 to $U_n^{-1}T_1(0)$, etc. We get in addition to α a finite sequence of simple loops $\beta_1, \beta_2, \ldots$ and a corresponding relation

$$(1) \qquad T_1 = U_n V_n \cdots .$$

If T_1 is not a power of U_n, one of the loops β_i, say $\beta = \beta_1$, will determine an element $V_n \neq U_n^k$ in G_n. Otherwise proceed to γ_2, etc. Since the transformations T_i generate G_n which is not cyclic by assumption, eventually we will find a simple loop β in R and a corresponding $V_n \neq U_n^k$ in G_n (then also $U_n \neq V_n^k$).

Case 1. For all n the subgroup $H_n = \langle U_n, V_n \rangle$ of G_n, which is not cyclic, is not generated by two elliptic elements of order 2. In this case an application of Lemma 8 gives a contradiction. Indeed for a subsequence perhaps $U = \lim U_n$, $V = \lim V_n$ exist, and since both U_n and V_n cannot have fixed points in D, U and/or $V = \mathrm{id}$.

Case 2. For all n, H_n is generated by two elements E_n, F_n of order 2. Because U_n and V_n are determined by simple loops on R and therefore on Δ/H_n, E_n and F_n can be chosen from the set U_n, V_n, U_nV_n. But $U_n(D_n) \cap D_n \neq \phi$, and $V_n(D_n) \cap D_n \neq \phi$ while D_n shrinks to $z = 0$. Hence for all large n (we will assume for all n) the fixed points of E_n and F_n must lie close to $z = 0$; in particular they must lie in D. By Corollary 1, there is a uniquely determined subgroup H_n^0 of Γ_n, generated by two elements of order 2, which contains H_n as a subgroup (in fact Lemma 10 below implies $H_n = H_n^0$). Consequently if Case 2 cannot be avoided by rechoosing U_n and V_n by the process above, the relations (1) show that G_n itself is a subgroup of H_n^0. But the only subgroups are cyclic or those again generated by two elliptic elements of order 2 contrary to assumption.

§8. *Proof of Theorem* 1, *part* (b)

We will use the notation introduced in Section 1. The first result is a direct consequence of the proof of part (a), since all the transformations constructed lie in $\mathcal{I}(0; r)$.

LEMMA 9. *$\mathcal{G}(p; r)$ is generated by transformations which are in $\mathcal{I}(p; r)$ and are determined by simple loops on Δ/Γ.*

LEMMA 10. *Suppose $\mathcal{G}(p; r)$ is generated by two transformations of order 2. Then for some $E, F \in \mathcal{G}$*

$$(1) \qquad \mathcal{G}(p; r) = \langle E, F;\ E^2 = F^2 = \text{id.} \rangle$$

where at least one of E, F has its fixed point in $D(p; r)$. Furthermore if $H \subset \Gamma$ is a subgroup generated by two elements of order 2 and $\mathcal{G}(p; r) \subset H$, then $\mathcal{G}(p; r) \equiv H$.

Proof. As a consequence of Lemma 9 we can find E and F so that two of the three elements E, F, EF lie in $\mathcal{I}(p; r)$. Assume $E \in \mathcal{I}(p; r)$ and let L be any line through the fixed point x of E. If H^+ is a component of $\Delta - L$ then $E(H^+) \cap H^+ = \phi$. So if $x \notin D(p;r)$ we can determine L so that $D(p;r) \subset H^+$ and hence $E(D(p;r)) \cap D(p;r) = \phi$ in contradiction to our assumption that $E \in \mathcal{I}(p;r)$.

If both E and F lie in $\mathcal{I}(p;r)$ the second assertion follows from the construction of Section 6, Case 3. If not, we can assume $H = \langle E, F_o; E^2 = F_o{}^2 = \mathrm{id} \rangle$ where F is conjugate in H to E or F_o. We must have $EF = (EF_o)^k$ for some k. But $EF \in \mathcal{I}(p;r)$ and therefore $EF(D(p;r)) \cap D(p;r) \neq \phi$. On the other hand elementary properties of hyperbolic transformations (or alternatively Lemma 9) insure that for $D = D(p;r)$, $(EF_o)^{\pm 1}(D) \cap D \neq \phi$ (consider transformations of the form $z \mapsto kz$).

LEMMA 11. *Given a compact set* $K \subset \Delta$, *there exists* $r = r(K) > 0$ *with the following property. Suppose* Γ *is a group such that for some* $p \in K$, $\mathcal{G}(p;r)$ *contains an element of infinite order. For each* $q \in K$, *if* $\mathcal{G}(q;r)$ *is isomorphic to* $\mathcal{G}(p;r)$ *then* $\mathcal{G}(q;r) = \mathcal{G}(p;r)$. *Otherwise either one of* $\mathcal{G}(p;r)$, $\mathcal{G}(q;r)$ *is a subgroup of the other or* $\mathcal{G}(p;r) \cap \mathcal{G}(q;r) = \phi$ *but together they generate a group of the form* (1).

Proof. Consider only those r no larger than that which satisfies Corollary 1 and part (a) of Theorem 1. Assume the conclusion of Lemma 11 is false. We can also assume that $p = 0$ and that for each n, there is a group Γ_n and a point $p_n \in K$ such that one of the two possibilities below occurs for all n. Set $G_{on} = \mathcal{G}(0; 1/n)$ and $G_n = \mathcal{G}(p_n; 1/n)$. In accordance with Lemma 9, generators will be taken so as to be induced by simple loops on Δ/Γ_n.

Case 1. G_{on} is infinite cyclic with generator $S_n \in \mathcal{I}(0; 1/n)$ and G_n is also cyclic with generator $T_n \in \mathcal{I}(p_n; 1/n)$ but $T_n \neq S_n{}^{\pm 1}$, id. (it then follows that $T_n \neq S_n{}^k$) and $\langle S_n, T_n \rangle$ is not of the form (1), that is, not generated by two elements of order 2. Passing to a subsequence if necessary, we can assume that $\lim S_n = \mathrm{id.}$, $\lim p_n = p \in K$, and $\lim T_n = T$ exists as a Möbius transformation. But once again we are in conflict with Lemma 8.

Case 2. $G_{on} = \langle E_n, F_n; E_n^2 = F_n^2 = \text{id.}\rangle$ with $E_n \in \mathcal{I}(0; 1/n)$ and $T_n \in \mathcal{I}(p_n; 1/n)$ a generator of G_n with $T_n \notin G_{on}$. Again we can assume $\lim E_n F_n = \text{id.}$, $\lim p_n = p \in K$, and $\lim T_n = T$. If we can show that the non-cyclic subgroup $\langle E_n F_n, T_n\rangle$ is not generated by two elements of order 2 we will again have a contradiction to Lemma 8 and the proof of Lemma 11 will be complete.

This assertion will be proved by drawing a contradiction from the assumption it is false. If it is false then T_n is of order 2 and $S_n = T_n E_n F_n$ must be as well. By Lemma 10 the fixed point of T_n lies in $D(p_n; 1/n)$ so if q_n is the fixed point of S_n, $\lim q_n = \lim p_n = p$.

Suppose first that $p = 0$. Then S_n and T_n as well as E_n and F_n have fixed points arbitrarily close to $z = 0$. In view of Corollary 1 and Lemma 10 $\langle S_n, T_n\rangle$ is a subgroup of $\langle E_n, F_n\rangle$. In particular $T_n \in G_{on}$ contrary to assumption.

So assume $p \neq 0$. The fact that S_n and T_n have their fixed points near $z = p$ implies that for a small fixed r the local group $\mathcal{G}_n = \mathcal{G}(p; r)$ for Γ_n contains S_n and T_n (for all large n) and is therefore of the form (1). Let U_n and V_n be elements of order 2 which generate $\mathcal{G}_n$ and have fixed points in $D(p; r)$. We want to show that neither U_n nor V_n (and hence neither S_n nor T_n) is conjugate in Γ_n to E_n or F_n. To see this suppose that $U_n = W E_n W_n^{-1}$ for example. Then by the uniqueness of $\mathcal{G}_n$ in Γ_n we can assume that $V_n = W F_n W^{-1}$. But for some k, $T_n S_n = (U_n V_n)^k$ and therefore $E_n F_n = T_n S_n = W(E_n F_n)^k W^{-1}$ which implies [5, Lemma 4], that $W \in \langle E_n, F_n\rangle$. That is $\mathcal{G}_n \subset G_{on}$ contrary to the assumption that $T_n \notin G_{on}$.

On the other hand, because $T_n S_n = E_n F_n$, the non-Euclidean line through the fixed points of S_n and T_n, which is also the line connecting the two fixed points of $S_n T_n$, is identical with the line through the fixed points of E_n and F_n. The analysis of Case 3, Section 6 shows that this occurs only if S_n and T_n are conjugate in $\langle E_n, F_n\rangle$ to E_n and F_n in some order.

Lemma 10 implies that if $\mathcal{G}(q; r)$ is of the form (1) and also is a subgroup of $\mathcal{G}(p; r)$, then there is actually equality.

Proof of part (b). Set $K = D(0; R) = \{z : d(0, z) \leq R\}$ and choose $r = r(K)$ in accordance with Lemma 11. We can find $0 < r_1 \leq r$ with the following property. If Γ is any group such that for some $p \in K$ there exists an element of infinite order $T \in \mathcal{G}(p; r_1)$ then $T \in \mathcal{G}(q; r)$ for all $q \in K$.

Now assume that for the group Γ and some $p \in D(0; R)$, the group $\mathcal{G}(p; r_1)$ contains an element of infinite order. Then so does each $\mathcal{G}(q; r)$, $q \in K$. Given $T \in \mathcal{I}(0; R)$ fix $x, y \in D(0; R)$ such that $T(x) = y$. Then T sends $\mathcal{I}(x; r)$ to $\mathcal{I}(y; r)$ and by Lemma 11, $\mathcal{G}(x; r) = \mathcal{G}(y; r)$.

If $\mathcal{G}(x; r)$ is infinite cyclic with generator S then $TST^{-1} = S^{\pm 1}$. Referring to [5] this implies that either T is a power of S in which case $T \in \mathcal{G}(x; r)$, or $T^2 = \mathrm{id}$. If the former case holds for all T then $\mathcal{G}(0; R)$ is cyclic and $\mathcal{G}(0; R) = \mathcal{G}(x; r)$. Otherwise we can assume the second possibility holds. But then by Lemma 10, the fixed point p of T lies in $D(0; R)$. Hence by Lemma 11, $\mathcal{G}(x; r)$ is a subgroup of $\mathcal{G}(p; r)$ which is generated by two elliptic elements of order 2. Using Lemma 11 we are led to the conclusion that $\mathcal{G}(0; R) = \mathcal{G}(p; r)$.

It remains to consider the case that $\mathcal{G}(x; r) = \langle E, F;\ E^2 = F^2 = \mathrm{id.} \rangle$ with the generators E, F determined by lifting from a point $O^* \in \Delta$ two simple loops from a point $O \in \Delta/\Gamma$ each of which is retractible to a branch point of order 2. Now for some $W \in \mathcal{G}(x; r)$, $TEFT^{-1} = WEFW^{-1}$ Again applying [5, Lemma 4], $W^{-1}T = \text{identity}$. This implies that $T \in \mathcal{G}(x; r)$ and in view of Lemma 11 shows that $\mathcal{G}(0; R) = \mathcal{G}(x; r)$.

§9. *Sequences of transformations converging to the identity*

In the statement of the following lemma, all Möbius transformations are assumed to preserve the unit disk.

LEMMA 12. *Suppose* $\{S_n\}$ *is a sequence of Möbius transformations with* $\lim S_n = \mathrm{id}$. *Then there is a subsequence, again labeled* $\{S_n\}$, *and another sequence* $\{T_n\}$ *such that* $\lim T_n S_n T_n^{-1}$ *is a parabolic transformation.*

Proof. We may divide the proof into three cases according to whether all of the transformations $\{S_n\}$ are (1) parabolic, (2) hyperbolic, or (3) elliptic. For the first two it is convenient to take the transformations as preserving the upper half plane but for the third we will return to the unit disk.

Case 1. This is trivial since every parabolic transformation S_n is conjugate to $z \mapsto z+1$.

Case 2. Let p_n, q_n be the attractive and repulsive fixed points of S_n respectively. We may assume the sequences $\{p_n\}, \{q_n\}$ converge. Then there is a sequence $\{U_n\}$ such that $U_n(p_n) = \infty$, $U_n(q_n) \neq 0$ yet $\lim U_n = U$ exists as a Möbius transformation. Consider then the transformations

$$U_n S_n U_n^{-1} : z \mapsto a_n z + b_n, \quad b_n \neq 0 .$$

Since $U_n \to U$ and $S_n \to \text{id.}$, we have $a_n \to 1$ and $b_n \to 0$. Set $V_n(z) = z/b_n$. Then

$$V_n U_n S_n U_n^{-1} V_n^{-1} : z \mapsto a_n z + 1 ,$$

and the sequence $\{T_n\}$ where $T_n = V_n U_n$ fulfills the requirements of the lemma.

Case 3. Let U_n map $z = 0$ to the fixed point of S_n. Then

$$U_n S_n U_n^{-1}(z) = e^{i\Theta_n} z, \quad -\pi < \Theta_n \leq \pi .$$

The trace of S_n is $2 \cos (\Theta_n/2)$ which by hypothesis must approach 2. Hence $\lim \Theta_n = 0$. By replacing S_n by S_n^{-1} if necessary we may assume that $\Theta_n > 0$. Set

$$x_n = (1 - \sin \Theta_n)^{\frac{1}{2}}$$

and

$$V_n(z) = (z + x_n)/(1 + x_n z) .$$

One finds that

$$V_n U_n S_n U_n^{-1} V_n^{-1}(z) = (a_n z + x_n)/(-x_n z - \bar{a}_n)$$

where

$$a_n = \left(e^{i\Theta_n} - x_n^2\right) \Big/ \left(1 - e^{i\Theta_n}\right).$$

Then $\lim a_n = (i-1)$, and setting $T_n = V_n U_n$, $\lim T_n S_n T_n^{-1}$ is the parabolic transformation

$$z \mapsto ((i-1)z+1)/(-z+i+1) .$$

§10. *Proof of Theorem* 2

LEMMA 13. *Suppose that* Γ *is a Fuchsian group with fundamental polygon* P *and that* $S \in \Gamma$ *pairs two sides of* P. *Denote by* M *the cyclic subgroup consisting of those* $T \in \Gamma$ *with the same fixed points as* S. *Then* S *is a generator of* M.

Proof. Let σ, σ' denote the sides which are paired by S and fix a simple arc γ^* in P joining a point $x \in \sigma$ to the point $S(x) \in \sigma'$. The projection γ of γ^* to $R = \Delta/M$ is a simple loop. But R is either doubly connected or simply connected with Δ branched over a point $b \in R$. In particular γ is a generator of $\pi_1(R-b)$. Because S is determined by a lift of γ the result follows.

Proof of Theorem 2. As is to be expected, we will prove the theorem by contradiction. Without changing notation we will also replace the original sequence by a subsequence whenever appropriate. Assume then that for each n there is a group Γ_n such that for no conjugate $T\Gamma_n T^{-1}$ of Γ_n is it true that the disk $D(0; 1/n)$ is contained in the Poincaré polygon $P(T\Gamma_n T^{-1})$ (if $z = 0$ is not a fixed point so that this can be formed).

First choose r in accordance with Theorem 1, part (a). If Γ_n has more than one fixed point in $D(0; r)$ there is a non-Euclidean line L_n

such that all the fixed points of Γ_n in $D(0;r)$ lie on $L_n \cap D(0;r)$. Hence in all cases there is a disk $D_n' \subset D(0;r)$ of radius $r/2$ which contains no fixed points of Γ_n at all. Let U_n be a transformation which maps $z=0$ to the center of D_n'. We can carry this out for each n and then assume that $U = \lim U_n$ exists as a Möbius transformation.

Now consider the groups $\Gamma_n' = U_n \Gamma_n U_n^{-1}$ and the Poincaré polygon $P(\Gamma_n')$ which we can form since $z = 0$ is not a fixed point of Γ_n'. In fact no Γ_n' has a fixed point in $D(0; r/2)$.

By assumption there is a side σ_n of $P(\Gamma_n')$ which intersects $D(0; 1/n)$. The "opposite" side σ_n' also intersects $D(0; 1/n)$. Consider the sequence $\{S_n\}$ where $S_n \epsilon \Gamma_n'$ maps σ_n to σ_n'. We may assume that $S = \lim S_n$ exists as a Möbius transformation. The point $z = 0$ is a fixed point of S so that S is either elliptic or the identity.

But S cannot be elliptic. For otherwise S_n would be elliptic for all large n. The fixed point of S_n would lie outside $D(0; r/2)$ while the sides σ_n and σ_n' become closer together. The only way this can happen is if $S = \lim S_n = \mathrm{id}$.

Now apply Lemma 12. We may assume that for a sequence $\{V_n\}$, $\lim V_n S_n V_n^{-1} = S'$ is parabolic. Correspondingly consider the groups $\Gamma_n'' = V_n \Gamma_n' V_n^{-1}$. Again for some W_n such that $\lim W_n = W$ exists as a Mobius transformation it is true that $W_n \Gamma_n'' W_n^{-1}$ has no fixed point in $D(0; r/2)$. It is also true that $\lim W_n V_n S_n V_n^{-1} W_n^{-1} = WS'W^{-1}$ remains parabolic. On the other hand our assumption on Γ_n (and a lot of relabelling) implies that there is also a side τ_n of the polygon $P(W_n \Gamma_n'' W_n^{-1})$ that intersects $D(0; 1/n)$. In other words as before there is a sequence $\{T_n\}$ with $T_n \epsilon W_n \Gamma_n'' W_n^{-1}$ and $\lim T_n = \mathrm{id}$.

Consider the groups

$$G_n = \langle T_n, W_n V_n S_n V_n^{-1} W_n^{-1} \rangle, \quad n = 1, 2, \ldots .$$

G_n is not cyclic by Lemma 13 since $W_n V_n S_n V_n^{-1} W_n^{-1} \neq T_n^{\pm 1}$. Again referring to the proof of Lemma 13 we see that both generators of G_n are induced by simple loops on Δ/G_n (this property is not destroyed by conjugation). Furthermore G_n is not generated by two elements of order 2 since neither T_n nor S_n for large n is elliptic of order 2. That is we are in a situation in which Lemma 8 applies and gives the desired contradiction.

UNIVERSITY OF MINNESOTA

REFERENCES

[1]. Bers, L., Inequalities for finitely generated Kleinian groups, J. d'Anal. Math. 18 (1967), pp. 23-41.

[2]. Chuckrow, V., On Schottky groups with applications to Kleinian groups, Annals of Math., 88 (1968), pp. 47-61.

[3]. Earle, C., and Marden, A., to appear.

[4]. Harvey, W. J., to appear.

[5]. Marden, A., On homotopic mappings of Riemann surfaces, Annals of Math. 90 (1969), pp. 1-8.

[6]. ————, The geometry of finitely generated Kleinian groups, Annals of Math, to appear 1974.

[7]. Purzitsky, N., All two-generator Fuchsian groups, to appear.

[8]. Sturm, J., and Shinnar, M., these proceedings.

[9]. Wang, H-C., Discrete nilpotent subgroups of Lie groups, J. Diff. Geom. 3 (1969), pp. 481-492.

REMARKS ON COMPLEX MULTIPLICATION AND SINGULAR RIEMANN MATRICES

Henrik H. Martens

According to Conforto ([1], p. 133) the existence of non-singular Riemann matrices which admit complex multiplication is one of the "tiefgründigsten" results of the investigations of Scorza. The present paper is a collection of observations on the situation where the complex multiplication is an automorphism of the associated polarized abelian variety. One of these observations shows that a period matrix of a closed Riemann surface of genus ≥ 2 is singular whenever the surface admits an automorphism which defines a quotient surface of positive genus. This was also shown by Lewittes [2] under the additional assumption that the automorphism is an involution.

§1. Let $n \geq 1$ be an integer, and consider a complex torus $T = \mathbb{C}^n/G$ where G is the group of translations of $\mathbb{C}^n$ generated by the $2n$ R-linearly independent column vectors of a period matrix Π. The torus is said to be an abelian variety, and Π is said to be a Riemann matrix, if there exists a skewsymmetric $2n \times 2n$ matrix P with rational entries such that

1.1 $$\Pi P {}^t\Pi = 0, \quad \text{and}$$

1.2 $$i\Pi P {}^t\overline{\Pi} > 0 \quad \text{(pos. def.)} .$$

Whenever these conditions are satisfied, P is said to be a principal matrix for Π. Two principal matrices are said to be equivalent if one is a multiple of the other. A Riemann matrix is said to be singular if it admits two non-equivalent principal matrices.

Let Π be a Riemann matrix and P a principal matrix for Π. To show that Π is singular, it suffices to produce a skewsymmetric $2n \times 2n$ matrix S with rational entries, not a multiple of P, such that $\Pi S {}^t\Pi = 0$. For then $P + \lambda S$ will not be a multiple of P, and will be principal for Π for all sufficiently small rational λ ([1], p. 131).

If Π is a Riemann matrix and P is a principal matrix for Π, then there exists an $n \times 2n$ matrix Λ such that

1.3 $$ {}^t\Pi\Lambda - {}^t\Lambda\Pi = P^{-1} $$

([1], pp. 83-88).

Since any skewsymmetric 2×2 matrix must be a multiple of

$$ \begin{pmatrix} 0 & 1 \\ -1 & 0 \end{pmatrix}, $$

it follows that there are no singular Riemann matrices in dimension 1. In what follows T will denote a complex torus of dimension $n \geq 2$, Π will denote a period matrix for T. If T is assumed to be an abelian variety, P will denote a principal matrix for Π, and Λ will denote the matrix of (1.3). T_1 will denote a complex torus of dimension $m \geq 1$, and the associated matrices will be denoted by Π_1, P_1, Λ_1 to the extent that they exist.

§2. Let $h: T \to T_1$ be a homomorphism of complex tori. Then h is represented by a uniquely determined linear map $\mathbf{C}^n \to \mathbf{C}^m$ which carries periods of T into periods of T_1. Hence h is represented by a uniquely determined $m \times n$ matrix H such that

2.1 $$ H\Pi = \Pi_1 M $$

where M is a $2m \times 2n$ matrix with integral entries. Conversely, any $m \times n$ matrix H which satisfies a relation (2.1) defines a homomorphism $T \to T_1$. The $\mathbf{R}$-linear independence of the column vectors implies that H and M determine each other uniquely.

Assume now that T is an abelian variety and let $S = MP^tM$. Then S is a skewsymmetric $2m \times 2m$ matrix with rational entries, and by (2.1) and (1.1) we have

$$\Pi_1 S^t\Pi_1 = \Pi_1 MP^tM^t\Pi_1 = H\Pi P^t\Pi^t H = 0 .$$

If v is a point of $\mathbb{C}^m$, then

$$i^tv\Pi_1 S^t\overline{\Pi}_1\overline{v} = i^tvH\Pi P^t\overline{\Pi}^t\overline{Hv} .$$

If h is surjective, then ${}^tvH \neq 0$ whenever $v \neq 0$, and then the right-hand side is strictly positive by (1.2). But this means that $i\Pi_1 S^t\overline{\Pi}_1$ is positive definite. Hence, if h is surjective, S will be a principal matrix for Π_1, and thus the homomorphic image of an abelian variety is an abelian variety.

§3. By a complex multiplication for a period matrix Π we shall mean a relation

3.1 $$A\Pi = \Pi R$$

where A is an $n \times n$ matrix and R is a $2n \times 2n$ matrix with rational entries, not a multiple of the unit matrix. Multiplying both sides of (3.1) by a suitable integer, we get a relation defining a non-trivial endomorphism of T.

Let Π be a Riemann matrix admitting a complex multiplication. Then, by the reasoning of (2), the matrix $S = RP^tR$ will be a skewsymmetric $2n \times 2n$ matrix with rational entries such that $\Pi S^t\Pi = 0$. It follows that Π is singular unless S is a multiple of P. Thus a complex multiplication of a non-singular Riemann matrix must be of the form (3.1) where $RP^tR = \lambda P$, whence R is non-singular. ($\lambda = 0$ implies $R = 0$ by the reasoning of (2)).

If T and T_1 are abelian varieties with dimensions $n > m \geq 1$, and if $h: T \to T_1$ is an epimorphism, then the period matrices of T are singular.

Proof. Let h be represented by matrices H and M. Let $R = P^tMP_1{}^{-1}M$. Then R is a $2n \times 2n$ matrix with rational entries and rank 2m. Moreover, by (1.3), (2.1) and (1.1)

$$\begin{aligned} \Pi R &= \Pi P^tM\,({}^t\Pi_1\Lambda_1 - {}^t\Lambda_1\Pi_1)\,M \\ &= \Pi P^t\Pi^tH\Lambda_1 - \Pi P^tM^t\Lambda_1 H\Pi \\ &= -\Pi P^tM^t\Lambda_1 H\Pi\ . \end{aligned}$$

Hence we have produced a complex multiplication for Π with $A = -\Pi P^tM^t\Lambda_1 H$, and RP^tR cannot be a multiple of P since R is of rank 2m with $2 \leq 2m < 2n$.

§4. Let X and X_1 be closed Riemann surfaces of genera g and g_1, such that $g > g_1 \geq 1$. Let $f: X \to X_1$ be a holomorphic map of X onto X_1. Then f induces a homomorphism of the Jacobian varieties $J(X) \to J(X_1)$ which is surjective, and by (3) it follows that the period matrices of X must be singular.

In particular, if X admits a non-trivial automorphism defining a quotient surface of positive genus, and if the genus of X is ≥ 2, then the above condition is satisfied and the period matrices of X must be singular.

§5. Let X be a closed Riemann surface of genus $g \geq 2$. The Jacobian variety J(X) of X is an abelian variety of dimension $n = g$, and if Π is taken as a canonical period matrix of X, a principal matrix for Π will be

$$J = \begin{pmatrix} 0 & E \\ -E & 0 \end{pmatrix}$$

where E is the $g \times g$ unit matrix.

Any automorphism of X gives rise to an automorphism of J(X) represented by matrices H and M where H is non-singular and M is unimodular. The matrix H represents the automorphism of X as a map of holomorphic differentials. The matrix M represents the automorphism of

X as a map of 1-cycles. Since the latter must preserve intersection products and canonical homology bases, we have $MJ^tM = J$.

Conversely, any automorphism h of J(X), given by matrices H and M such that $MJ^tM = J$, will map theta-divisors on theta-divisors (by elementary transformation theory) and hence, by Torelli's theorem, will map W^1 on a translate of $\pm W^1$, where W^1 denotes a canonical image of X in J(X). It follows that either h or −h preserves the family of translates of W^1, and whichever does will induce and be induced by an automorphism of X. In the hyperelliptic case, and only then, will h and −h both represent automorphisms of X.[1]

If the automorphism does not satisfy the condition $MJ^tM = J$, it will not be induced by an automorphism of X, and will not preserve theta-divisors. In this case the abelian variety has two distinct structures as canonically polarized Jacobian variety of X.

(For background on Jacobian varieties see, e.g., [3]. The observations of this section are valid, mutatis mutandis, in the abstract case. Note that the hyperelliptic involution is always represented by the involution $u \to -u$ on J(X).)

§6. Conforto ([1], pp. 131-132) shows that a Riemann matrix Π admits a complex multiplication if and only if there is a $2n \times 2n$ matrix S, not a multiple of P, with rational entries, such that $\Pi S^t\Pi = 0$. (S is not assumed to be skewsymmetric.) Indeed, if Π admits a complex multiplication (2.2) we may take $S = RP$ and get

$$\Pi S^t\Pi = \Pi RP^t\Pi = A\Pi P^t\Pi = 0 .$$

On the other hand, if $\Pi S^t\Pi = 0$, then set $R = SP^{-1}$ and get

1 Thus the group of automorphisms of X is isomorphic to that of the canonically polarized Jacobian variety in the hyperelliptic case, and to its quotient modulo $\{e, -e\}$ otherwise.

$$\Pi R = \Pi SP^{-1} = \Pi S({}^t\Pi\Lambda - {}^t\Lambda\Pi)$$
$$= -\Pi S^t\Lambda\Pi \ ,$$

and observe that this gives a complex multiplication with $A = -\Pi S^t\Lambda$. For $R = SP^{-1}$ cannot be a multiple of the identity unless S is a multiple of P, contrary to hypothesis.

This shows in particular that singular Riemann matrices admit complex multiplication.

If the complex multiplication defines an automorphism of the polarized variety, i.e., an automorphism given by H and M such that $MP^tM = P$, then the above construction yields a matrix $S = MP$ which is skewsymmetric if and only if

$$-{}^tP^tM = P^tM = MP$$

or

$$P = MP^tM = M^2P$$

or

$$M^2 = E \ .$$

The last condition means that the automorphism is an involution, and this is the case considered by Lewittes [2].

Let Π be a canonical period matrix of a closed Riemann surface X. Then J is a principal matrix for Π. Note that J is unimodular. If P is another unimodular principal matrix for Π, not a multiple of J, then PJ^{-1} defines an automorphism of the Jacobian variety. This will induce an automorphism of the Riemann surface if and only if

$$(PJ^{-1})\, J^t(PJ^{-1}) = J$$

or, since $J^{-1} = {}^tJ = -J$ and ${}^tP = -P$,

$$PJ^tP = J = -PJP \ ,$$

whence $PJ^{-1}PJ^{-1} = PJPJ = E$ and the automorphism must be an involution other than the hyperelliptic involution. In this case a quotient surface of positive genus must exist.[2] If $PJ^tP \neq J$, then the abelian variety carries several distinct structures as canonically polarized Jacobian variety of X. This may be worth a closer look.

§7. Let Γ be a finite group of automorphisms of T. One might define a quotient torus of T by a homomorphism $h : T \to T : \Gamma$ such that $h \cdot g = h$ for all $g \in \Gamma$, and satisfying a universal mapping property. If T is the Jacobian variety of a closed Riemann surface, such a quotient torus exists as the Jacobian variety of the quotient surface when Γ corresponds to a group of automorphisms of the Riemann surface. This would lead to a generalization of the criterion of (4) to arbitrary abelian varieties. The quotient torus should be the Albanese variety of the quotient space of T by Γ. If Γ contains $-e$, where e is the identity, no non-trivial quotient torus can exist. This should imply that the Albanese variety of a Kummer variety is trivial.

UNIVERSITY OF TRONDHEIM
TRONDHEIM, NORWAY

REFERENCES

[1]. Conforto, F., Abelsche Funktionen und Algebraische Geometrie, Springer, 1956.

[2]. Lewittes, J., *Singular Riemann Matrices*, Advances in the Theory of Riemann Surfaces, Ann. of Math. Studies No. 66, pp. 287-294, Princeton University Press, 1971.

[3]. Martens, H., *Torelli's Theorem and a Generalization for Hyper-Elliptic Surfaces*, Comm. Pure and Appl. Math. XVI (1963), pp. 97-110.

---

[2] This gives a characterization of those Riemann surfaces which are p-hyperelliptic for some $p > 0$ as those whose canonical period matrices admit a skewsymmetric symplectic integral matrix, other than J, satisfying (1.1).

INTERSECTIONS OF COMPONENT SUBGROUPS OF KLEINIAN GROUPS

Bernard Maskit*

The basic geometry of a finitely generated Kleinian group with an invariant component is now fairly well understood, but very little is known about Kleinian groups which do not have an invariant component. In this paper we investigate the geometric relations among the different components of a Kleinian group, and the algebraic relations among the corresponding component subgroups.

We start with a Kleinian group G, a component Δ_0 of G (i.e., Δ_0 is a connected component of $\Omega(G)$, the set of discontinuity of G), and the corresponding component subgroup G_0 (i.e., $G_0 = \{g \in G | g\Delta_0 = \Delta_0\}$). We assume that Δ_0/G_0 is a finite Riemann surface; using Ahlfor's finiteness theorem [1], this assumption is equivalent to assuming that Δ_0 is a component of G_0 and that G_0 is finitely generated. In particular, this assumption is satisfied if G is finitely generated.

In order to state our main result, we need the following definition. For any Kleinian group G and subgroup H, a subset A of the Riemann sphere is said to be *precisely invariant under* H *in* G if

i) $gA = A$, $g \in H$, and

ii) $gA \cap A = \phi$, $g \in G - H$.

We let $p : \Omega(G) \to \Omega(G)/G$ be the natural projection. If Y is a connected subset of $\Omega(G)/G$, if A is a connected component of $p^{-1}(Y)$,

* Research supported by National Science Foundation Grant: P019572000.

and if H is the stabilizer of A in G (i.e., $H = \{g \in G | gA = A\}$, then A is precisely invariant under H in G. In this case we say that H is a *covering subgroup* of G, that H *stabilizes* A, and that H *covers* Y.

In the special case of the above that A is a component of G, we also say that H is a *component subgroup* of G.

THEOREM 1. *Let* G *be a Kleinian group, let* Δ_0 *be a component of* G, *and let* G_0 *be the stabilizer of* Δ_0 *in* G. *Assume that* $\Delta_0/G_0 = X_0$ *is a finite Riemann surface. Then there is a finite set of simple disjoint loops on* X_0, *which divide* X_0 *into a finite number of subsurfaces* $Y_1, \ldots, Y_s$, *so that if* G_1 *is some other component subgroup of* G, *then* $G_0 \cap G_1$ *is either trivial, or cyclic, or is a covering subgroup of some* Y_i.

As an immediate corollary to the above, we obtain

THEOREM 2. *Let* G_1 *and* G_2 *be component subgroups of the finitely generated Kleinian group* G. *Then* $G_1 \cap G_2$ *is finitely generated.*

The proof of Theorem 1 appears in Section 4.

Our first example, given in Section 5.1, is of a finitely generated purely loxodromic (including hyperbolic) Kleinian group with infinitely many components where every component is a circular disc, all the components have disjoint closures, and the intersection of any two distinct component subgroups is trivial.

We also have some results concerning the boundaries of components. We note that if G_i is the stabilizer of the component Δ_i, and Δ_i/G_i is a finite surface, then $\Lambda(G_i)$, the limit set of G_i, is the boundary of Δ_i.

THEOREM 3. *Let* G_i *be the stabilizer of the component* Δ_i, *of the Kleinian group* G, $i = 1, 2$. *Assume that* Δ_i/G_i *is a finite Riemann surface,* $i = 1, 2$. *Then*

$$\Lambda(G_1 \cap G_2) = \Lambda(G_1) \cap \Lambda(G_2) = \overline{\Delta}_1 \cap \overline{\Delta}_2 \ .$$

THEOREM 4. *Let* G_0 *be the stabilizer of the component* Δ_0, *of the Kleinian group* G. *Assume that* Δ_0/G_0 *is a finite Riemann surface. Let* g_0 *be a loxodromic element of* G *with at least one fixed point lying in* $\overline{\Delta}_0$. *Then for some positive integer* n, $g_0^n \in G_0$.

THEOREM 5. *Let* G_0, Δ_0, G *be as in Theorem* 4. *Let* g_0 *be a parabolic element of* G *whose fixed point* z *lies on the boundary of* Δ_0. *Then there is a parabolic element* $g_1 \in G_0$ *which has* z *as a fixed point.*

The proofs of Theorems 3, 4 and 5 appear in Section 3.

The conclusions of Theorems 4 and 5 are the best possible. In Section 5.2 we produce for every positive integer n, a finitely generated, purely loxodromic Kleinian group G, with a particular element g_0 lying on the boundary of 2n distinct components of G, where g_0 cyclically permutes these components. In the special case $n = 1$, we get a purely loxodromic finitely generated Kleinian group which contains a Fuchsian group of the first kind as a subgroup of index 2.

In Section 5.3, we give a simple example of a finitely generated Kleinian group G with a component Δ_0 and a parabolic element $h \in G$, where the fixed point of h lies on the boundary of Δ_0, but $h(\Delta_0) \neq \Delta_0$.

The theorems listed above are all fairly easy consequences of this author's Decomposition theorem [11], some known but deep results about Fuchsian groups, and a version of Koebe's theorem, stated below.

For any set S, we denote the spherical diameter of S by dia(S).

THEOREM 6. *Let* G_0 *be the stabilizer of the component* Δ_0 *of the Kleinian group* G. *Assume that* Δ_0/G_0 *is a finite Riemann surface. Let* $\Delta_1, \Delta_2, \ldots,$ *be the transforms of* Δ_0 *in* G. *Then* $\Sigma_i \operatorname{dia}^4(\Delta_i) < \infty$.

One would like to know that the sum of the squares of the diameters converges, but results of this sort are known only in very special cases, [2], [7], [8], [16].

The hypothesis that Δ_0/G_0 be a finite Riemann surface is essential. One easily constructs a Kleinian group – necessarily infinitely generated – with a sequence of equivalent components where the spherical diameters are bounded away from zero. This construction is given in Section 5.4.

§1. *Fuchsian and quasi-Fuchsian groups*

1.1. Let F be a Fuchsian or quasi-Fuchsian group operating on the (quasi-) discs U and L. These two regions are separated by a (quasi-) circle C. We regard U and L as being endowed with the Poincaré metric so that there is a unique (non-Euclidean) straight line in U, and in L, connecting any two points of C. If g is a loxodromic element of F, then the straight line in U(L) is called the *axis* of g in U(L). The union of these two axes, together with the fixed points of g is a simple closed curve called the *complete axis* of g.

Looking now only in U, the *Nielsen convex region* K is the non-Euclidean convex hull of the limit set of F; i.e., K is the smallest convex region containing all lines joining limit points of F. Using Lemma 1 below, we see that K can equivalently be defined as the convex hull of the set of axes of F in U.

LEMMA 1. *Let* G *be a non-elementary Kleinian group, and let* $z \neq z'$ *be limit points of* G. *Then there is a sequence* $\{g_n\}$ *of loxodromic elements of* G, *where* g_n *has fixed points* z_n, z_n', *so that* $z_n \to z$, $z_n' \to z'$.

Proof. Normalize G so that $\infty \in \Omega(G)$, and choose sequences $\{h_n\}$, $\{h_n'\}$ of elements of G, so that $h_n(\infty) \to z_n$, $h_n'(\infty) \to z_n'$. Let g_0 be some fixed loxodromic element of G, and set $j_n = h_n g_0 h_n^{-1}$, $j_n' = h_n' \cdot g_0 \cdot (h_n')^{-1}$. Then both fixed points of j_n converge to z, and both fixed points of j_n' converge to z'. One easily sees that $g_n = j_n' \cdot j_n$ has the desired properties.

1.2. We will need the following known facts about Fuchsian and quasi-Fuchsian groups. For the reader's convenience, we include outlines of proofs, where there are no ready references.

LEMMA 2. *The following statements are equivalent*

i) F *is finitely generated.*

ii) *The Dirichlet fundamental polygon for* F *in* U *has finitely many sides.*

iii) K/F *has finite (non-Euclidean) area.*

Proof. All of these statements are by definition invariant under conformal maps; hence, we can assume that U is the upper half plane and F is Fuchsian. In this case the equivalence of i) and ii) can be found in Marden [10], and there is a proof of the equivalence of i) and iii) in the Fenchel-Nielsen manuscript [6]. We actually need the equivalence of ii) and iii); we outline a proof below.

If F has a finite sided fundamental polygon P then one sees at once that $P \cap K$ is, except for a finite number of parabolic cusps, relatively compact in U. Each of these parabolic cusps is isometric to a region of the form $\{z \mid \text{Im } z > 1, \ |\text{Re } z| < \frac{1}{2}\}$, hence $P \cap K$ has finite area.

We next assume that P has infinitely many sides. Let $P' = P \cap K$, then P' is again a generalized polygon with sides $s_1, s_2, \ldots,$ and vertices $v_1, v_2, \ldots,$ where some of the v_i might lie on the boundary of U; the vertices of P' lying in U are called finite vertices, the others on the boundary of U are called infinite vertices. Each vertex is an intersection of two sides; the angle at the vertex is the angle of intersection of these two sides measured from inside P'.

The equivalence classes of the finite vertices are called cycles; each cycle contains only finitely many vertices, and the sum of the angles of the vertices in each cycle is a submultiple of 2π. If a cycle has exactly one vertex in it, then the angle at that vertex is at most $2\pi/3$; we do not consider "vertices" of angle π to be vertices. If a cycle has exactly two vertices, then the sum of the two angles is at most π. Regarding the infinite vertices as having angle 0, we conclude that P' has infinitely many vertices with angle at most $2\pi/3$.

Let P^* be the convex hull of the vertices of P with angle at most $2\pi/3$, intersected with K. Pick an interior point O in P^* and draw lines from O to the vertices of P^*. This divides P^* into infinitely many triangles, $T_1, T_2, \ldots$. For n sufficiently large the angle of T_n at O is less than $\pi/18$. For any three consecutive triangles, one of them has angle sum, exclusive of the angle at O, at most $8\pi/9$. Hence, there are infinitely many triangles with angle sum less than $17\pi/18$, and so P^* has infinite area. Since $P^* \subset P'$, P' also has infinite area.

1.3. For our next lemma, we need the following definition. A sequence of simple closed curves $\{\gamma_n\}$ *nests about* the point z, if γ_{n+1} separates z from γ_n, and γ_n converges to z.

LEMMA 3. *Let F be a finitely generated Fuchsian or quasi-Fuchsian group of the first kind. Let $z_0 \in \Lambda(F)$, where z_0 is not a fixed point of any parabolic element of G. Then there is a loxodromic element $g_* \in F$, with complete axis A, and there is a sequence $\{g_n\}$ of elements of F so that $\{g_n(A)\}$ nests about z_0.*

Proof. It obviously suffices to assume that F is a Fuchsian group operating on U. Passing to a subgroup of finite index we can assume that F is tension free.

On U/F we can find a finite number of closed geodesics $w_1, \ldots, w_n$ so that each connected component of $U/F - \{w_1, \ldots, w_n\}$ is conformally either a disc or a punctured disc.

Let L be the straight line from some point O in U to z_0. Since z_0 does not project to one of the punctures on U/F, the projection of L necessarily crosses some w_i infinitely often. We then choose g_* to be that element of F which identifies the endpoints of some lifting of w_i.

1.4. In general, a limit point z_0 of a Kleinian group G is called a *point of approximation* for G if there is a point $y_0 \in \Omega(G)$ and a sequence $\{g_n\}$ of distinct elements of G, so that the spherical distance from

$g_n(z_0)$ to $g_n(y_0)$ is bounded away from zero. Other equivalent definitions are given in [2].

It was shown in [2] that if z_0 is a point of approximation for the Kleinian group G, then z_0 is not a fixed point of any parabolic element of G.

LEMMA 4. *Let* H *be a finitely generated Fuchsian or quasi-Fuchsian group. Then every point* $z_0 \in \Lambda(H)$ *is either a fixed point of a parabolic element of* H *or is a point of approximation for* H.

Proof. Let $\{g_n(A)\}$ be the sequence of axes nesting about z_0 given by Lemma 3. One sees at once that A, the axis of g_* separates $g_n^{-1}(A)$ from $g_n^{-1}(z_0)$. For each n, there is an appropriate power α_n of g_*, so that $j_n(z_0) = g_*^{\alpha_n} g_n^{-1}(z_0)$ is bounded away from A. Then for any y_0 on A, the distance between $j_n(z_0)$ and $j_n(y_0)$ is bounded from below by the distance of $j_n(z_0)$ to A.

§2. *Koebe's theorem*

2.1. In this section we prove Theorem 6. Assume that $\infty \in \Delta_0$, and let $G = \Sigma_i g_i G_0$ be a coset decomposition.

Let $S = gG_0$ be a particular coset. For any $h \in G_0$ the center of the isometric circle of gh is the point $(gh)^{-1}(\infty) = h^{-1}(g^{-1}(\infty))$. Hence, the set of centers of isometric circles of elements of S is invariant under G_0. We conclude that we can choose the coset representative g so that the center of the isometric circle of g lies in a particular fundamental set for G_0.

2.2. We have assumed that G_0 is finitely generated, hence $\Omega(G_0)/G_0$ is a finite union of finite Riemann surfaces. Then we can find a fundamental set D, for the action of G_0 on $\Omega(G_0)$, where D is relatively compact in $\Omega(G_0)$ except for a finite number of cusped regions.

A set R is a *cusped region* for G_0 if there is a fractional linear transformation ζ, and there is a parabolic element j in G_0 as follows.

The set

$$\{z \mid \operatorname{Im}(z) > 0,\ 0 \le \operatorname{Re}(z) < 1\}$$

is the image of R under ζ, and $\zeta^{-1} j \zeta(z) = z + 1$.

Each cusped region R then has associated with it a parabolic element j, which identifies the sides of R and a vertex z, the fixed point of j.

2.3. If now $g_i \in G - G_0$, then we can estimate the diameter of $g_i(\Delta_0)$ by inverting in the isometric circle of g_i. One easily sees that

2.3.1) $$\operatorname{dia} g_i(\Delta_0) \le |c_i|^{-2} \delta_i^{-1}$$

where δ_i is the distance from $g_i^{-1}(\infty)$ to Δ_0, and we have written $g_i = \begin{pmatrix} a_i & b_i \\ c_i & d_i \end{pmatrix}$ so that $|c_i|^{-1}$ is the radius of the isometric circle of g_i.

Using the well-known fact that

2.3.2) $$\Sigma\, |c_i|^{-4} < \infty ,$$

we see that

2.3.3) $$\Sigma \operatorname{dia}^2 g_i(\Delta_0) < \infty ,$$

where the sum is taken over all cosets where $g_i^{-1}(\infty)$ is bounded away from Δ_0.

2.4. In order to complete the proof of our theorem, it suffices to assume that the points $g_i^{-1}(\infty) = -d_i/c_i$ all lie in a cusped region R, and converge to the vertex. Conjugating G by a translation, we can further assume that the vertex of R is at the origin.

Let δ_i be the distance from $g_i^{-1}(\infty)$ to Δ_0. One easily sees that for $g_i^{-1}(\infty)$ in R, there is a constant $K > 0$, so that

2.4.1) $$\delta_i \ge K|g_i^{-1}(\infty)| = K|d_i/c_i| .$$

The existence of a cusped region as part of a fundamental domain near the origin shows that the subgroup J, of G, keeping the origin fixed is either cyclic, or contains a cyclic subgroup of index 2. Hence, we can assume that the elements $g_i \cdot j \cdot g_i^{-1}$ are all distinct. We compute the entries in the matrix $g_i \cdot j \cdot g_i^{-1}$ and use (2.3.2) to obtain

2.4.2) $$\Sigma_i \, |d_i|^{-8} < \infty \, .$$

Finally, combining (2.3.1), (2.4.1) with the fact that $|d_i/c_i| \to 0$ we obtain

2.4.3) $$\text{dia}^4 \, g_i(\Delta_0) \leq |c_i|^{-8} \, \delta_i^{-4} \leq K^{-4} \, |c_i d_i|^{-4} \leq K^{-4} \, |d|^{-8}.$$

Theorem 6 now follows from (2.4.2) and (2.4.3).

§3. *Boundaries of components*

3.1. Let G_0 be a finitely-generated Kleinian group with an invariant component Δ_0. Every subgroup H of G_0 has an invariant component $\Delta(H)$ containing Δ_0. If $\Delta(H)$ is simply-connected and hyperbolic, then there is a schlicht function $\zeta : \Delta(H) \to U$, the unit disc. A parabolic element g in H is called *accidental* if $\zeta g \zeta^{-1}$ is hyperbolic.

A subgroup H of G_0 is called a *factor subgroup* if H is a maximal subgroup of G_0 having the following three properties.

i) $\Delta(H)$ is simply-connected;

ii) H contains no accidental parabolic elements,

iii) H contains every parabolic element of G_0 whose fixed point lies on $\Lambda(H)$.

It was shown in [11] that every factor subgroup is finitely generated.

It is known (see Maskit [12], Bers [3], and Kra-Maskit [9]) that every factor subgroup is either quasi-Fuchsian (i.e., a quasi-conformal deformation of a Fuchsian group of the first kind), degenerate (i.e., $\Omega(H)$ is both connected and simply connected), or elementary (i.e., $\Lambda(H)$ contains at most one point).

LEMMA 5. *Let* z_0' *be a point on the boundary of* Δ_0, *where* z_0 *also lies on the boundary of some other component* Δ_1 *of* G. *Then there is a quasi-Fuchsian factor subgroup* H *of* G_0, *so that* $z_0 \in \Lambda(H)$.

Proof. It was shown in [11] that every component of $\Omega(G_0)$ other than Δ_0 is a component of some quasi-Fuchsian factor subgroup of G_0. The result follows from the fact that Δ_1 must lie in a component of $\Omega(G_0)$ other than Δ_0.

3.2. *Proof of Theorem* 5. If z_0 is a fixed point of a parabolic element g_0 in G, and g_0 is not in G_0, then $g_0(\Delta_0) \neq \Delta_0$, and so by Lemma 5, z_0 lies in the limit set of some quasi-Fuchsian subgroup H of G_0. Since z_0 is the fixed point of g_0, z_0 is not a point of approximation of H; hence, by Lemma 4, there is a parabolic $g_1 \in H$ with fixed point at z_0.

3.3. *Proof of Theorem* 4. If z_1, z_2 are the fixed points of a loxodromic g_0 in G, where z_1 and z_2 both lie on the boundary of Δ_0 then the spherical diameter of $g_0{}^n(\Delta_0)$ is not less than the spherical distance from z_1 to z_2, for every n. Theorem 4 now follows from Theorem 6.

3.4. *Proof of Theorem* 3. The only thing that needs proving is that if $z_0 \in \Lambda(G_1) \cap \Lambda(G_2)$, then $z_0 \in \Lambda(G_1 \cap G_2)$. By Lemma 5, it suffices to assume that G_1 and G_2 are both quasi-Fuchsian.

If z_0 is a fixed point of a parabolic element g_1 in G_1, then by Theorem 5, there is a parabolic g_2 in G_2 with fixed point z_0. Now g_1 and g_2 must both belong to the same cyclic parabolic subgroup of G, for otherwise an invariant curve under g_1 would necessarily intersect an invariant curve under g_2.

By Lemma 4, the only other possibility is that z_0 is a point of approximation for G_1. Then let A be the complete axis of g_* in G_1, and let $\{g_n\}$ be the sequence of elements of G_1 given by Lemma 2. For n sufficiently large, $g_n(A)$ intersects Δ_2. For each n, there is a power α_n so that $g_*{}^{\alpha_n} \cdot g_n{}^{-1}(\Delta_2)$ intersects an arc of A that is bounded away from $\bar{\Delta}_1$. Hence by Theorem 6, there are only finitely many distinct regions of the form $g_*^{\alpha_n} \cdot g_n^{-1}(\Delta_2)$; i.e., after passing to a subsequence, we can write $g_*^{\alpha_n} \cdot g_n^{-1} = \tilde{g} \cdot j_n$ where $\tilde{g}$ is a fixed element of G, and the j_n lie in G_2. We set $h_n = j_n^{-1} \cdot j_1 \in G_1 \cap G_2$, and observe that $h_n(g_1(A)) = g_n(A) \to z_0$. Hence, $z_0 \in \Lambda(G_1 \cap G_2)$.

§4. *Proof of Theorem* 1

4.1. We need the following part of the decomposition theorem [11].

THEOREM. *Let* G_0 *be a finitely generated Kleinian group with an invariant component* Δ_0. *There is a finite set* $w_1, \ldots, w_n$ *of disjoint simple closed curves on* $X = \Delta_0/G_0$, *which divide* X *up into subsurfaces* $Y_1, \ldots, Y_s$ *so that the following holds. Let* H *be a factor subgroup of* G_0, *and let* $\Delta(H)$ *be the component of* H *containing* Δ_0. *Then* H *is a covering subgroup of some* Y_i, *and* $\Delta(H)/H - Y_i$ *is a finite union of disjoint discs and punctured discs.*

Combining the above theorem with Lemma 5, we see that it suffices to prove Theorem 1 for the case that G_0 is quasi-Fuchsian.

4.2. We now assume that G_0 is quasi-Fuchsian and we set $H = G_0 \cap G_1$. Let K be the Nielsen convex region of H in Δ_0; we need to show that the interior of K is precisely invariant under H in G_0.

If z is an interior point of K, then by Lemma 1, we can find an axis of K arbitrarily close to z. Hence, if z is an interior point of $K \cap gK$, $g \notin H$, there are axes A_1, A_2 in K, so that A_1 intersects $g(A_2)$. Regarding A_1 and A_2 as complete axes, A_1 and $g(A_2)$ must have a second point of intersection. Now A_1 lies, except for two fixed points in $\Delta_0 \cup \Delta_1$, and similarly A_2 lies in $\Delta_0 \cup g(\Delta_1)$. This second point of intersection must be a point of intersection of Δ_1 and $g\Delta_1$.

We have shown that for $g \in G_0$, if $gK \cap K$ has non-empty interior, then $g \in G_1$, and so $gK = K$; i.e., the interior of K is precisely invariant under H.

We remark that gK and K might have a common axis, in which case, H is loxodromic cyclic or a Z_2 extension of a loxodromic cyclic group.

4.3. We have shown that the interior of the Nielsen convex region K for $H = G_0 \cap G_1$ is precisely invariant under H in G_0. This means that K/H is conformally embedded as a subsurface of Δ_0/G_0, hence K/H has finite non-Euclidean area. By Lemma 2, H is finitely generated; we have proven Theorem 2.

4.4. To conclude the proof of Theorem 1, we need to observe that if K_i is the Nielsen convex region of $H_i = G_0 \cap G_i$, $i = 1, 2$, then $K_1 = K_2$, or $K_1 \cap K_2 = \phi$, or K_1 and K_2 have exactly one axis in common. Since K_1 and K_2 are closed and convex, it suffices to show that the interiors of K_1 and K_2 are either equal or disjoint.

If z is an interior point of K_1 and K_2, then by Lemma 1, there are axes A_1 in H_1, A_2 in H_2, which intersect in the interior of $K_1 \cap K_2$. The complete axes must have a second point of intersection, necessarily a point of $\Delta_1 \cap \Delta_2$. Of course, if $\Delta_1 \cap \Delta_2 \neq \phi$, then $\Delta_1 = \Delta_2$, so that $K_1 = K_2$.

Each of the regions $K_1, K_2, \ldots,$ projects to a region on X_0 bounded by a finite number of sides. Area considerations show that there can be only a finite number of distinct such regions.

In order to complete the proof of Theorem 1, we need to observe that if H in G has a Nielsen convex region with empty interior, then H is either trivial, cyclic, or is a Z_2 extension of a loxodromic cyclic group. This last case can occur for only finitely many conjugacy classes and in each of them, the axis can be thickened, to give a region on X_0 which is a neighborhood of the line joining the two branch points of order 2.

§5. *Examples*

5.1. For our first example, we let H and H^* be finitely generated purely hyperbolic Fuchsian groups of the first kind operating on the unit disc U. For each complex number $t \neq 0$ let a_t be the transformation

$$a_t(z) = tz \ ,$$

let $H_t = a_t H a_t^{-1}$, and let G_t be the group generated by H_t and H^*.

For $|t|$ sufficiently small G_t can be formed from H_t and H^* via Klein's combination theorem; for these values of t, we know that G_t is a Kleinian group with an invariant component, that G_t is the free product of H_t and H^*, and that G_t contains no parabolic elements [16].

For each fixed t, the construction above defines a homomorphism Φ_t from the abstract free product of H and H^* into PSL(2; C) raised to an appropriate power. For any fixed element $g \in H * H^*$, the square of the trace of $\Phi_t(g)$ is a holomorphic function of t. It follows that for a fixed non-trivial $g \in H * H^*$, $\Phi_t(g)$ can be parabolic for at most countably many points t. We conclude that there is a ray $\arg t = \theta_0$, where no element of G_t on this ray is parabolic; it is no loss of generality to assume that $\theta_0 = 0$.

We now consider G_t only for t real and positive. Let T be the set of points t with the property that $G_{t'}$ is a free product of $H_{t'}$ and H^* in the sense of Combination Theorem I [13], for every t' with $0 < t' \leq t$. One sees at once from the definition in [13] that T is open. Let t_0 be the lub of T and set $G = G_{t_0}$.

We already know that G contains no parabolic elements. Then by Chuckrow's theorem [5], Φ_{t_0} is an isomorphism onto a purely loxodromic group.

For every $t \in T$, the disc $U^* = \{|z| > 1\}$ is precisely invariant under $H^* = \Phi_t(H^*)$ in G_t. By continuity, U^* is precisely invariant under H^* in G. In particular we conclude that G is Kleinian and that U^* is a component of G.

As above, for each $t \in T$, the disc $U_t = \{|z| < t\}$ is precisely invariant under $\Phi_t(H)$ in G_t, and so by continuity, $U_0 = \{|z| < t_0\}$ is precisely invariant under $H_0 = H_{t_0}$ in G.

We also observe that no element of G_t maps U_t onto U^*, hence no element of G maps U_0 onto U^*; i.e., $\Omega(G)/G$ contains at least the 2 components, U^*/H^* and U_0/H_0.

We next show that the translates of U^* and U_0 are the only components of G. Suppose Δ were some other components of G, with component subgroup J. There are two possibilities to consider: either U^* and U_0 lie in the same component of the complement of Δ, or they lie in different components.

If there is a g in G so that U^* and $g(U_0)$ lie in the same component of the complement of Δ, then there is a quasi-Fuchsian factor subgroup J_1 of J, where the limit set of J_1 separates U^* and $g(U_0)$ from the interior of Δ [11]. Since J_1 is isomorphic to the fundamental group of a closed orientable surface, it cannot be written as a non-trivial free product. Using the Kurosh subgroup theorem, we see that J_1 is a subgroup of a conjugate of either H_0 or H^*; we assume it is H^*, the argument is essentially the same in the other case. Then there is a $g^* \epsilon G$, where g^* maps $\Lambda(J_1)$, the limit set of J_1 onto the unit circle. Since $\Lambda(J_1)$ separates Δ from both $g(U_0)$ and U^*, and since $g^* \cdot g(U_0)$ must be contained in $|z| < 1$, $g^*(\Delta) = U^*$.

The other possibility is that $g(U_0)$ and U^* lie in different components of the complement of Δ for every g in G. Then we can find quasi-Fuchsian factor subgroups J_1 and J_2 of J, so that $\Lambda(J_1)$ separates Δ from U^* and $\Lambda(J_2)$ separates Δ from U_0. In fact U^* must be the full complementary component of $\Lambda(J_1)$, for if not, there would be a translate of U_0 also separated from Δ by $\Lambda(J_1)$. Similarly, U_0 is the full complementary component of $\Lambda(J_2)$. We conclude that $J_1 = H^*$, $J_2 = H_0$.

H_0 and H^* are now non-conjugate factor subgroups in the purely loxodromic group G, where G has an invariant component. Under these circumstances, it was shown in [11] that G is the free product of H_0 and H^* in the sense of Combination Theorem I.

We have shown that every component of G is a translate of either U_0 or U^*; in particular, every component of G is a circular disc.

Finally, we remark that for $t \epsilon T$, the translates of U_t and U^* all have pairwise disjoint closures, hence the corresponding component subgroups have trivial pairwise intersections. By Chuckrow's theorem [5], the component subgroups of G have trivial pairwise intersections, and so by Theorem 3 above, the components of G have pairwise disjoint closures.

5.2. For our next example we start by constructing a Fuchsian group H, acting on the upper half plane U, and depending on a real parameter $\varepsilon > 0$. We require H to be finitely generated of the first kind, and purely hyperbolic; then U/H is a closed Riemann surface of genus $g = g(\varepsilon) \geq 2$, and H can be generated by elements $A_1, B_1, \ldots, A_g, B_g$. We require further that A_1 be the transformation $z \to 2z$, and that there be a closed fundamental polygon P for H which contains the region $D = \{z \mid 1 \leq |z| \leq 2, \varepsilon \leq \arg z \leq \pi - \varepsilon\}$.

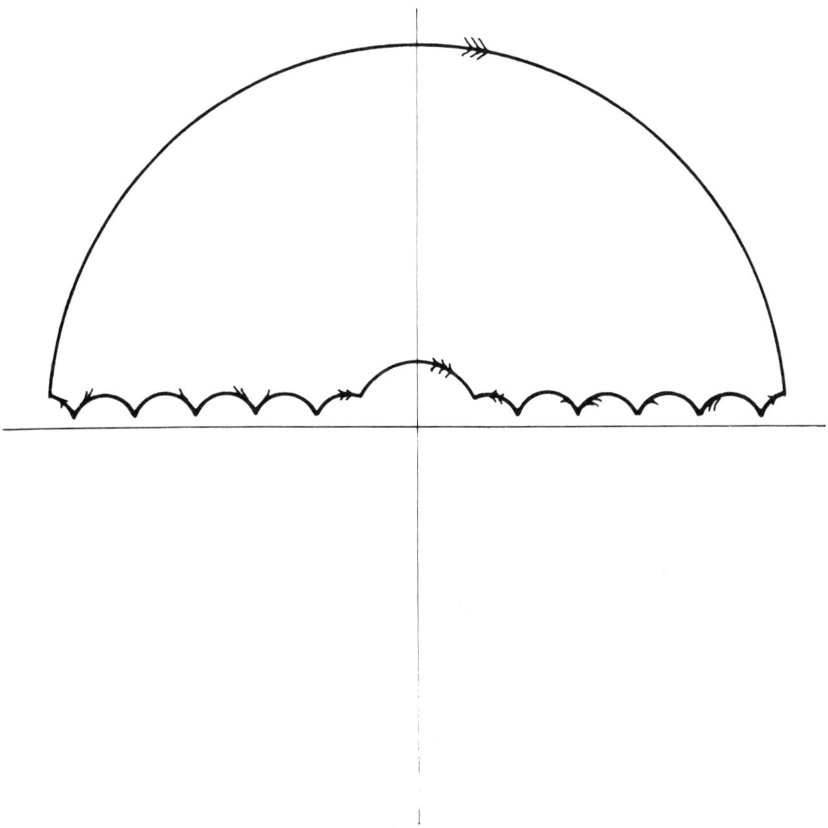

Fig. 1

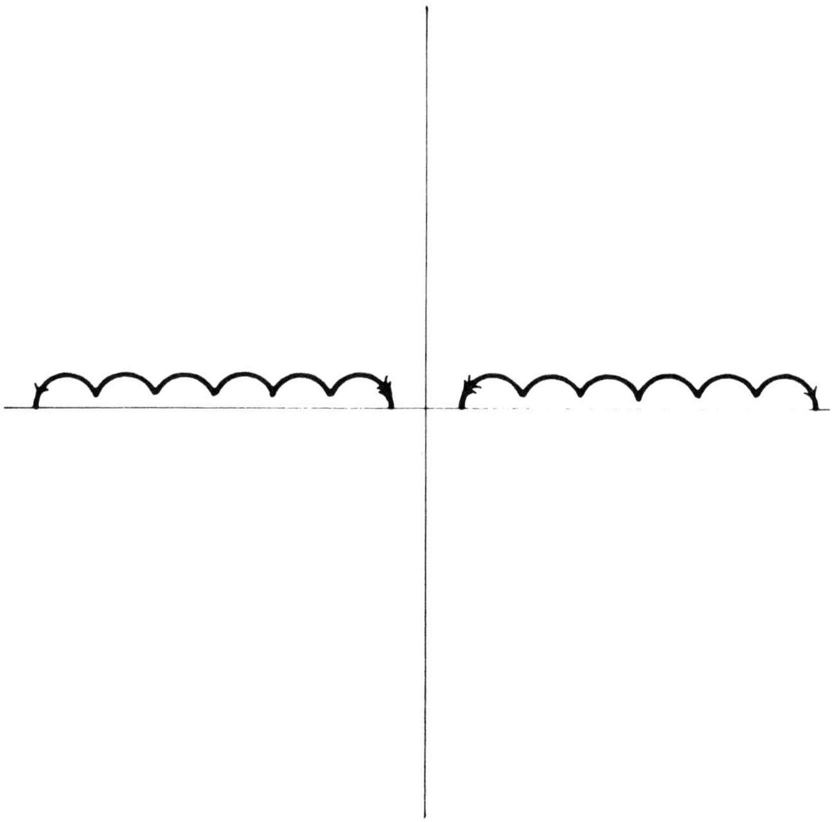

Fig. 2

We construct this auxiliary group H by constructing its fundamental polygon P; this construction is a variant of a construction due to Bers [4]. We will make P symmetric with respect to the imaginary axis. We pick $g' \geq 1$ and place $4g' + 2$ circles, each of equal radius δ, and each centered on the positive real axis, so that the first side is orthogonal to $|z| = 1$, the last side is orthogonal to $|z| = 2$, and all the other angles are equal. For appropriate choice of $\delta = \delta(g')$ each of these other angles measures $2\pi/(4g' + 1)$. For g' sufficiently large, $\delta(g') < \varepsilon$. Our group H is then the group generated by the obvious identification of the sides of P (see Figure 1, for $g' = 1$). By Poincaré's polygon theorem [15], H is Fuchsian and has the desired properties.

The group H above has a distinguished subgroup H_0 generated by the elements $B_1, A_2, \ldots, B_g$. One easily sees that a fundamental polygon $\tilde{P}_0$ for H_0 can be obtained from P by deleting the sides $|z| = 1$, $|z| = 2$, and extending the other four sides (see Figure 2).

We want to regard H_0 as a Kleinian group, and so we get a fundamental domain P_0 for H_0 by reflecting $\tilde{P}_0$ in the real axis.

Let r be the rotation $z \to e^{2\pi i/n}z$, and let $H_i = r^n H_0 r^{-n}$. Then $P_i = r^i(P_0)$ is a fundamental domain for H_i. If we choose $\varepsilon < \pi/2n$, then the regions P_i all have disjoint boundaries, and so by Combination Theorem I [13], the group G_1 generated by $H_0, H_1, \ldots, H_{n-1}$ is Kleinian, purely loxodromic, and has $\cap P_i$ as a fundamental domain.

Since we started with the Fuchsian group H, we could apply Combination Theorem II and adjoin $z \to 2z$ to G_1 to get a Kleinian group. We

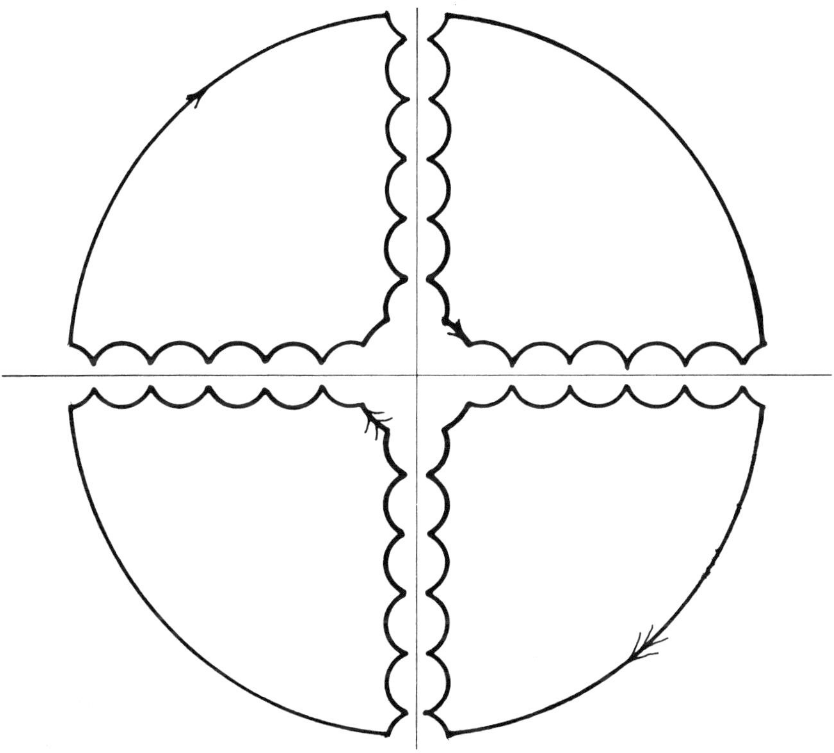

Fig. 3

instead adjoin the element $g_0 : z \to 2\, e^{\pi i/n} z$ to G_1; Combination Theorem II [16] is still applicable, and one easily sees that the resulting group has the desired properties (see Figure 3 for the case $n = 2$).

We remark that the element g_0 cyclically permutes an even number of components of G. It is not known whether or not this restriction to an even number of components is necessary.

5.3. For our next example, we start with the classical unimodular group G_0, and adjoin to it the element $h : z \to z + 4i$. The weak form of Combination Theorem II given in [14], shows that G, the group generated by G_0 and h is Kleinian. One sees at once that there are infinitely many components of G having ∞ as a boundary point, and that h keeps no component of G invariant.

5.4. For our last example, we pick a sequence of disjoint rectangles $R_1, R_2, \ldots,$ in the plane, where the R_n all lie on the boundary of a common region, they are all bounded, they all have height one, and width tending to zero. For each n, we pick a finite number of tangent circles which cover R_n. These circles are chosen so that they all bound a common region, and any two circles intersect only at points of tangency. These circles are chosen so that they separate the plane into an unbounded region D_0, and a sequence of regions $D_1, D_2, \ldots,$ where R_n separates D_n from ∞. We choose the circles sufficiently small, so that the diameter of D_n does not converge to zero. Let $\tilde{G}_1$ be the group generated by the reflections in all these circles. Then $\tilde{G}_1$ is discontinuous and $\cap D_i$ is at least part of a fundamental domain for $\tilde{G}_1$.

We choose a small circle C_i inside each D_i, $i = 1, 2, \ldots,$ and a circle C'_i of the same radius as C_i, inside D_0, where the C'_i are all disjoint and all lie on the boundary of a common region. Let g_i be a loxodromic transformation mapping the outside of C_i onto the inside of C'_i.

One easily sees that G, the orientation preserving half of the group generated by $\tilde{G}_1$ together with $g_1, g_2, \ldots,$ has the desired properties.

REFERENCES

[1]. Ahlfors, L. V., Finitely generated Kleinian groups, Am. J. Math. 86 (1964), pp. 413-429.

[2]. Beardon, A., and Maskit, B., Limit points of Kleinian groups and finite-sided fundamental polyhedra, Acta Math., (to appear).

[3]. Bers, L., On boundaries of Teichmüller spaces and on Kleinian groups: I, Ann. of Math. 91 (1970), pp. 570-600.

[4]. Bers, L., Riemann Surfaces, Lecture notes, N.Y.U. (1958).

[5]. Chuckrow, V., On Schottky groups with applications to Kleinian groups, Ann. of Math. (2), 88 (1968), pp. 47-61.

[6]. Fenchel, W., and Nielsen, J., Discontinuous groups of non-Euclidean motions, (unpublished manuscript).

[7]. Koebe, P., Über die Uniformisierung der Algebraischen Kurven II, Math. Ann. 69 (1910), pp. 1-81.

[8]. Koebe, P., Über die Uniformisierung der Algebraischen Kurven III, Math. Ann. 72 (1912), pp. 437-516.

[9]. Kra, I., and Maskit, B., Involutions on Kleinian groups, Bull. Amer. Math. Soc. 78 (1972), pp. 801-805.

[10]. Marden, A., On finitely generated Fuchsian groups, Comm. Math. Helv. 42 (1967), pp. 81-85.

[11]. Maskit, B., Decomposition of certain Kleinian groups , Acta Math. 130 (1973), pp. 243-263.

[12]. Maskit, B., On boundaries of Teichmüller spaces and on Kleinian groups: II, Ann. of Math. 91 (1970), pp. 607-639.

[13]. Maskit, B., On Klein's Combination Theorem, Trans. Amer. Math. Soc. 120 (1965), pp. 499-509.

[14]. Maskit, B., Construction of Kleinian groups, Proc. of the Conf. on Complex Anal., Minneapolis, 1964, pp. 281-296, Springer-Verlag, 1965.

[15]. Maskit, B., On Poincaré's Theorem for Fundamental Polygons, Advances in Math. 7 (1971), pp. 219-230.

[16]. Maskit, B., On Klein's Combination Theorem III, Advances in the Theory of Riemann Surfaces, pp. 297-316, Princeton University Press, 1971.

POLYNOMIAL APPROXIMATION IN THE BERS SPACES

Thomas Metzger* and Mark Sheingorn**

Introduction

In his 1965 paper [2], Bers began the study of his Banach spaces of automorphic forms. In that paper he defines for any region D and discontinuous group Γ on $D: A_q(D,\Gamma) = \{f | f$ is a holomorphic automorphic form of degree $-2q$ with respect to Γ on D, and $\int_R |f| \lambda_D^{2-q} \, dxdy < \infty\}$. (R is a standard fundamental region and λ_D is the Poincaré metric for D.) With the integral providing the norm, this is a Banach space independent of R. Bers develops its fundamental properties in that paper. Among the most interesting is

THEOREM A. *If* D *is simply connected and not the whole plane and* $q > 1$, *then the space of Poincaré series of* $A_q(D,\langle 1\rangle)$ *is dense in* $A_q(D,\Gamma)$ ([2], p. 96).

REMARKS. (a) $A_q(D,\langle 1\rangle)$ is just the set of analytic functions f on D with $\int_D |f| \lambda_D^{2-q} \, dxdy < \infty$. We denote it by $A_q(D)$.

(b) $\theta(f)(z)$, the Poincaré series of the function f, is defined as:

$$\theta(f)(z) = \sum_{A \in \Gamma} f(A_z) A'^q(z) \overline{\mu(A)} ,$$

where μ is the multiplier system.

---

* Texas A and M University, College Station, Texas 77843

** This work was done while the author was a National Academy of Sciences-National Research Council Postdoctoral Research Associate at the National Bureau of Standards, Washington, D. C. 20234.

(c) In a subsequent paper ([1], p. 197), Bers was able to show that

$$\theta : A_q(D) \to A_q(D, \Gamma)$$

as defined above is a continuous, onto mapping.

Thus, by remark (c), questions of density in $A_q(D, \Gamma)$ can be settled by considering the corresponding problem in $A_q(D)$. However, Bers points out that "The usefulness of (Theorem A) is limited by the difficulty of describing explicity those holomorphic functions (which are in $A_q(D)$)" ([2], p. 96).

In another subsequent paper [3], Bers resolves this question for $q \geq 2$ by showing

THEOREM B. *If* D *is any simple connected region and* $q \geq 2$ *then the set of polynomials is dense in* $A_q(D)$.

As we said above, this immediately implies that Poincaré series of polynomials are dense in $A_q(D, \Gamma)$, if $q \geq 2$ and D is simply connected. This interesting fact allows us to construct non-vanishing Poincaré series (see Knopp, [8], p. 601).

The situation for $1 < q < 2$ is much more difficult. For example, it is not obvious that bounded functions are in $A_q(D)$ if $1 < q < 2$. I.e., in general we cannot decide if $\int_D \lambda_D^{2-q}\, dxdy < \infty$. Certain special cases have been handled: the case of the unit disc as well as regions arising in the work of Earle and Marden [7] have been resolved in Sheingorn [11], where he showed that Theorem B held in those cases.

It is the purpose of this paper to present two large classes of regions for which Theorem B holds:

THEOREM 1. *Let* D *be simply connected, and* $\psi : U \to D$ *a conformal map of the unit disc onto* D. *Then*

(a) $\psi' \in H^1 \implies$ *bounded functions (and thus polynomials) are in* $A_q(D)$.

(b) $\psi' \in H^{1+\delta}$, *some* $\delta > 0 \implies$ *polynomials are dense in* $A_q(D)$

THEOREM 2. *Let* D *be a Smirnov region. Then the polynomials are dense in* $A_q(D)$.

REMARKS. (1) In Theorem 1, the H spaces are Hardy spaces. (2) $\psi' \in H^1$ is known to be equivalent to D having rectifiable boundary.

These theorems may be substantially stronger than Bers anticipated when he wrote in a similar context, "The case $1 < q < 2$ could be included under certain smoothness conditions on the boundary of D, but the domains occurring in applications are either discs or bounded by non-differentiable Jordan curves." [2], p. 97.

§1. In this section we prove Theorem 1.

Proof of (a). We will in fact show that $\int_D \lambda_D^{2-q}(z)\, dxdy < C(\|\psi'\|_1)^q$ where $\|\ \|_1$ is the H^1 norm. By change of variables, this is the same as showing

$$(*) \qquad \int_U \frac{|\psi'(z)|^q}{(1-|z|^2)^{2-q}}\, dxdy < C(\|\psi'\|_1)^q .$$

We need Duren's extension of a lemma of Carleson:

THEOREM C ([5], p. 163). *Let* μ *be a finite measure on* $|z| < 1$ *and suppose* $0 < p \le q < \infty$. *Then in order that there exist a constant* C *such that for all* $f \in H^p$:

$$\left\{ \int_U |f(z)|^q\, d\mu(z) \right\}^{\frac{1}{q}} \le C\, \|f\|_p ,$$

it is necessary and sufficient that $\mu(S) \le A\, h^{q/p}$ *for every set* S *of the form*

$$(1) \qquad \{ re^{i\theta} \mid 1 - h \le r < 1,\ \theta_0 \le \theta \le \theta_0 + h \} .$$

Here, of course, A is a constant independent of h. If we can apply Theorem C with

$$\mu(E) = \int_E \frac{dxdy}{(1-|z|^2)^{2-q}}$$

we will have shown (*). We calculate $\mu(s)$ with any S of form (1):

$$\mu(s) = \int_S \frac{dxdy}{(1-|z|^2)^{2-q}} = \int_{\theta_0}^{\theta_0+h} \int_{1-h}^{1} \frac{r\,drd\theta}{(1-r^2)^{2-q}} = \int_{1-h}^{1} \int_{\theta_0}^{\theta_0+h} \frac{r\,d\theta}{(1-r^2)^{2-q}}\,dr$$

$$= h\left(\int_{1-h}^{1} \frac{r\,dr}{(1-r^2)^{2-q}}\right) \frac{h}{2}\left(\int_{(1-h)^2}^{1} \frac{dm}{(1-m)^{2-q}}\right) = \frac{h}{2}\left(\int_{(1-h)^2}^{1} (1-m)^{q-2}\,dm\right)$$

$$= -\frac{h}{2} \cdot \frac{(1-m)^{q-1}}{q-1}\Bigg|_{(1-h)^2}^{1} = -\frac{h}{2} \cdot \frac{-(2h-h^2)^{q-1}}{q-1} = h^q \cdot \frac{(2-h)^{q-1}}{2(q-1)} \le A\,h^q.$$

Thus the condition of Theorem C (with $q = q$, $p = 1$) is verified and (*) follows.

Proof of (b). The proof proceeds in four lemmas, the first three of which have appeared elsewhere.

LEMMA 1. *Suppose there exist polynomials* P_n *such that*

$$\int_U \frac{|P_n(z)\,\psi'^q(z) - 1|}{(1-|z|^2)^{2-q}}\,dxdy \to 0 .$$

Then the polynomials are dense in $A_q(D)$.

This lemma is stated in H. Shapiro [9]. A proof appears in Sheingorn [11].

LEMMA 2. *Let* μ *be a positive measure on* U. *Suppose* $f \in H(U)$, f *non-vanishing, and for some* $\varepsilon > 0$, *we have*:

(1) $$\int_U |f(z)|\, m(r)^{-\varepsilon}\, d\mu < \infty$$

(2) $$\int_U m(r)^{-\varepsilon}\, d\mu < \infty$$

where

$$m(r) = \min_\theta |f(re^{i\theta})| \ .$$

Then there exist polynomials P_n *such that* $\int_U |P_n(z) f(z) - 1|\, d\mu \to 0$.

This lemma appears in H. Shapiro [10].

LEMMA 3. *Let*

$$\mu(E) = \int_E \frac{dxdy}{(1-|z|^2)^{2-q}} \ .$$

Let ψ *be as in Theorem* 1. *Suppose that*

(**) $$\int_U \frac{|\psi'(z)|^q}{(1-|z|^2)^{2-q+\varepsilon}}\, dxdy < \infty, \quad \textit{some} \ \ \varepsilon > 0 \ .$$

Then there exists $\varepsilon_0 > 0$ *such that:*

(1) $$\int_U |\psi'(z)|^q\, m(r)^{-\varepsilon_0}\, d\mu < \infty$$

(2) $$\int_U m(r)^{-\varepsilon_0}\, d\mu < \infty$$

where

$$m(r) = \min_\theta |\psi'(re^{i\theta})| \ .$$

The proof is contained in Sheingorn [11], but it is not state there explicitly. We give the proof now. Since $\psi : U \to D$ conformally,

$|\psi'(re^{i\theta})| \geq K(1-r)$, where K depends only on ψ. Thus $m(r) \geq K(1-r)$ and $m(r)^{-\varepsilon} \leq K^{-\varepsilon}(1-r)^{-\varepsilon}$. Thus

$$\int_U m(r)^{-\varepsilon}\, d\mu \leq K^{-\varepsilon} \int_U (1-|z|^2)^{q-2}(1-r)^{-\varepsilon}\, dxdy$$

$$\leq K^{-\varepsilon} \int_U (1-|z|^2)^{q-2-\varepsilon}\, dxdy$$

since $(1+r)^{-\varepsilon} < 1$. The last integral is finite if $\varepsilon < q-1 = \varepsilon_1$ so we have shown (2).

In precisely the same way we get

$$\int_U |\psi'^q(z)|\, m(r)^{-\varepsilon}\, d\mu \leq K^{-\varepsilon} \int_U |\psi'^q(z)|(1-|z|^2)^{q-2-\varepsilon} dxdy\ .$$

The last integral is finite by (**) so we have (1). We get (1) and (2) with the same ε_0 by choosing $\varepsilon_0 < \min(\varepsilon, \varepsilon_1)$.

LEMMA 4. *Assume* $\psi' \in H^{1+\delta}$, *some* $\delta > 0$. *Then* (**) *holds for some* $\varepsilon > 0$.

Proof. We proceed as before, with Theorem C. Define

$$\mu(E) = \int_E \frac{dxdy}{(1-|z^2|)^{2-q+\varepsilon}}\ .$$

This will be finite if $\varepsilon < q-1$; we will specify ε exactly later. We must compute $\mu(S)$ again:

$$\mu(S) = \int_S \frac{dxdy}{(1-|z|^2)^{2-q+\varepsilon}} = h \cdot \int_{1-h}^{1} \frac{r\, dr}{(1-r^2)^{2-q+\varepsilon}} \leq A\, h^{q-\varepsilon}\ ,$$

Apply Theorem C with $q = q$, $p = \frac{q}{q-\varepsilon}$ now choosing ε so small that $p \leq q$ and $p \leq 1+\delta$. This can be done since $q > 1$ and $\delta > 0$. We get that

$$\int_U \frac{|\psi'(z)|^q}{(1-|z^2|^{2-q+\varepsilon}} \, dxdy \leq C \, \|\psi'\|_{1+\delta} < \infty \, .$$

This completes the proof of Lemma 4.

The proof of Theorem 1, (b) follows easily from the four lemmas taken in reverse order.

Theorem 1 has the following:

COROLLARY 1. *Under the condition of Theorem* 1 (b), *the set* $\{\theta(p) | p$ *a polynomial*$\}$ *is dense in* $A_q(D,\Gamma)$, $q > 1$.

§2. The space $E^p(D)$, $0 < p < \infty$, D a simply connected region, is defined as follows: A function f analytic in D is of class $E^p(D)$ if there exists a sequence of rectifiable Jordan curves C_n tending to the boundary of D (i.e., any compact $K \subset D$ is contained in int C_n for large n) such that

$$\int_{C_n} |f(z)|^p \, |dz| \leq m < \infty \, .$$

Smirnov first defined these spaces. It has been shown that the polynomials are dense in $E^p(D)$ if and only if D is a Smirnov domain. In our notation a Smirnov domain is one for which ψ' is an inner function (Duren [5], p. 173).

It is not too difficult to see that $f \in E^p(D)$ if $(f \circ \psi)(\psi')^{\frac{1}{p}} \in H^p$ and that

$$\|f\|_{E^p(D)} \geq \left\| f \circ \psi \cdot \psi'^{\frac{1}{p}} \right\|_{H^p} \tag{1}$$

(Duren [5], pp. 168-169).

Duren, Romberg, and Sheilds have established a connection between the Hardy spaces and the spaces $A_q(U)$ ([6]). We state this as

THEOREM D. *Let* $q > 1$. *Then*

$$\|f\|_{H^{\frac{1}{q}}} \geq \|f\|_{A_q(U)} . \tag{2}$$

In addition, $\overline{H^{\frac{1}{q}}} = A_q(U)$, *i.e.,* $H^{\frac{1}{q}}$ *is dense in* $A_q(U)$.

We now proceed with the proof of Theorem 2. Choose $\varepsilon > 0$ and $f \in A_q(D)$. Then as before by change of variable, $(f \circ \psi) \cdot \psi'^q \in A_q(U)$. Since $H^{\frac{1}{q}}$ is dense in $A_q(U)$, we can find $G \in H^{\frac{1}{q}}$ such that

$$\|(f \circ \psi) \cdot \psi'^q - G\|_{A_q(U)} < \varepsilon/2 .$$

But by the above, $G = (g \circ \psi) \cdot \psi'^q$ where $g \in E^{\frac{1}{q}}(D)$. Hence

$$\frac{\varepsilon}{2} > \|(f \circ \psi) \cdot \psi'^q - (g \circ \psi) \cdot \psi'^q\|_{A_q(U)} = \|f - g\|_{A_q(D)} . \tag{\#}$$

Since D is a Smirnov domain, there exist a polynomial P such that $\|g - P\|_{E^{\frac{1}{q}}(D)} < \varepsilon/2$. By (1) this implies that:

$$\|(g - P) \circ \psi \cdot \psi'^q\|_{H^{\frac{1}{p}}} < \varepsilon/2 .$$

But then (2) implies that

$$\|(g - P) \circ \psi \cdot \psi'^q\|_{A_q(U)} < \varepsilon/2 .$$

Changing variables again,

$$\varepsilon/2 > \|g - P\|_{A_q(D)} . \tag{\#\#}$$

Now (#) and (##) imply Theorem 2.

COROLLARY 2. *Let* $q > 1$ *and* D *be a Smirnov domain. Then the set* $\{\theta(p) | P$ *is a polynomial*$\}$ *is dense in* $A_q(D, \Gamma)$.

REFERENCES

[1]. Bers, L., *Automorphic Forms and Poincaré Series for infinitely generated Fuchsian groups*, American Journal of Mathematics, Vol. 87 (1965), pp. 196-214.

[2]. ———, *Completeness theorems for Poincaré series in one variable*, Proceedings of the International Symposium on Linear Spaces, Jerusalem, Jerusalem Academic Press - Pergamon Press, (1961), pp. 88-100.

[3]. ———, *A non-standard integral equation with applications to quasi-conformal mappings*, Acta Mathematica, vol. 116 (1966), pp. 113-134.

[4]. Carleson, L., *Interpolations by bounded analytic functions and the corona problem*, Annals of Mathematics, vol. 76 (1962), pp. 547-559.

[5]. Duren, P., *Theorem of* H^p *Spaces*, New York, Academic Press, 1970.

[6]. Duren, P., Romberg, B., and Sheilds, A., *Linear functionals on* H^p *spaces with* $0 < p < 1$, J. Reine Angew. Math., vol. 238 (1969), pp. 32-60.

[7]. Earle, C., and Marden, A., *Projections to automorphic forms*, Proc. Amer. Math. Soc., vol. 19 (1968), pp. 274-278.

[8]. Knopp, M., *A corona theorem for automorphic forms and related results*, American Journal of Mathematics, vol. 91 (1969), pp. 599-618.

[9]. Shapiro, H., *Some remarks on weighted polynomial approximations of holomorphic functions*, Math. USSR. Sbornik, vol. 2 (1967), pp. 285-294.

[10]. ———, *Weighted polynomial approximation and boundary behavior of analytic functions*, Contemporary Problems in Theory of Analytic Functions, Moscow, Nauka 1966, pp. 326-335. (Russian)

[11]. Sheingorn, M., *Poincaré series of polynomials bounded away from zero on a fundamental region*, American Journal of Mathematics, (to appear).

SIMPLE ILLUSTRATIONS OF THE USES OF EXPLICIT COMPUTATION OF THETA CONSTANTS

Harry E. Rauch

§1. *Introduction*

In deriving relations among g-theta constants for a given $g \geq 1$, in particular, identities and linear transformation formulas, and in applying these to particular cases, one often finds that general algorithms lead quickly to results which are well determined up to signs or, more generally, roots of unity. A technique that is often useful for determining these roots of unity is to specialize the theta matrix (period matrix) in the relation and then use the remarkably rapid convergence of the theta series for explicit numerical computation. With any luck, the "origin" and "first layer" of terms suffice. Examples of this technique for $g = 1$ are given in the book by Rauch and Lebowitz (see references). In the present manuscript a particularly cogent (because it seems the only available technique) example for $g = 2$ is given, one based on the paper by Rauch listed in the references. Section 2 below is a modification of some material to appear in the book "Theta Functions with Applications to Riemann Surfaces" by H. Rauch and H. Farkas which will be published by Williams & Wilkins, Baltimore, in January, 1974, and which they have copyrighted.

§2. *The example for* $g = 2$

We are going to exhibit a particular element M of $Sp(2, Z)$ and discuss its action on the theta constants by means of Theorems A, B, C, D in reference [1] below. The origin of M is in Riemann surface theory, see same reference, but we do not need to know this here. What we do need to know from that source is that there is a unique theta matrix π (to be distinguished from the number π by the context) such that $M \circ \pi = \pi$, and it is instructive, if not essential, to know that $M^5 = I_4$, which alerts us to the possibility that the fifth roots of unity may play a role. We give here a reworking and elaboration of parts of the paper by Rauch, "Theta Constants on a Riemann Surface...," listed in the references.

Let

$$1) \qquad M = \begin{pmatrix} 1 & 0 & 1 & 0 \\ -1 & 0 & -1 & 1 \\ -1 & -1 & 0 & 0 \\ 0 & -1 & 0 & 0 \end{pmatrix} = \begin{pmatrix} D & C \\ B & A \end{pmatrix} .$$

We verify that $M \in Sp(2, Z)$.

We want to find the fixed matrix π. Using $A = 0_2$ and $\det|B| = 1$, we find

$$2) \qquad \pi = M \circ \pi = (A\pi + B)(C\pi + D)^{-1} = B(C\pi + D)^{-1}$$

and thus

$$3) \qquad \Delta = \pi_{11}\pi_{22} - \pi_{12}^2 = \det|\pi| = \frac{1}{\det|C\pi + D|} .$$

But from 1) we find

$$4) \qquad (C\pi + D) = \begin{pmatrix} \pi_{11} + 1 & \pi_{12} \\ -\pi_{11} + \pi_{12} - 1 & -\pi_{12} + \pi_{22} \end{pmatrix}$$

so that, using 3), one finds

$$\frac{1}{\Delta} = \det|C\pi + D| = \pi_{11}\pi_{22} - \pi_{12}^2 + \pi_{22} = \Delta + \pi_{22}$$

or

$$5) \qquad \pi_{22} = \frac{1}{\Delta} - \Delta .$$

From 2), 3), 4) we derive

$$\pi = \frac{1}{\det|C\pi+D|}\begin{pmatrix} 2\pi_{12}-\pi_{11}-\pi_{22}-1 & \pi_{12}-\pi_{11}-1 \\ \pi_{12}-\pi_{11}-1 & -\pi_{11}-1 \end{pmatrix}$$

from which we obtain

6) a) $\pi_{11} = (2\pi_{12}-\pi_{11}-\pi_{22}-1)\Delta$

b) $\pi_{12} = (\pi_{12}-\pi_{11}-1)\Delta$

c) $\pi_{22} = (-\pi_{11}-1)\Delta$.

We solve 6c) for π_{11}, using 5) to obtain

7) $$\pi_{11} = \left(\Delta - \frac{1}{\Delta}\right)/\Delta - 1 = -\frac{1}{\Delta^2} ,$$

and we solve 6b) for π_{12}, using 6c) and 5),

$$\pi_{12}(1-\Delta) = (-\pi_{11}-1)\Delta = \pi_{22} = \frac{1-\Delta^2}{\Delta}$$

or

8) $$\pi_{12} = \frac{1+\Delta}{\Delta} = 1 + \frac{1}{\Delta} .$$

Substituting 5), 7), 8), we obtain

$$\Delta = \pi_{11}\pi_{22} - \pi_{12}^2 = \frac{-1-\Delta-\Delta^2-\Delta^3}{\Delta^3}$$

or

9) $$\Delta^4 + \Delta^3 + \Delta^2 + \Delta + 1 = 0$$

so that $\Delta \neq 1$ is a fifth root of unity.

To determine Δ we set $\varepsilon = e^{2\pi i/5}$ and $\Delta = \varepsilon^{\ell}$, $1 \leq \ell \leq 4$ so that

$$\pi = \begin{pmatrix} -\varepsilon^{-2\ell} & 1+\varepsilon^{-\ell} \\ 1+\varepsilon^{-\ell} & \varepsilon^{-\ell}-\varepsilon^{\ell} \end{pmatrix} .$$

We want π to be a theta matrix so that $\operatorname{Im} \pi$ must be positive-definite, in particular,

$$\operatorname{Im} \pi_{11} = \sin(4\pi\ell/5) > 0$$
$$\operatorname{Im} \pi_{22} = -2 \sin(2\pi\ell/5) > 0 \ ,$$

both of which hold if and only if $\ell = 3$, when $1 \leq \ell \leq 4$. Thus

10) a) $\Delta = \varepsilon^3$

b) $$\pi = \begin{pmatrix} -\varepsilon^4 & 1+\varepsilon^2 \\ 1+\varepsilon^2 & \varepsilon^2-\varepsilon^3 \end{pmatrix} .$$

It will be useful to compute $\varepsilon, \varepsilon^2, \varepsilon^3, \varepsilon^4$ explicitly. We use Gauss's device and set

$$\eta_1 = \varepsilon^2 + \varepsilon^3, \quad \eta_2 = \varepsilon + \varepsilon^4$$

and note that

$$\eta_1 \rightarrow \eta_2 = \varepsilon^3 + \varepsilon + \varepsilon^4 + \varepsilon^2 = -1$$
$$\eta_1 + \eta_2 = \varepsilon^2 + \varepsilon^3 + \varepsilon + \varepsilon^4 = -1 \ .$$

Thus η_1, η_2 are the roots of

$$(x-\eta_1)(x-\eta_2) = x^2 - (\eta_1+\eta_2) + \eta_1 \eta_2 = x^2 + x - 1 = 0$$

whose roots are also $(-1 \pm \sqrt{5})/2$. A diagram of the fifth roots of unity on the unit circle will convince the reader that

$$\varepsilon^2 + \varepsilon^3 = \eta_1 = \frac{-1-\sqrt{5}}{2}$$
$$\varepsilon + \varepsilon^4 = \eta_2 = \frac{-1+\sqrt{5}}{2} \ .$$

The first of these equations can be written

$$\varepsilon^2 + \frac{1}{\varepsilon^2} = \frac{-1-\sqrt{5}}{2}$$

from which one derives, after consulting the diagram of fifth roots of unity,

11) $$\varepsilon^2 = \frac{-1-\sqrt{5}+i\sqrt{10-2\sqrt{5}}}{4}$$

$$\varepsilon^3 = \frac{-1-\sqrt{5}-i\sqrt{10-2\sqrt{5}}}{4} .$$

Similarly from the expression for $\varepsilon + \varepsilon^4$, since $\varepsilon^4 = 1/\varepsilon$, we derive

12) $$\varepsilon = \frac{-1+\sqrt{5}+i\sqrt{10+2\sqrt{5}}}{4}$$

$$\varepsilon^4 = \frac{-1+\sqrt{5}-i\sqrt{10+2\sqrt{5}}}{4} .$$

We now apply Theorem A to the ten $= 2^{g-1}(2^g+1) = 2(2^2+1)$ 2-theta constants, with π defined by 10b), M given by 1), $\sqrt{\det|C\pi+D|} = \sqrt{1/\Delta} = \sqrt{\varepsilon^2} = \varepsilon$ by 3), 10a), and a convenient choice of square root. Since $M \circ \pi = \pi$ we find the remarkable fact that *except for multiplication by roots of unity the theta constants in question are simply permuted among themselves and, more than that, are equal in each set of transitivity.* Let us carry out the calculations, to see this. We find

$$\mathrm{dv}(CD^T) = \mathrm{dv}\begin{pmatrix} 1 & -1 \\ -1 & 1 \end{pmatrix} = \begin{pmatrix} 1 \\ 1 \end{pmatrix}$$

$$\mathrm{dv}(AB^T) = \mathrm{dv}\begin{pmatrix} 0 & 0 \\ 0 & 0 \end{pmatrix} = \begin{pmatrix} 0 \\ 0 \end{pmatrix}$$

and thus that

$$\tilde{\varepsilon} = \begin{pmatrix} 1 & 0 \\ -1 & 0 \end{pmatrix}\begin{pmatrix} \varepsilon_1 \\ \varepsilon_2 \end{pmatrix} - \begin{pmatrix} 1 & 0 \\ -1 & 1 \end{pmatrix}\begin{pmatrix} \varepsilon_1' \\ \varepsilon_2' \end{pmatrix} + \begin{pmatrix} 1 \\ 1 \end{pmatrix} = \begin{pmatrix} \varepsilon_1 - \varepsilon_1' + 1 \\ -\varepsilon_1 + \varepsilon_1' - \varepsilon_2' + 1 \end{pmatrix}$$

$$\tilde{\varepsilon}' = -\begin{pmatrix} -1 & -1 \\ 0 & -1 \end{pmatrix}\begin{pmatrix} \varepsilon_1 \\ \varepsilon_2 \end{pmatrix} = \begin{pmatrix} \varepsilon_1 + \varepsilon_2 \\ \varepsilon_2 \end{pmatrix} .$$

We also find that

$$\emptyset\left(M,\begin{bmatrix}\varepsilon\\ \varepsilon'\end{bmatrix}\right)=\frac{1}{4}\left\{-(\varepsilon_1,\varepsilon_2)\begin{pmatrix}-1 & 0\\ 0 & 0\end{pmatrix}\begin{pmatrix}\varepsilon_1\\ \varepsilon_2\end{pmatrix}+2(\varepsilon_1',\varepsilon_2')\begin{pmatrix}-1 & 0\\ 0 & -1\end{pmatrix}\begin{pmatrix}\varepsilon_1\\ \varepsilon_2\end{pmatrix}\right\}$$

$$=\frac{1}{4}\left\{\varepsilon_1^2-2(\varepsilon_1\varepsilon_1'+\varepsilon_2\varepsilon_2')\right\} .$$

Thus, applying Theorem A in succession to $\theta\begin{matrix}0 & 0\\ 0 & 0\end{matrix}$ and its transforms while always using the reduced representation of any characteristic which occurs and the reduction formula, we obtain

$$\theta\begin{bmatrix}1 & 1\\ 0 & 0\end{bmatrix}=\kappa(M)\varepsilon\,\theta\begin{bmatrix}0 & 0\\ 0 & 0\end{bmatrix}$$

$$\theta\begin{bmatrix}0 & 0\\ 0 & 1\end{bmatrix}=\kappa(M)\varepsilon\,e^{\pi i/4}\,\theta\begin{bmatrix}1 & 1\\ 0 & 0\end{bmatrix}$$

$$\theta\begin{bmatrix}1 & 0\\ 0 & 0\end{bmatrix}=\kappa(M)\varepsilon\,\theta\begin{bmatrix}0 & 0\\ 0 & 1\end{bmatrix}$$

$$\theta\begin{bmatrix}0 & 0\\ 1 & 0\end{bmatrix}=\kappa(M)\varepsilon\,e^{\pi i/4}\,\theta\begin{bmatrix}1 & 0\\ 0 & 0\end{bmatrix}$$

$$\theta\begin{bmatrix}0 & 0\\ 0 & 0\end{bmatrix}=\kappa(M)\varepsilon\,\theta\begin{bmatrix}0 & 0\\ 1 & 0\end{bmatrix} .$$

We make an immediate deduction from 13) by successive substitution of the first four lines into the fifth to obtain

$$\kappa^5(M)\varepsilon^5\,e^{\pi i/2}=1$$

$$\kappa^5(M)=-i .$$

But $\kappa(M)=e^{2\pi i\ell/8}$ for some ℓ by the last assertion of Theorem A so we must have

$$e^{10\pi i\ell/8}=e^{3\pi i/2}$$

or

$$5\ell\equiv 6 \pmod 8$$

whose solution is $\ell\equiv 6 \pmod 8$ so that

14) $$\kappa(M)=-i .$$

We are fortunate in this case not to have to use Theorems B and D to compute $\kappa(M)$ even though it is easy to find, using the generators and method of proof of Theorem C, that

$$M\ {}^{-}C_1\ {}^{-}A_{21}\ {}^{+}B_2\ {}^{+}A_{21}\ {}^{-}C_2\ {}^{+}B_2 = I_2$$

or

$$M = {}^{-}B_2\ {}^{+}C_2\ {}^{-}A_{21}\ {}^{-}B_2\ {}^{+}A_{21}\ {}^{+}C_1\ .$$

We should also observe that, for example, $M \circ \begin{bmatrix} 1 & 1 \\ 0 & 0 \end{bmatrix} = \begin{bmatrix} 2 & 0 \\ 2 & 1 \end{bmatrix}$ and that the reduction formula implies $\theta\begin{bmatrix} 2 & 0 \\ 2 & 1 \end{bmatrix} = (-1)^2\,\theta\begin{bmatrix} 0 & 0 \\ 0 & 1 \end{bmatrix}$. In 15) below, on the other hand, one finds $M \circ \begin{bmatrix} 1 & 1 \\ 1 & 1 \end{bmatrix} = \begin{bmatrix} 1 & 0 \\ 2 & 1 \end{bmatrix}$ and the reduction formula implies $\theta\begin{bmatrix} 1 & 0 \\ 2 & 1 \end{bmatrix} = -\,\theta\begin{bmatrix} 1 & 0 \\ 0 & 1 \end{bmatrix}$.

The five 2-characteristics in 13) form a set of transitivity of the permutation representation induced by the action $M\circ$ of the cyclic subgroup of $Sp(2, Z)$ generated by M. We take a 2-characteristic not in that transitivity set, say, $\begin{matrix} 0 & 1 \\ 1 & 0 \end{matrix}$ and repeat the process to obtain, in view of 14) and the remarks on the reduction formula,

$$15)\qquad \theta\begin{bmatrix} 0 & 0 \\ 1 & 1 \end{bmatrix} = -\,i\varepsilon\,\theta\begin{bmatrix} 0 & 1 \\ 1 & 0 \end{bmatrix}$$

$$\theta\begin{bmatrix} 0 & 1 \\ 0 & 0 \end{bmatrix} = -\,i\varepsilon\,\theta\begin{bmatrix} 0 & 0 \\ 1 & 1 \end{bmatrix}$$

$$\theta\begin{bmatrix} 1 & 1 \\ 1 & 1 \end{bmatrix} = -\,i\varepsilon\,\theta\begin{bmatrix} 0 & 1 \\ 0 & 0 \end{bmatrix}$$

$$\theta\begin{bmatrix} 1 & 0 \\ 0 & 1 \end{bmatrix} = i\varepsilon\, e^{-3\pi i/4}\,\theta\begin{bmatrix} 1 & 1 \\ 1 & 1 \end{bmatrix}$$

$$\theta\begin{bmatrix} 0 & 1 \\ 1 & 0 \end{bmatrix} = -\,i\varepsilon\, e^{\pi i/4}\,\theta\begin{bmatrix} 1 & 0 \\ 0 & 1 \end{bmatrix}\ .$$

From 13) and 15) we derive

16)
$$\theta\begin{bmatrix}1 & 1\\ 0 & 0\end{bmatrix} = -i\,\varepsilon\,\theta\begin{bmatrix}0 & 0\\ 0 & 0\end{bmatrix}$$

$$\theta\begin{bmatrix}0 & 0\\ 0 & 1\end{bmatrix} = -\,\varepsilon^2\, e^{\pi i/4}\,\theta\begin{bmatrix}0 & 0\\ 0 & 0\end{bmatrix}$$

$$\theta\begin{bmatrix}1 & 0\\ 0 & 0\end{bmatrix} = i\,\varepsilon^3\, e^{\pi i/4}\,\theta\begin{bmatrix}0 & 0\\ 0 & 0\end{bmatrix}$$

$$\theta\begin{bmatrix}0 & 0\\ 1 & 0\end{bmatrix} = i\,\varepsilon^4\,\theta\begin{bmatrix}0 & 0\\ 0 & 0\end{bmatrix}$$

and

17)
$$\theta\begin{bmatrix}0 & 0\\ 1 & 1\end{bmatrix} = -\,i\,\varepsilon\,\theta\begin{bmatrix}0 & 1\\ 1 & 0\end{bmatrix}$$

$$\theta\begin{bmatrix}0 & 1\\ 0 & 0\end{bmatrix} = -\varepsilon^2\,\theta\begin{bmatrix}0 & 1\\ 1 & 0\end{bmatrix}$$

$$\theta\begin{bmatrix}1 & 1\\ 1 & 1\end{bmatrix} = i\,\varepsilon^3\,\theta\begin{bmatrix}0 & 1\\ 1 & 0\end{bmatrix}$$

$$\theta\begin{bmatrix}1 & 0\\ 0 & 1\end{bmatrix} = -\,\varepsilon^4\, e^{-3\pi i/4}\,\theta\begin{bmatrix}0 & 1\\ 1 & 0\end{bmatrix} .$$

We now substitute from 16) and 17) into the identity on p. 311 of Rauch, "Theta Constants...," to obtain

$$\theta^4\begin{bmatrix}0 & 0\\ 0 & 0\end{bmatrix}(1+\varepsilon^3) - \theta^4\begin{bmatrix}0 & 1\\ 1 & 0\end{bmatrix}(\varepsilon^3+\varepsilon^2) = 0 .$$

We shall establish below that $\theta\begin{bmatrix}0 & 0\\ 0 & 0\end{bmatrix}$ is not zero. This also follows from the fact, established in the reference [1], that π is a period matrix of a Riemann surface of genus 2. Thus we obtain

18)
$$\frac{\theta^4\begin{bmatrix}0 & 1\\ 1 & 0\end{bmatrix}}{\theta^4\begin{bmatrix}0 & 0\\ 0 & 0\end{bmatrix}} = \frac{1+\varepsilon^3}{\varepsilon^2+\varepsilon^3}$$

$$= \frac{\varepsilon^3(1+\varepsilon^3)}{1+\varepsilon}$$

$$= -(\varepsilon+\varepsilon^3)\, \varepsilon^3(1+\varepsilon^3)$$

$$= -(\varepsilon+\varepsilon^3)^2$$

$$= -\varepsilon^4\left(\varepsilon+\frac{1}{\varepsilon}\right)^2 ,$$

where we have used $\varepsilon^2 = 1/\varepsilon^3$ and

$$\frac{1}{1+\varepsilon} = -(\varepsilon+\varepsilon^3) .$$

While this last is easily verified *a posteriori* we can derive it *ab initio* as follows. Write

$$\frac{1}{1+\varepsilon} = a + b\varepsilon + c\varepsilon^2 + d\varepsilon^3 ,$$

multiply both sides by $1 + \varepsilon$, use $1 + \varepsilon + \varepsilon^2 + \varepsilon^3 + \varepsilon^4 = 0$, and compare coefficients of powers of ε on both sides to obtain $1 = a - d$, $0 = a+b-d$, $0 = (b+c-d)$, $c = 0$, from which one derives $b = -1 = d$, $a = 0$. Taking the fourth root of both sides of 18), we obtain

$$\text{19)} \qquad \frac{\theta\begin{bmatrix}0 & 1\\ 1 & 0\end{bmatrix}}{\theta\begin{bmatrix}0 & 0\\ 0 & 0\end{bmatrix}} = (-1)^{\frac{1}{4}} \sqrt{\varepsilon+1/\varepsilon}\ \varepsilon .$$

We see from 11) that $\varepsilon + 1/\varepsilon = \varepsilon + \varepsilon^4$ is positive and we take the positive square root. There remains only the problem of determining which fourth root of (-1) to take. This can only be decided by reference to the left side of 19), and we use numerical computation.

We have

$$\text{20)} \quad \theta\begin{bmatrix}0 & 1\\ 1 & 0\end{bmatrix} = \sum_{n_1,n_2} \exp \pi i\left\{\pi_{11}n_1^2 + 2\pi_{12}n_1\left(n_2+\frac{1}{2}\right) + \pi_{22}\left(n_2+\frac{1}{2}\right)^2 + n_1\right\}$$

$$= \sum_{n_1,n_2} \exp \pi i\left\{\left(\frac{1-\sqrt{5}}{4} + i\,\frac{\sqrt{10+2\sqrt{5}}}{4}\right)n_1^2\right.$$

$$+ \left(1 + \frac{-1-\sqrt{5}+i\sqrt{10-2\sqrt{5}}}{4}\right)(2n_1)\left(n_2+\frac{1}{2}\right)$$

$$\left. + \frac{i\sqrt{10-2\sqrt{5}}}{2}\left(n_2+\frac{1}{2}\right)^2 + n_1\right\}$$

and

21)
$$\theta\begin{bmatrix}0 & 0\\ 0 & 0\end{bmatrix} = \sum_{n_1,n_2} \exp \pi i\left\{\pi_{11}n_1^2 + 2\pi_{12}n_1n_2 + \pi_{22}n_2^2\right\}$$
$$= \sum_{n_1,n_2} \exp \pi i\left\{\left(\frac{1-\sqrt{5}}{4} + \frac{i\sqrt{10+2\sqrt{5}}}{4}\right)n_1^2\right.$$
$$+ \left(1 + \frac{-1-\sqrt{5}+i\sqrt{10-2\sqrt{5}}}{4}\right)(2n_1)(n_2)$$
$$\left. + \frac{i\sqrt{10-2\sqrt{5}}}{2}\, n_2^2\right\},$$

where the explicit periods are computed from 10b), 11), and 12). Experience has shown (see Rauch and Lebowitz, Chapter IV, pp. 269-270) that very good accuracy in computing 1-theta constants is achieved by using only the terms at the "origin" and in the "first layer," i.e., altogether the terms for which $-1 \leq n \leq 1$, where n is the single summation index. The analogue for the double series of the 2-theta constants is the "origin and the first square layer," i.e., the nine terms for which $-1 \leq n_1$, $n_2 \leq 1$. Using the values

22)
$$\sqrt{5} = 2.236 \qquad \pi = 3.142$$
$$\sqrt{10-2\sqrt{5}} = 2.36 \qquad \sqrt{10+2\sqrt{5}} = 3.81$$

in 20) and 21) and the notation (n_1, n_2) for the indicated term, we find

$$\theta\begin{bmatrix}0 & 1\\ 1 & 0\end{bmatrix} = (0,0) + (0,-1) + (-1,0) + (1,-1) + (1,0) + (-1,-1)$$
$$+ (-1,1) + (0,1) + (1,1) + \ldots$$
$$= 2(.395) + 2(.127i) + 3(-.003 + .0011i)$$
$$+ .00024 + \text{negligible} + \ldots$$
$$= .787 + .257i + \ldots ,$$

where $(0,0) = (0,-1) = .395$, etc., in the order written. We also have

$$\theta\begin{bmatrix}0 & 0\\ 0 & 0\end{bmatrix} = (0,0) + \{(1,0) + (-1,0)\} + \{(1,-1) + (-1,1)\}$$
$$+ \{(0,1) + (0,-1)\} + \{(1,1) + (-1,-1)\} + \ldots$$
$$= 1 + (.0565 - .0825i) + (-.0565 - .0825i) + .049$$
$$+ \text{negligible} + \ldots$$
$$= 1.049 - .165i + \ldots\ ,$$

where the terms in each brace are equal and each is half the indicated numerical value in the succeeding line. Consequently one has

$$23)\qquad \frac{\theta\begin{bmatrix}0 & 1\\ 1 & 0\end{bmatrix}}{\theta\begin{bmatrix}0 & 0\\ 0 & 0\end{bmatrix}} \sim \frac{.787 + .257i}{1.049 - .165i}$$
$$= .695 + .354i\ .$$

Thus the argument of the left side of 23) is $\sim$ arctan $.509 = 27°$. On the other hand, from 19) we see that that same argument, the argument of $\varepsilon +$ argument of $(-1)^{\frac{1}{4}}$, $= 72° + 45° + k90°$, with $-2 \leq k \leq 1$, which is $27°$ precisely for $k = -1$ so that we must choose $(-1)^{\frac{1}{4}} = (\sqrt{2}/2) - (\sqrt{2}/2)i$. Using this value, 11), 12), and 23), one gets

$$24)\qquad \frac{\theta\begin{bmatrix}0 & 1\\ 1 & 0\end{bmatrix}}{\theta\begin{bmatrix}0 & 0\\ 0 & 0\end{bmatrix}} = \left(\frac{\sqrt{2}}{2} - \frac{\sqrt{2}}{2}i\right)\left(\frac{\sqrt{-1+\sqrt{5}}}{\sqrt{2}}\right)\left(\frac{-1+\sqrt{5}+i\sqrt{10+2\sqrt{5}}}{4}\right)$$
$$= \left(\frac{1-i}{8}\right)\left(\left(-1+\sqrt{5}\right)^{\frac{3}{2}} + i\sqrt{(10+2\sqrt{5})(-1+\sqrt{5})}\right)$$
$$= \left(\frac{1-i}{8}\right)\left(\left(-1+\sqrt{5}\right)^{\frac{3}{2}} + i\sqrt{8\sqrt{5}}\right)$$
$$= \left(\frac{1-i}{8}\right)\left((1.236)^{\frac{3}{2}} + i\sqrt{17.9}\right)$$
$$\sim .701 + .359i\ ,$$

so that 23) is off approximately one digit in the second place in real and imaginary parts from the value in 24).

REMARK. We append a list of the more significant typographical errors in Rauch, "Theta constants... ." On p. 307, lines 16, 22, ε and ε' should be exchanged; on p. 308, line 8, i should be a subscript; p. 309, lines 24, 25 should read "and the side on the axis directed vertically as δ. Now if $2\Omega_1$ and $2\Omega_2$ are the periods of du, ..." Line 28 should have δ and γ exchanged. Page 310, line 11, the square root should be a fourth root. Page 313, line 4, no prime on γ_2. Page 318, line 5, A and C should be exchanged; line 9, the second C should be deleted. Page 319, line 1, "expect" should be except; line 7, "no" should be not. Page 321, line 13, ε and ε' should be exchanged; lines 3 and 4 from bottom, $\begin{bmatrix} 0 & 0 \\ 0 & 1 \end{bmatrix}$ should read $\begin{bmatrix} 0 & 0 \\ 1 & 0 \end{bmatrix}$. Page 322, line 8, the number .4 should be replaced by the value given here in equation 24).

THE GRADUATE SCHOOL AND UNIVERSITY CENTER OF
THE CITY UNIVERSITY OF NEW YORK

REFERENCES

[1]. Rauch, Harry E., Theta Constants on a Riemann Surface with Many Automorphisms, Symposia Mathematica, vol. III, 1970, Istituto Nazionale di Alta Matematica, Academic Press, N. Y.

[2]. Rauch, Harry E., and Lebowitz, Aaron, Elliptic Functions, Theta Functions, and Riemann Surfaces, Williams and Wilkins, Baltimore, 1973.

ON THE RELATION BETWEEN LOCAL AND GLOBAL PROPERTIES OF BOUNDARY VALUES FOR EXTREMAL QUASI-CONFORMAL MAPPINGS

Edgar Reich*

§1. *Introduction*

A one-to-one quasi-conformal mapping $w = f(z)$ of $E = \{|z| < 1\}$ onto $E = \{|w| < 1\}$ will be said to be *extremal* (for its boundary values) if for every quasi-conformal mapping $g(z)$ of E onto E for which

$$g|_{\partial E} = f|_{\partial E} \tag{1.1}$$

the relation

$$\|\kappa_g\|_\infty \geq \|\kappa_f\|_\infty, \quad (\kappa_g = g_{\bar{z}}/g_z,\ \kappa_f = f_{\bar{z}}/f_z) \tag{1.2}$$

is satisfied. ($\|\ \|_\infty$ denotes the essential sup over E.) Given a boundary correspondence $f|_{\partial E}$ of ∂E onto ∂E for which a quasi-conformal extension to the closed disk $\bar{E}$ exists, there will always exist an extremal quasi-conformal extension f^*. The quantity

$$k^* = k^*(f|_{\partial E}) = \|\kappa_{f^*}\|$$

denotes the extremal dilatation for the given boundary values $f|_{\partial E}$.

To allow for consideration of local effects let Γ be a closed arc on $\{|z| = 1\}$ consisting of more than a single point, and consider the problem of minimizing the maximal dilatation for mappings of E onto E whose

* Work done with partial support from the National Science Foundation.

boundary values on Γ are specified while those on the complementary arc $\tilde{\Gamma} = \partial E \setminus \Gamma$ are *free*. We call the mapping f of E onto E *extremal* (Γ) if (1.2) holds for every quasi-conformal mapping g of E onto E for which

$$g|_\Gamma = f|_\Gamma \ .$$

Given Γ, and a mapping $f|_\Gamma$ of Γ into ∂E for which a quasi-conformal extension to $\bar{E}$ exists there will also, by normality, exist an extremal (Γ) quasi-conformal extension f_Γ^*. We will write

$$k_\Gamma^* = k^*(f|_\Gamma) = \|\kappa_{f_\Gamma^*}\| \ .$$

Given $f|_{\partial E}$ and Γ, it is obvious that $k_\Gamma^* \le k^*$. If it turns out that $k_\Gamma^* = k^*$ we will say that Γ is *substantial*. This nomenclature is intended to convey the idea that the restriction of f to Γ plays a substantial role in determining the value of k^*, in spite of the fact that this latter quantity is *a priori* determined only by a global knowledge of f on all of ∂E. For given $f|_{\partial E}$ shortening Γ evidently does not increase k_Γ^*. For a fixed $\zeta \in \partial E$, let

$$k_\zeta^* = \inf k_\Gamma^* \ ,$$

where the inf is over all arcs Γ of which ζ is an interior point. The quantity k_ζ^* is determined from a specification of f on ∂E in an arbitrarily small neighborhood of ζ. Given $f|_{\partial E}$ it may turn out that $k_\zeta^* = k^*$. In this case we will say that the point ζ is *substantial*;[1] that is, ζ is substantial if and only if each Γ of which ζ is an interior point is substantial.

In Section 3 we will obtain a necessary and sufficient condition for a boundary point E to be substantial as related to the class $\mathfrak{b}$ of functions $\phi(z)$ holomorphic in E with

[1] The foregoing formulation was developed during a conversation with K. Strebel.

$$\|\phi\| = \iint_E |\phi(z)|\, dxdy \leq 1 \ ,$$

and the extremum problem

$$(1.3) \qquad \sup_{\phi \in \mathfrak{b}} \left| \iint_E \kappa(z)\phi(z)\, dxdy \right| \ .$$

A boundary correspondence may or may not give rise to substantial boundary points. This is illustrated explicitly in Section 4 for boundary correspondences of certain Teichmüller mappings. Finally, in Section 5 we find that we are able to give a simple new example of what might at first appear to be an anomalous behavior of extremal sequences for the extremal problem (1.3).

§2. *Characterization of extremal mappings for the free boundary problem*

Let Γ be a closed arc on ∂E containing more than a single point. Let $\mathfrak{b}_\Gamma$ be the family of functions $\phi(z)$ holomorphic in E, continuous on $E \cup \tilde{\Gamma}$, such that

$$(2.1) \qquad \operatorname{Im}\{\phi(z)\, dz^2\} = 0, \qquad z \in \tilde{\Gamma} = \partial E \setminus \Gamma \ ,$$

and with

$$(2.2) \qquad \|\phi\| = \iint_E |\phi(z)|\, dxdy \leq 1 \ .$$

Condition (2.1) is equivalent to

$$(2.3) \qquad \operatorname{Im}\{z^2\phi(z)\} = 0, \qquad z \in \tilde{\Gamma} \ .$$

Hence, by the reflection principle, and condition (2.2) it follows that $\mathfrak{b}_\Gamma$ is a compact normal family.

THEOREM 2.1. *Let* f *be a quasi-conformal mapping of* E *onto* E, $\kappa = f_{\bar{z}}/f_z$. *Let*

$$k = \|\kappa\|_\infty , \quad H_\Gamma = \sup_{\phi \epsilon \mathfrak{b}_\Gamma} \left| \iint_E \kappa(z)\phi(z)\,dxdy \right| ,$$

$$I_\Gamma = \sup_{\phi \epsilon \mathfrak{b}_\Gamma} \left| \iint_E \frac{\kappa(z)}{1-|\kappa(z)|^2}\, \phi(z)\,dxdy \right| .$$

Then the following three conditions are equivalent:

(i) f *is extremal* (Γ), *i.e.,* $k_\Gamma^* = k$

(ii) $H_\Gamma = k$

(iii) $I_\Gamma = \dfrac{k}{1-k^2}$.

Proof. Let

$$\Delta_\Gamma = \sup_{\phi \epsilon \mathfrak{b}_\Gamma} \iint_E \frac{|\kappa(z)|^2}{1-|\kappa(z)|^2}\, |\phi(z)|\,dxdy .$$

Suppose (i) holds. Then

$$\Delta_\Gamma \le \frac{k^2}{1-k^2} = \frac{k_\Gamma^{*2}}{1-k_\Gamma^{*2}} .$$

By [2], Theorem 7 (forming the polygons such that the maximal side length between vertices on Γ shrinks to zero)

$$\frac{k_\Gamma^*}{1-k_\Gamma^*} \le I_\Gamma + \Delta_\Gamma .$$

Therefore

$$I_\Gamma \ge \frac{k_\Gamma^*}{1-k_\Gamma^{*2}} = \frac{k}{1-k^2} ,$$

and (iii) follows.

Suppose (iii) holds. By [2], Theorem 2,

$$\frac{k_\Gamma^*}{1+k_\Gamma^*} \geq I_\Gamma - \Delta_\Gamma \ .$$

Hence, using (iii),

$$\frac{k_\Gamma^*}{1+k_\Gamma^*} \geq \frac{k}{1-k^2} - \frac{k^2}{1-k^2} = \frac{k}{1+k} \ .$$

Therefore, $k_\Gamma^* \geq k$. Since $k_\Gamma^* \leq k$, (i) follows.

The equivalence of conditions (ii) and (iii) follows as in [2] (there only the case $\Gamma = \partial E$ is treated in detail, but formally this is no different from the the present more general case) from the easily verifiable fact that $\phi_n \in \mathfrak{b}_\Gamma$ is an extremal sequence for the functional H_Γ if and only if it is an extremal sequence for I_Γ.

§3. *A characterization of substantial boundary points*

We first require a lemma dealing with the possibility of approximating functions of class $\mathfrak{b}$ that are small on $\tilde{\Gamma}$ by functions from $\mathfrak{b}_\Gamma$.

LEMMA 3.1. *Suppose* $\phi \in \mathfrak{b}$, *and there exist* $\delta > 0$ *and* $\varepsilon > 0$ *such that* $|\phi(z)| < \delta$ *whenever* $\text{distance}(z, \tilde{\Gamma}) < \varepsilon$. *Then there exists a function* $\tilde{\phi} \in \mathfrak{b}_\Gamma$ *such that*

$$\|\tilde{\phi} - \phi\| < A\delta \ .$$

The constant A *depends only on the length of* Γ.

Proof. (i) Without loss of generality $\|\phi\| = 1$, and $\tilde{\Gamma}$ is the arc $-\tau < \theta < \tau$, $z = e^{i\theta}$. We assume first that $\tau = \frac{\pi}{2}$, and that ϕ is holomorphic in $\bar{E}$.

Let $f(z) = z^2\phi(z)$, $v(\theta) = \operatorname{Im} f(e^{i\theta})$.

We define a real valued, piecewise continuous function $\tilde{v}(\theta)$, $0 \leq \theta < 2\pi$, as follows

$$\tilde{v}(\theta) = \begin{cases} 0 \quad , & e^{i\theta} \in \tilde{\Gamma} \\ v(\theta) + h(\theta + \pi), & e^{i\theta} \in \Gamma \end{cases} \tag{3.1}$$

where $h(\theta)$, $e^{i\theta} \in \Gamma$ will be determined so that

$$\int_0^{2\pi} \tilde{v}(\theta)\, d\theta = 0, \quad \int_0^{2\pi} e^{-i\theta}\tilde{v}(\theta)\, d\theta = 0 \ . \tag{3.2}$$

From the Poisson integral representation of f,

$$f(z) = \frac{i}{2\pi} \int_0^{2\pi} \frac{e^{i\theta}+z}{e^{i\theta}-z}\, v(\theta)\, d\theta + \text{real const.}, \qquad z \in E \ ,$$

and the fact that $f(0) = f'(0) = 0$, it follows that

$$\int_0^{2\pi} v(\theta)\, d\theta = 0, \quad \int_0^{2\pi} e^{-i\theta} v(\theta)\, d\theta = 0 \ . \tag{3.3}$$

Using (3.1), (3.3), we find that the requirement (3.2) is equivalent to

$$\begin{cases} \int_{\tilde{\Gamma}} h(\theta)\, d\theta = \int_{\tilde{\Gamma}} v(\theta)\, d\theta = W_1 \\ \int_{\tilde{\Gamma}} h(\theta)\cos\theta\, d\theta = -\int_{\tilde{\Gamma}} v(\theta)\cos\theta\, d\theta = W_2 \\ \int_{\tilde{\Gamma}} h(\theta)\sin\theta\, d\theta = -\int_{\tilde{\Gamma}} v(\theta)\sin\theta\, d\theta = W_3 \ . \end{cases} \tag{3.4}$$

These relations will be satisfied if we, e.g., define $h(\theta)$ as

$$h(\theta) = A + B\cos\theta + C\sin\theta \ , \tag{3.5}$$

where

$$A = \frac{\pi W_1 - 4W_2}{\pi^2 - 8}, \quad B = \frac{2\pi W_2 - 4W_1}{\pi^2 - 8}, \quad C = \frac{2}{\pi} W_3 \ .$$

Next, let
$$\tilde{f}(z) = \frac{i}{2\pi}\int_0^{2\pi} \frac{e^{i\theta}+z}{e^{i\theta}-z}\,\tilde{v}(\theta)\,d\theta \ .$$

Equation (3.2) implies that $\tilde{f}(0) = \tilde{f}'(0) = 0$. Thus, the function

$$\tilde{\phi}(z) = \frac{\tilde{f}(z)}{z^2}, \qquad z \in E$$

is holomorphic in E. Im $(z^2\tilde{\phi}(z))$ has boundary values $\tilde{v}(\theta)$, as $z \to e^{i\theta}$, $\theta \neq \pm\frac{\pi}{2}$. It follows, by the reflection principle, that $z^2\tilde{\phi}(z)$ is continuous across $\tilde{\Gamma}$, and that

$$-\operatorname{Im}(z^2\tilde{\phi}(z)) = \operatorname{Im}(\tilde{\phi}(z)\,dz^2) = 0, \qquad z \in \tilde{\Gamma} \ .$$

By considering the power series we find, using (3.1), (3.4) and (3.5),

$$\text{(3.6)} \quad \int_0^{2\pi} |\tilde{f}(re^{i\theta}) - f(re^{i\theta})|^2\,d\theta = 2\int_0^{2\pi} \{\operatorname{Im}[\tilde{f}(re^{i\theta}) - f(re^{i\theta})]\}^2\,d\theta$$

$$\leq 2\left[\int_{\tilde{\Gamma}} v(\theta)^2\,d\theta + \int_{\tilde{\Gamma}} h(\theta+\pi)^2\,d\theta\right]$$

$$= 2\int_{\tilde{\Gamma}} [v(\theta)^2 + h(\theta)^2]\,d\theta$$

$$= 2\int_{\tilde{\Gamma}} v(\theta)^2\,d\theta + 2\pi A^2 + \pi B^2 + \pi C^2 + 8AB$$

$$\leq 65\int_{\tilde{\Gamma}} |\phi(e^{i\theta})|^2\,d\theta < 65\pi\delta^2 \ .$$

One also verifies, using power series, that

$$\text{(3.7)} \quad \iint_E |\tilde{\phi}(z) - \phi(z)|^2\,dxdy \leq \frac{1}{2}\,\overline{\lim_{r\to 1}}\int_0^{2\pi} |\tilde{f}(re^{i\theta}) - f(re^{i\theta})|^2\,d\theta \ .$$

Hence, by (3.6), (3.7) and Schwarz' inequality,

$$\|\tilde{\phi} - \phi\| \leq \sqrt{\pi}\,\sqrt{\tfrac{1}{2}\,65\pi\delta^2} < 20\delta \ .$$

To insure that $\|\tilde{\phi}\| \leq 1$ we take

$$\tilde{\phi}_1 = \frac{\tilde{\phi}}{1+20\delta} .$$

Thus

$$\|\tilde{\phi}_1 - \phi\| \leq \left\| \frac{\tilde{\phi}}{1+20\delta} - \frac{\phi}{1+20\delta} \right\| + \left\| \frac{\phi}{1+20\delta} - \phi \right\| \leq \frac{40\delta}{1+20\delta} < 40\delta .$$

Dropping the subscript 1, we can summarize as follows:

$$\tilde{\phi} \in \mathfrak{b}_{\Gamma}, \quad \|\tilde{\phi} - \phi\| < 40\delta . \tag{3.8}$$

(ii) To dispose of the assumption that ϕ is holomorphic on ∂E, consider $\phi(Rz)$, $|z| \leq 1$, where $1-\varepsilon < R < 1$. By part (i), there exists a function $\tilde{\phi}_R(z) \in \mathfrak{b}_{\Gamma}$, $\|\tilde{\phi}_R - \phi(Rz)\| < 40\delta$. Since

$$\|\tilde{\phi}_R\| \leq \|\tilde{\phi}_R - \phi(Rz)\| + \iint_E |\phi(Rz)|\, dxdy < 40\delta + 2$$

it follows that the family

$$\{\tilde{\phi}_R(z)\}_{1-\varepsilon<R<1}$$

is normal over E, and, in fact, by reflection, across $\tilde{\Gamma}$. Choosing $\tilde{\phi}(z)$ as the limit as $R = 1 - \frac{1}{n} \to 1$ of a subsequence converging uniformly on all compact subsets of $E \cup \tilde{\Gamma}$ we obtain (3.8) again.

(iii) If τ is arbitrary, $0 < \tau < \pi$, consider the linear transformation

$$w = T(z) = \frac{z-a}{1-az}, \quad a = \frac{\cos \tau}{1+\sin \tau}$$

which maps E onto E in such a manner that Γ is mapped onto the arc $\frac{\pi}{2} \leq \theta \leq \frac{3\pi}{2}$. Define

$$\psi(w) = \phi(z)\left(\frac{dz}{dw}\right)^2, \quad (w = T(z)) \tag{3.9}$$

and construct $\tilde{\psi}(w)$ as in parts (i), (ii); then define

$$\tilde{\phi}(z) = \tilde{\psi}(w)\left(\frac{dw}{dz}\right)^2, \quad (w = T(z)) . \tag{3.10}$$

Let $M = \sup |T'(z)|^2$. Then, by (3.9), we have

$$|\psi(w)| \le M^2\delta, \quad (|w| = 1, |\arg w| \le \frac{\pi}{2} \quad ,$$

and therefore, $\|\tilde{\psi} - \psi\| < 40M^2\delta$.

Since $\|\tilde{\phi} - \phi\| = \|\tilde{\psi} - \psi\|$ we have

$$\|\tilde{\phi} - \phi\| < A\delta, \quad (A = 40M^2) \ .$$

The condition $\tilde{\phi} \in \mathfrak{b}_\Gamma$ is insured by the invariance condition (3.10).

THEOREM 3.1. *Let* $f(z)$ *be a quasi-conformal mapping of* E *onto* E, $\zeta \in \partial E$. *A necessary and sufficient condition that* f *is extremal for its boundary values, and that* ζ *is a substantial boundary point is that there exists a sequence* $\phi_n \in \mathfrak{b}$ *satisfying the following two properties:*

(3.11) $\lim_{n\to\infty} \phi_n(z) = 0$, *uniformly on every compact subset of* $\overline{E}\setminus\{\zeta\}$

and

$$\lim_{n\to\infty} \iint_E \kappa(z)\phi_n(z)\,dxdy = k \quad (\kappa = f_{\bar{z}}/f_z,\ k = \|\kappa\|_\infty) \ . \tag{3.12}$$

Proof. (i) Assume that (3.11) and (3.12) hold. By (3.12) it follows ([2], [3]) that f is extremal (this is our present Theorem 2.1 when $\Gamma = \partial E$), $k = k^*$.

Let Γ be a closed arc on $\partial E\setminus\{\zeta\}$. Choose $\phi_n \in \mathfrak{b}$ so that

$$|\phi_n(z)| < \frac{1}{n}$$

for all z in a neighborhood of $\tilde{\Gamma}$. By Lemma (3.1) there exists a function $\tilde{\phi} \in \mathfrak{b}_\Gamma$ with

$$\|\tilde{\phi}_n - \phi_n\| < \frac{A}{n} \ .$$

Since

$$\left| \iint_E \kappa(z)\tilde{\phi}_n(z)\,dxdy - \iint_E \kappa(z)\phi_n(z)\,dxdy \right| \le k\|\tilde{\phi}_n - \phi_n\|$$

we evidently obtain

$$H = \lim_{n\to\infty} \iint_E \kappa(z)\tilde{\phi}_n(z)\,dxdy = k .$$

By Theorem 2.1, we conclude that Γ is substantial. Since this is true for arbitrary Γ, ζ is substantial.

(ii) Conversely, suppose f is extremal and ζ is substantial. Let $\{\Gamma_n\}$ be closed arcs centered at ζ with arc length shrinking to zero as $n \to \infty$. Since Γ_n is substantial,

$$\sup_{\phi \in \mathfrak{B}_{\Gamma_n}} \left| \iint_E \kappa(z)\phi(z)\,dxdy \right| = k .$$

Therefore there is an element $\phi_n \in \mathfrak{B}_{\Gamma_n}$ such that

$$\iint_E \kappa(z)\phi_n(z)\,dxdy > k - \frac{1}{n} .$$

Choose a subsequence of $\{\phi_n(z)\}$ such that

$$\lim_{n\to\infty} \phi_n(z) = \phi(z)$$

exists, uniformly on compact subsets of $\overline{E} \setminus \{\zeta\}$. Evidently

$$\|\phi\| \le 1, \quad \operatorname{Im}\{\phi(z)\,dz^2\} = 0, \quad z \in \partial E \setminus \{\zeta\} . \tag{3.13}$$

Condition (3.13) implies that $\phi(z) \equiv 0$. (Transfer ϕ to $\operatorname{Im} w > 0$ by means of $\psi(w)\,dw^2 = \phi(z)\,dz^2$. We get $\iint_{\operatorname{Im} w > 0} |\psi(w)|\,dudv \le 1$, and since $\psi(w)$ is real on the real axis it can be reflected to give an entire function with $\iint_{|w|<\infty} |\psi(w)|\,dudv \leq 2$. Hence, by integrating the Cauchy coefficient estimates, $\psi(w) \equiv 0$.) This completes the proof.

The following is an immediate corollary.

THEOREM 3.2. *For the Teichmüller mapping* f *of* E *with complex dilatations*

$$\kappa(z) = k\,\frac{\overline{\phi_0(z)}}{|\phi_0(z)|}, \qquad \phi_0 \in \mathfrak{b} \tag{3.14}$$

the boundary point $\zeta \in \partial E$ *is substantial if and only if there exists a sequence* $\phi_n \in \mathfrak{b}$, $n = 1, 2, \ldots$, *such that*

$$\lim_{n\to\infty} \iint \frac{\overline{\phi_0(z)}\,\phi_n(z)}{|\phi_0(z)|}\,dxdy = 1\ ,$$

for which (3.11) *holds.*

§4. *Example: Affine mappings*

In this section we restrict ourselves to boundary correspondences induced by affine mappings

$$w = F_K(z) = Kx + iy \qquad (K > 1) \tag{4.1}$$

of a Jordan domain Ω. Although the definitions and theorems of the preceding sections have always been stated for the unit disk E they immediately extend to mappings of Ω provided the holomorphic functions involved are transformed as quadratic differentials. We will calculate k^*_ζ explicitly in some cases of interest, attempting, however, to avoid complication at the expense of some generality. Since Ω has finite area the mapping (4.1) is extremal. Thus $z_0 \in \partial\Omega$ will be a substantial boundary point if and only if

$$K^*_{z_0} = \frac{1 + k^*_{z_0}}{1 - k^*_{z_0}} = K\ .$$

THEOREM 4.1. *Suppose* $\partial\Omega$ *has a tangent with tangent angle* $\tau(s)$ (s = *arc length*) *in a neighborhood of* $z_0 \epsilon \partial\Omega$. *Suppose* $\tau(s)$ *has modulus of continuity* $\mu(s)$, *where* $\int_0 \frac{\mu(t)}{t} < \infty$. *Then* $k^*_{z_0} = 0$.

Proof.[2] Put $w_0 = F_K(z_0)$, $\Omega' = F_K(\Omega)$. Let $\zeta = \xi + i\eta = \Phi(z)$ map Ω conformally onto the strip $\Sigma = \{\zeta \mid 0 < \eta < \pi\}$, and let $\omega = \alpha + i\beta = \Psi(w)$ map Ω' conformally onto the strip $\Sigma' = \{\omega \mid 0 < \beta < \pi\}$, so that z_0 and w_0 map onto $\xi = -\infty$ and $\alpha = -\infty$ respectively. If we write $\Phi = \log g$, then we know by Warschawski's theorem [5] that $g'(z)$ has a continuous non-vanishing extension up to the boundary of Ω in a neighborhood of z_0. The same holds for the derivative of e^{Ψ} up to the boundary of Ω' in a neighborhood of w_0. Let θ_0 denote the tangent angle of $\partial\Omega$ at z_0. The boundary correspondence (4.1) between $\partial\Omega$, $\partial\Omega'$ satisfies

$$\left|\frac{dw}{dz}\right|_{z=z_0} = K_1 = \left|\frac{K+1}{2} + \frac{K-1}{2} e^{-2i\theta_0}\right| .$$

Let $\alpha = A_1(\xi)$, $\alpha = A_2(\xi)$ denote the correspondences induced between $\partial\Sigma$ and $\partial\Sigma'$ at $\eta = 0, \pi$ respectively. In view of the above, we obtain the following asymptotic estimates:

$$(4.2) \qquad A_1(\xi) = \xi + \log K_1 + o(1), \quad A_2(\xi) = \xi + \log K_1 + o(1), \quad \xi \to -\infty ,$$

and

$$(4.3) \qquad A_1'(\xi) = 1 + o(1), \quad A_2'(\xi) = 1 + o(1), \quad \xi \to -\infty .$$

We will now proceed to replace this correspondence between $\partial\Sigma$ and $\partial\Sigma'$ by a new one,

$$(4.4) \qquad \alpha = R_1(\xi), \ (-\infty<\xi<\infty, \ \eta = 0), \ \alpha = R_2(\xi), \ (-\infty<\xi<\infty, \ \eta = \pi)$$

---

[2] For the case when z_0 is an interior point of a side of a rectangle, Dr. A. Sontag has indicated to the author that she independently obtained a proof by a different method.

as follows. Given δ, $0 < \delta < 1$, choose $M > 0$ so large that

$$(4.5) \quad 1-\delta \le A_j'(\xi) \le 1+\delta, \quad (j=1,2), \quad |A_2(\xi) - A_1(\xi)| < \delta, \text{ for } \xi \le -M .$$

Let

$$(4.6) \qquad R_j(\xi) = \begin{cases} A_j(\xi), & \xi \le -M \\ A_j(-M) + \xi + M, & \xi \ge -M \end{cases} \qquad (j = 1,2) .$$

By (4.5), (4.6) we see that $R_j(\xi)$ are continuous and strictly increasing, and satisfy

$$(4.7) \qquad 1-\delta \le R_j'(\xi) \le 1+\delta, \quad |R_2(\xi) - R_1(\xi)| < \delta, \text{ for } all \ \xi ,$$

(except at $\xi = -M$ where $R_j'(\xi)$ is possibly undefined). The formula

$$(4.8) \qquad \alpha = \left(1-\frac{\eta}{\pi}\right) R_1(\xi) + \frac{\eta}{\pi} R_2(\xi), \quad \beta = \eta, \quad (-\infty<\xi<\infty,\ 0<\eta<\pi)$$

defines a quasi-conformal mapping

$$\omega = g(\zeta), \quad \zeta \in \Sigma$$

of Σ onto Σ' with boundary correspondence $R_j(\xi)$ and complex dilatation

$$\frac{g_{\bar\zeta}}{g_\zeta} = \frac{\left[(1-\frac{\eta}{\pi}) R_1'(\xi) + \frac{\eta}{\pi} R_2'(\xi) - 1\right] + \frac{i}{\pi}[R_2(\xi) - R_1(\xi)]}{\left[(1-\frac{\eta}{\pi}) R_1'(\xi) + \frac{\eta}{\pi} R_2'(\xi) + 1\right] - \frac{i}{\pi}[R_2(\xi) - R_1(\xi)]} .$$

By (4.7)

$$\left|\frac{g_{\bar\zeta}}{g_\zeta}\right| \le \frac{(1+\delta-1) + \frac{\delta}{\pi}}{(1-\delta+1) - \frac{\delta}{\pi}} = \frac{(1+\frac{1}{\pi})\delta}{2-(1+\frac{1}{\pi})\delta} .$$

Thus, if $\delta > 0$ is chosen sufficiently small,

$$G_\delta(z) = \Psi^{-1} \circ g \circ \Phi(z), \quad z \in \Omega$$

is a $(1+\varepsilon)$-quasi-conformal mapping of Ω onto Ω' whose boundary correspondence agrees with $F_K(z)$ in a neighborhood of z_0. This completes the proof.

To find out what happens if Ω has a "corner" consider first the case of a straight sector

$$V = \{\theta_1 < \arg z < \theta_2\} .$$

If ∂V is subjected to the mapping induced by (4.1) the result is to map ∂V onto $\partial V'$,

$$V' = \{\theta_1' < \arg w < \theta_2'\} .$$

There exists [4] an extremal quasi-conformal mapping $f(K, \theta_1, \theta_2; z)$ of V onto V′ which assumes this boundary correspondence, and which has dilatation

$$K^* = S(K, \theta_1, \theta_2) < K .$$

(The function $S(K, \theta_1, \theta_2)$ can be calculated almost explicitly [4].)

THEOREM 4.2. *Suppose $\partial\Omega$ has a corner at $z = z_0$ produced by two arcs meeting at z_0 with tangent angles $\theta_1, \theta_2, (\theta_1 < \theta_2)$ and suppose that the mapping $(z-z_0)^{\frac{\pi}{\theta_2-\theta_1}}$ straightens the corner out locally into an arc satisfying the hypothesis of Theorem* 4.1. *Then* $K^*_{z_0} = S(K,\theta_1,\theta_2) < K$.

Outline of proof. Let Σ be the strip $\{\zeta \mid \theta_1 < \eta < \theta_2\}$. In place of (4.2) we get

$$A_1(\xi) = \xi + a_1 + o(1), \quad A_2(\xi) = \xi + a_2 + o(1), \quad \xi \to -\infty ,$$

where $a_j = \frac{1}{2} \log (\sin^2\theta_j + K^2\cos^2\theta_j)$, $j = 1, 2$. A mapping g can be constructed analogously to (4.8). This finally leads to an $[S(K,\theta_1,\theta_2)+\varepsilon]$-quasi-conformal mapping of Ω onto Ω' with the same boundary correspondence as F_K in a neighborhood of the corner z_0. Thus,

$$K^*_{z_0} \leq S(K, \theta_1, \theta_2) .$$

To prove that $K^*_{z_0} \geq S(K, \theta_1, \theta_2)$ one can, e.g., use an extremal length argument as in [4].

Since $S(K, \theta_1, \theta_2) < K$ we see that neither smooth arcs nor corners give rise to substantial boundary points. The latter will however occur when the Jordan domain Ω has "cusps." We will consider here only the following special type.

$$\Omega = \{z = x + iy \mid 0 < y < x^{\alpha},\ 0 < x < 1\} \cup \Omega_1 \ , \tag{4.9}$$

where $\alpha > 1$, and where

$$\inf \{\operatorname{Re} z \mid z \in \Omega_1\} = h > 0 \ .$$

The region Ω has a cusp at $0 \in \partial\Omega$. To simplify matters we will assume that $\partial\Omega$ is sufficiently nice that at points other than $z = 0$ the conformal mapping onto E has a continuous non-vanishing derivative.

As a corollary of Theorem 3.1 we will show that *the point* 0 *is a substantial boundary point of* Ω. By Theorem 3.2 it will be sufficient to find a sequence of functions $g_n(z)$ holomorphic in $\bar{\Omega} \setminus \{0\}$ with the following three properties (4.10) - (4.12):

$$\lim_{n\to\infty} g_n(z) = 0, \quad \text{locally uniformly in } \bar{\Omega} - \{0\} \ , \tag{4.10}$$

$$\overline{\lim_{n\to\infty}} \iint_{\Omega} |g_n(z)| \, dxdy > 0 \ , \tag{4.11}$$

$$\lim_{n\to\infty} \frac{\displaystyle\iint_{\Omega} g_n(z) \, dxdy}{\displaystyle\iint_{\Omega} |g_n(z)| \, dxdy} = 1 \ . \tag{4.12}$$

In fact, let

$$g_n(z) = c_n e^{-nz}, \qquad n = 1, 2, \dots \ ,$$

where

$$c_n = \frac{n^{\alpha+1}}{\int_0^\infty t^\alpha e^{-t} dt} .$$

It is easy to verify by routine estimation methods that

$$\iint_\Omega |g_n(z)|\, dxdy = 1 + o(1), \quad \iint_\Omega g_n(z)\, dxdy = 1 + o(1), \quad \text{as} \quad n \to \infty .$$

Therefore (4.10) - (4.12) hold.

§5. *A counterexample*

Suppose f is a quasi-conformal mapping of E onto E which is extremal for $f|_{\partial E}$. Then, by Hamilton's theorem, there exists an extremal sequence $\phi_n \in \mathfrak{b}$ for the functional (1.3) such that

$$(5.1) \qquad \lim_{n\to\infty} \iint_E \kappa(z)\phi_n(z)\, dxdy = \|\kappa\|_\infty , \quad (\kappa = f_{\bar{z}}/f_z) .$$

Even if we assume that f is a Teichmüller mapping with finite norm

$$\kappa(z) = k \frac{\phi_0(z)}{|\phi_0(z)|} , \quad \|\phi_0\| = 1$$

and $\lim_{n\to\infty} \phi_n(z)$ converges locally uniformly in E, it does not necessarily follow that

$$\lim_{n\to\infty} \phi_n(z) = \phi_0(z) .$$

This was first shown in ([2], p. 26), but the concept of a substantial boundary point now allows for a simpler and more intuitively understandable example as follows.

Let Ω be the Jordan domain with the cusp as given by (4.9). Let $\Phi(z)$, $\Phi(-1) = 0$, map $\{|z| < 1\}$ conformally onto Ω. For $K = \frac{1+k}{1-k} > 1$

let $\Psi(w)$ map $\{|w| < 1\}$ conformally onto $F_K(\Omega)$. Then $f = \Psi^{-1} \circ F_K \circ \Phi$ is a Teichmüller mapping with complex dilatation

$$\kappa(z) = k \frac{\overline{\phi_0(z)}}{|\phi_0(z)|}, \quad \phi_0(z) = \Phi'(z)^2, \quad \|\phi_0\| = \text{Area}(\Omega) .$$

Since 0 is a substantial boundary point for $F_K|_{\partial\Omega}$ it follows that $-1 = \Phi^{-1}(0)$ is a substantial boundary point of ∂E for $f|_{\partial E}$. By Theorem 3.1 there therefore exists a sequence $\phi_n(z) \in \mathfrak{b}$ for which (5.1) holds, but for which

$$\lim_{n\to\infty} \phi_n(z) = 0, \qquad z \in E.$$

In fact, the sequence

$$\phi_n(z) = g_n(\Phi(z))\Phi'(z)^2 = \frac{n^\alpha}{\int_0^\infty t^\alpha e^{-t} dt} e^{-n\Phi(z)} \phi_0(z)$$

is precisely an example of such a sequence.

UNIVERSITY OF MINNESOTA

REFERENCES

[1]. Hamilton, Richard S., "Extremal quasi-conformal mappings with prescribed boundary values," Trans. Am. Math. Soc. 138 (1969), pp. 399-406.

[2]. Reich, Edgar and Strebel, Kurt, "Extremal quasi-conformal mappings with given boundary values," Notes, Forschungsinstitut für Mathematik, E.T.H., Zürich, September, 1972.

[3]. Reich, Edgar and Strebel, Kurt, "Extremal quasi-conformal mappings with given boundary values," Bull. Am. Math. Soc. 79 (1973), pp. 488-490.

[4]. Strebel, Kurt, "Zur Frage der Eindeutigkeit extremaler quasikonformer Abbildungen des Einheitskreises," Comment. Math. Helv. 36 (1962), pp. 306-323.

[5]. Warschawski, S. E., "On differentiability at the boundary in conformal mapping," Proc. Am. Math. Soc. 12 (1961), pp. 614-620.

SYMMETRIC EMBEDDINGS OF RIEMANN SURFACES

Reto A. Rüedy

§1. *Embedded surfaces*

1.1. If the boundary S of an open subset of E^3, the three-dimensional Euclidean space, is a smooth (C^∞) surface, it can locally be mapped into $\mathbb{C}$ by orientation- and angle-preserving mappings. Using these mappings as local parameters, we obtain a complex analytic structure, the natural structure of S, and we will call the resulting Riemann surface a (complete) *classical surface*. Riemann surfaces of this type appeared already in Riemann's thesis.

1.2. If there is a conformal mapping of a given Riemann surface onto a classical surface, we call the Riemann surface *embeddable* and the mapping a conformal embedding. If the mapping is only topological or a diffeomorphism we speak of a topological or differentiable embedding.

1.3. It is well known that every compact Riemann surface can be topologically embedded and it is also a classical result that topologically equivalent Riemann surfaces are diffeomorphic. However the following result was established only recently ([3], [10]):

EMBEDDING THEOREM. *Every Riemann surface is (conformally) embeddable.*

This is the solution to a problem that was stated for the first time by Felix Klein in 1882 ([7]).

§2. *Symmetric embeddings*

2.1. If the Riemann surface R allows a non-trivial conformal automorphism A, we will call the topological or conformal embedding $e: R \to E^3$ symmetric (with respect to A) if $e \circ A \circ e^{-1}$ is (the restriction to e(R) of) a rotation. If a conformal symmetric embedding exists we will call A embeddable. Our aim is to characterize in an intrinsic way the embeddable automorphisms of finite order if the Riemann surface R is compact. (If the genus of the surface is ≥ 2, then every automorphism is of finite order.)

2.2. We denote by $<A>$ the group generated by A, O(A) denotes the order of the group $<A>$, F(A) the set of fixed points of A and N(A) the number of fixed points of A. Finally we put $F(<A>) = \cup F(A^m)$ where m goes from 1 to O(A)–1. If p is a fixed point of A then we can find a local coordinate system (U, ϕ) such that $\phi(p) = 0$ and $\phi \circ A \circ \phi^{-1}(z) = z \cdot \exp(i\alpha)$ for $z \in \phi(U)$. $\alpha = \alpha(A, p)$ is uniquely determined up to multiple of 2π, independent of the local coordinate system and we have $\alpha(A^m, p) = m\alpha(A, p)$. Usually we will normalize α by assuming that $-\pi < \alpha \leq \pi$.

2.3. Using the notation introduced in 2.2., we can characterize the embeddable automorphisms in the following way:

THEOREM. *Let* R *be a compact Riemann surface and* A *an automorphism of* R *of finite order.* A *is embeddable exactly if*

1. $F(A) = F(<A>)$,
2. N(A) *is even, i.e.,* $N(A) = 2n(A)$, $n(A) \in Z$,
3. $|\alpha(A, p)| = |\alpha(A, q)|$ *if* $p, q \in F(A)$,
4. *either* $\alpha(A, p) = \pi$ (*i.e.,* $A^2 = id$) *or* $\Sigma_{p \in F(A)} \alpha(A, p) = 0$.

In conditions 3. and 4. $\alpha(A, p)$ is normalized.

The first condition is automatically satisfied if O(A) is prime. All conditions are (trivially) satisfied if $F(<A>) = \emptyset$. The same is true if A is an involution since the Riemann-Hurwitz relation (see 3.1) implies that an involution has an even number of fixed points.

COROLLARY 1. *Every involution* $(A^2 = \mathrm{id})$ *is embeddable.*

COROLLARY 2. *If* $F(<A>) = \emptyset$ *then* A *is embeddable.*

COROLLARY 3. *A fixed point free automorphism of prime order is embeddable.*

2.4. If S is a classical surface and $A : S \to S$ a rotation then F(A) is the intersection of S and the axis of the rotation. Since this axis is perpendicular to S at all points of intersection, N(A) is even if S is compact. The other conditions of the theorem are obviously satisfied which shows that the four conditions are necessary. It remains to be shown that they are also sufficient.

§3. *Symmetric topological embeddings*

3.1. If A is an automorphism of finite order of a compact Riemann surface R of genus g, then the orbit space $R' = R/<A>$ is a compact Riemann surface of genus g′. If $F(A) = F(<A>)$, we find an explicit expression for g′ by the Riemann-Hurwitz relation:

$$g-1 = O(A) \cdot (g'-1) + (O(A)-1) \cdot N(A)/2 = O(A) \cdot (g'-1) + (O(A)-1) \cdot n(A).$$

If $F(<A>) = \emptyset$, O(A) divides $g-1$ and $g = O(A) \cdot (g'-1) + 1$. If A has an even number of fixed points O(A) divides $g-1+n(A)$ and we have $g = g' \cdot O(A) + (O(A)-1)(n(A)-1)$.

3.2. R′ may be obtained by choosing a fundamental domain for $<A>$ on R and identifying certain boundary points of this fundamental domain. In order to find a symmetric topological embedding we reverse this process: We cut R' along certain curves in such a way that the canonical projection $\pi: R \to R'$ has a continuous inverse on the complement of these curves and paste together in a symmetric way $O(A)$ copies of the image of R'. The main problem is to find the right cuts. These are described in Lemma 1 and Lemma 2.

3.3. In the following lemmas A is an automorphism of finite order of a compact Riemann surface R.

LEMMA 1. *Assume that* A *is fixed point free.*

a) *For some simple closed curve* γ *on* $R' = R/\langle A\rangle$, $\pi^{-1}(\gamma)$ *consists of* $O(A)$ *cycles which divide* R *into* $O(A)$ *parts.*

b) *The canonical projection* π *maps each part topologically onto* $R' - \gamma$ *and each part is bounded by two of the* $O(A)$ *cycles.*

c) *Some element* $B \in \langle A\rangle$ *identifies the two boundary components simultaneously for each part.*

LEMMA 2. *Assume that* $N(A) = 2n \neq 0$.

a) *The branch points* $p_1, \ldots, p_{2n}$ *of* $\pi: R \to R'$ *can be arranged in such a way that there are* n *disjoint simple curves* γ_j *joining* p_{2j-1} *and* p_{2j}, $j = 1, \ldots, n$, *whose preimages divide* R *into* $O(A)$ *parts.*

b) π *maps each part topologically onto* $R' - \cup \gamma_j$ *and each part is bounded by* n *disjoint simple closed curves each containing two fixed points.*

c) *The fixed points divide these curves into two halves and some* $B \in \langle A\rangle$ *maps one half of the curve onto the other one simultaneously for all these curves and for all parts.*

3.4. Before we prove these lemmas, we will show that A can be topologically embedded if it satisfies our conditions. The embedded surface will even be invariant under a dihedral group of rotations. If the additional rotations correspond to mappings of R onto itself which are homotopic to automorphisms, we may assume that $e^{-1} \circ r \circ e$ is a conformal automorphism for each element r of the dihedral group. In this case the arguments in Section 4 can be modified in an obvious way in order to prove that the full dihedral group is simultaneously conformally embeddable.

LEMMA 3. *Under the assumptions of the theorem,* A *is topologically embeddable.*

We will prove this lemma in 3.5. for the case $F(\langle A \rangle) = \emptyset$ and in 3.6. for the remaining cases.

3.5. Lemma 1 implies that R may be obtained by pasting together O(A) spheres with 2 holes and $g'-1 = (g-1)/O(A)$ handles. B cyclically permutes these parts mapping each part into an adjacent one. Let R'' be the closure of one of these parts. R'' can be mapped topologically into a classical surface S in $\mathbb{C} \times R$ $(= E^3)$ which is invariant under the rotation $r(z, x_3) = (z \exp(i\alpha), x_3)$, $\alpha = 2\pi/O(A)$, in such a way that $e(R'') \subset \{(z,x_3) \mid 0 \le \arg z \le \alpha\}$, and $e \circ B \circ e^{-1} = r$ if $\arg z = 0$. This embedding has a unique extension for which $e \circ B \circ e^{-1} = r$ holds on all of S. If we choose S as in Figure 1, the group of rotations leaving S invariant is dihedral. The numbers k and m (see Figure 1) have to be chosen in such a way that $k + 2m = g'-1$.

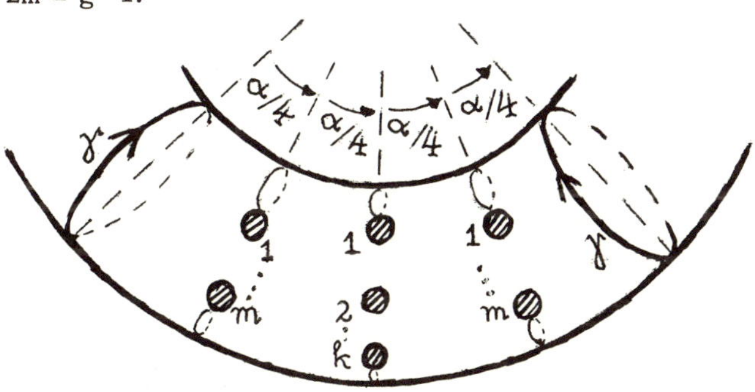

Fig. 1

3.6. If $F(A) \neq \emptyset$ Lemma 2 implies that R is obtained by pasting together O(A) spheres with n(A) holes and g′ handles. The embedding can be achieved in the same way as before (see Figure 2). Again $e \circ B \circ e^{-1} = r$ where $r(z, x_3) = (z \exp(i\alpha), x_3)$. In this case $k + 2m = g'$ and the embedded surface is again invariant under a dihedral group of rotations.

3.7. *Proof of Lemma* 1. Accola showed in [1] that the lemma is true under the additional assumption that O(A) is prime. In the general case we write O(A) in the form pq where p is prime and apply Accola's theorem

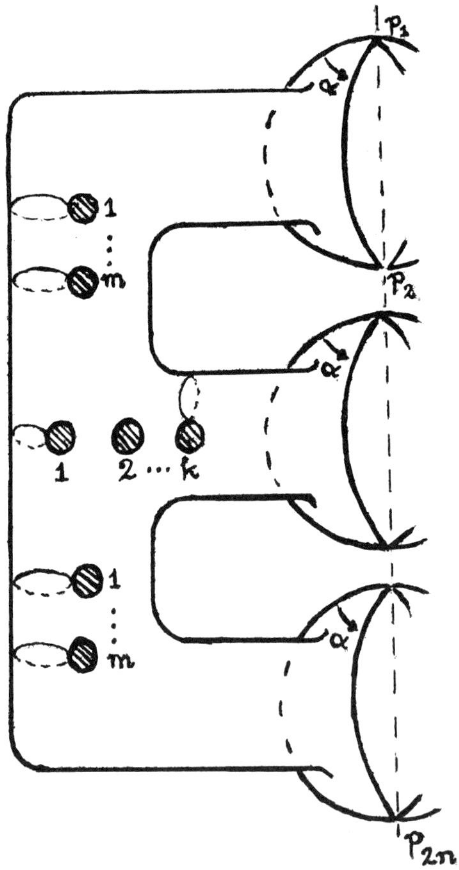

Fig. 2

to $<A^q>$ and $\pi_q : R \to R/<A^q>$. In this way we obtain a curve γ on $R/<A^q>$ such that the p curves forming $\pi_q^{-1}(\gamma)$ divide R into p parts $R_1, \ldots, R_p$. Let γ_0 be a lifting of γ with respect to π_q. Because $A^m \circ \gamma_0$ and $A^n \circ \gamma_0$ are liftings of $\pi \circ \gamma_0$ and because $\pi : R \to R'$ is an unbranched covering, $A^m \circ \gamma_0$ and $A^n \circ \gamma_0$ are disjoint unless $A^m = A^n$. $\pi^{-1}(\gamma) = \cup A^n \circ \gamma_0$ consists of $pq = O(A)$ disjoint curves $(q-1)$ of which lie in the interior of each R_j. Each of these $(q-1)$ curves γ_{jk} divides R_j since γ_{jk} and $A^q \circ \gamma_{jk}$ divide R and $(A^q \circ \gamma_{jk}) \cap R_j = \emptyset$. Therefore $\pi^{-1}(\gamma)$ divides R into O(A) parts each of which is bounded by two images of γ_0 and Lemma 1 follows easily.

3.8. *Proof of Lemma* 2. Let p^* be a fixed point of $<A>$ and choose $B \in <A>$ such that $0 < \alpha(B, p^*) \leq |\alpha(A^n, p^*)|$, $1 \leq n < O(A)$. If $p_0 \in R' = R/<A>$ is not a branch point, there is on R' a family of $2g'$ simple closed curves through p_0 which have no other points in common, which contain no branch points and whose homology classes form a canonical homology basis. The complement R'' of these curves is simply connected. We will show that we can modify this system without destroying any of the above properties such that the lifting of each of the modified curves to R is a cycle. If γ is an element of the original system of curves and if σ is a loop around the branch point $\pi(p^*)$, $\sigma \subset R'' \cup \{p_0\}$, then the lifting of one of the curves $\gamma\sigma^n$, $0 \leq n < O(A)$, is a cycle. If this is true for $\gamma\sigma^m$ we replace γ by a simple closed curve γ' which is homotopic to $\gamma\sigma^m$ in $R' - \pi(F(A))$ and which intersects the other curves of the system only in p_0 (see Figure 3 where R' is replaced by a canonical dissection of R').

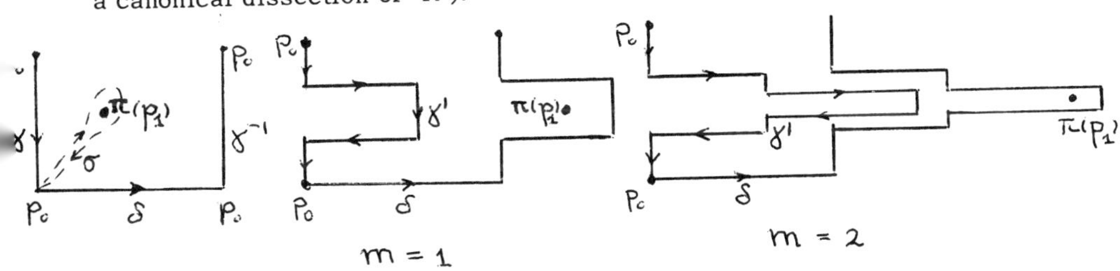

Fig. 3

In this way we obtain the modified homology basis Γ'. We label the branch points $p_1, \ldots, p_{N(A)}$ in such a way that $\alpha(B,p_i)+\alpha(B,p_{i+1})=0$ and choose a simple path $\gamma:[O, N(A)] \to R'' \cup \{p_0\}$ with the property $\gamma(n) = p_n$, $O \leq n \leq N(A)$.

Since the complement of $\gamma \cup \Gamma'$ is simply connected we can find on it O(A) continuous branches of π^{-1}. These branches have a continuous extension to Γ' since the liftings of all curves of Γ' are cycles. For the same reason they can also be extended continuously to $\gamma|_{[0,1]}$. The same is true for $\gamma|_{[2n,2n+1]}$, $1 \leq n \leq n(A)-1$, because $\alpha(B, p_j) = -\alpha(B, p_{j+1})$ whereas an extension across $\gamma|_{[2n-1,2n]} = \gamma_n$ is impossible except if we replace γ_n by a 2-sheeted covering.

If π_1 is a branch of π^{-1} then $\pi_1(R' - \cup\, \gamma_n)$ is a fundamental domain for $\langle A \rangle$ and is bounded on R by the two images of $\pi_1^{-1}(\cup\, \gamma_n)$. Since $|\alpha(B, p_n)| = 2\pi/O(A)$ it follows that B maps one of these two images onto the other and all statements in Lemma 2 are easily verified.

§4. *Conformal symmetric embeddings*

4.1. If we view the compact Riemann surface R as a differentiable surface, let $\mathfrak{A}(R, A)$ be the set of analytic structures on R such that A is homotopic to an automorphism with respect to this analytic structure. Two such structures are equivalent if the corresponding Riemann surfaces can be mapped onto each other by a conformal mapping which is homotopic to the identity. The set of equivalence classes T(R, A) is a metric space with respect to the Teichmüller metric and is homeomorphic to a finite dimensional Euclidean space. The dimension is positive except if $g = 0$. For further information about T(R, A) see [8], [2] and [4].

4.2. If $e: R \to S$ is a symmetric topological embedding, we deform S symmetrically in such a way that the induced mapping is K-quasi-conformal and $(1+\delta)$-quasi-conformal except on a small part of R, where δ is small, too. In order to preserve the symmetry the constructions described in [10] may first be carried out on a compact subset of $S \cap \{(z,x_3) | 0 < \arg z < \alpha$

and extended such that $(z, x_3) \mapsto (z \exp(i\alpha), x_3)$ maps the deformed surface onto itself. Since the deformed surface is an element of $T(R, A)$ by abuse of language, and because $T(R, A)$ is a finite dimensional Euclidean space, the continuity argument used in [9], p. 387, may be employed again to prove our theorem.

4.3. One might ask under what conditions a general group of automorphisms is embeddable. A group of rotations leaving a compact surface invariant is necessarily a group of motions of a regular solid, i.e., it is either cyclic or dihedral except for three isolated cases (the groups of the regular tetrahedron, octohedron and icosahedron). Therefore an embeddable group must be isomorphic to one of these groups.

On the other hand by a theorem of Hurwitz, every finite group is isomorphic to the group of automorphisms of some compact Riemann surface ([6], [5]). This alone excludes in many cases the possibility of embedding simultaneously the full group of automorphisms of a given surface.

It would be interesting to find intrinsic conditions for a cyclic group and an involution to be simultaneously embeddable. Partial answers are contained in 3.4., 3.5., 3.6 and the problem is of course again purely topological.

Acknowledgments. The author wishes to express his thanks to Lipman Bers who suggested the problem and to Jane Gilman, Joseph Lewittes, William Abikoff and Dave Patterson for helpful comments.

FORDHAM UNIVERSITY

REFERENCES

[1]. Accola, R. D. M., Automorphisms of Riemann surfaces, J. Analyse Math. 18 (1967).

[2]. Earle, C. J., On the moduli of closed Riemann surfaces with symmetries, Ann. of Math. Studies, #66, Advances in the theory of Riemann surfaces, Princeton (1971).

[3]. Garsia, A. M., An Embedding of Closed Riemann Surfaces in Euclidean Space, Comment. Math. Helv. 35 (1961).

[4]. Gilman, J. P., Relative Modular Groups in Teichmüller Space Theory, Doctoral dissertation, Columbia University, New York, 1971.

[5]. Greenberg, L., Maximal groups and signatures, these proceedings.

[6]. Hurwitz, A., Über diejenigen Gebilde, welche eindeutige Transformationen in sich zulassen, Math. Ann. 32 (1888), p. 306.

[7]. Klein, F., Über Riemann's Theorie der algebraischen Funktionen und ihrer Integrale, (1882).

[8]. Kuribayashi, A., On analytic families of compact Riemann surfaces with non-trivial automorphisms, Nagoya Math. J., 28 (1966), pp. 119-165.

[9]. Rüedy, R. A., Deformations of embedded Riemann surfaces, Ann. of Math. Studies, #66, Princeton (1971).

[10]. Rüedy, R. A., Embeddings of open Riemann surfaces, Comm. Math. Helv. 46 (1971).

ON THE TRAJECTORY STRUCTURE OF QUADRATIC DIFFERENTIALS

Kurt Strebel

§1. *Introduction*

1. In this paper the trajectory structure of holomorphic quadratic differentials of finite norm on a finite Riemann surface is investigated. The norm of a quadratic differential ϕ on a Riemann surface R is the integral $\iint_R |\phi(z)|\, dxdy$, and a finite Riemann surface is a compact surface minus finitely many connected closed sets. By the Riemann mapping theorem one can always choose local parameters in the neighborhood of the boundary components in such a way that they become analytic Jordan curves or points.

The zeroes of ϕ are called critical points. At each critical point a finite number of trajectories starts (or ends). These are called critical trajectories, no matter how their other end behaves. They are denumerable and can be left aside for double integrals as they only cover a set of measure zero. The other trajectories are called regular.

There are three types of regular trajectories: The closed ones, the spirals (almost closed ones) and the cross cuts. With each of these types we associate characteristic point sets on R, namely respectively ring domains, spiral sets (cluster sets of spirals) and strips.

For some of the results the finiteness of the norm or the finiteness of the Riemann surface or both are not needed.

2. The investigation is justified not so much by the interest into the trajectory structure itself as by the purpose to provide the basic facts for the application of the length area method to quadratic differentials. This

technique uses, besides the Schwarz inequality, two things: The metric induced by ϕ, namely $|\phi(z)|^{\frac{1}{2}}|dz|$, and the mapping $\Phi = \int\sqrt{\phi(z)}\,dz$ along the trajectories in the large.

The Riemann surface is partitioned into sets of trajectories or trajectory intervals. These sets have to be chosen in such a way that they are mapped by a branch of Φ in a one-one way onto some plane sets consisting of intervals on horizontal straight lines. The latter are then introduced as parameter domains with Φ^{-1} as conformal mapping into the surface. On the other hand the trajectories are geodetic lines in the ϕ-metric. Their continuous deformations therefore have a length $\int|\phi(z)|^{\frac{1}{2}}|dz|$ which is not smaller than the length of the corresponding trajectory. The deformation might be given by a quasi-conformal mapping or by other prescriptions. It is at this point where one needs that the parameter domains contain entire trajectories or something like that. One now computes the length of a deformed curve, integrates over the parameter domain and then sums up to get a double integral over the whole surface. To this integral the Schwarz inequality is applied.

3. The easiest to deal with are the closed trajectories. They cover finitely many disjoint ring domains. If we cut such a ring domain along an orthogonal trajectory the resulting quadrilateral is mapped by Φ onto a horizontal rectangle which is one of our parameter domains.

To deal with the spirals one uses the fact that they are almost closed. Their cluster sets, again finite in number on a finite Riemann surface, can be covered by arbitrarily long and thin rectangles, i.e., domains which are mapped by Φ onto horizontal rectangles in the ordinary sense. We use these as parameter domains. But the length inequality for closed trajectories now has to be generalized to closed curves composed of intervals on trajectories and orthogonal trajectories. It is then only approximately true and needs a passage to the limit.

What remains are trajectories which run from the boundary of R to the boundary (cross cuts). They can be gathered into denumerably many horizontal strips of a somewhat more complicated form than rectangles. Homotopic cross cuts with the same boundary points have at least the same length.

4. The length area method is then applied in the Appendix to generalize a theorem by E. Reich and the present author (see [1], Section 1) from the disk to finite Riemann surfaces. Its setting is the following: One considers two quasi-conformal mappings f and $\tilde{f}$ of a finite Riemann surface R onto another such surface R^* which are homotopic modulo the boundary. Let $g = \tilde{f}^{-1}$. The composition $g \circ f$ then is a quasi-conformal self-mapping of R which can be deformed into the identity, the deformation being constant on the boundary. An inequality can be derived which holds for all holomorphic quadratic differentials of finite norm on R (inequality (1.2.8) in [1]). The converse of R. S. Hamilton's theorem as well as the uniqueness theorem for Teichmüller mappings associated with quadratic differentials of finite norm are rather easy consequences of this theorem.

Hamilton's theorem itself is then proved for finite Riemann surfaces. Using Teichmüller's theorem on extremal quasi-conformal mappings of compact Riemann surfaces it is a consequence of the same inequality.

§2. *The length inequality for closed curves*

5. In this section ϕ is a holomorphic quadratic differential, of finite or infinite norm, on an arbitrary Riemann surface R. We will have to deal with the following three types of closed curves γ:

(1) $\gamma = \alpha$, α a closed trajectory

(2) $\gamma = \alpha_1 + \beta_1$. Here α_1 is an interval on a trajectory and β_1 is an interval on an orthogonal trajectory. They have their endpoints and only these in common, but initial point and endpoint of α_1 lie on different borders, say on the upper respectively the lower border, of β_1. (See Figure 1; the drawings are in the Φ-plane, trajectories horizontal.)

In both cases (1) and (2) γ evidently is a Jordan curve.

(3) $\gamma = \alpha_1+\beta_1+\alpha_2+\beta_2$. The situation is as follows: Given an interval β on an orthogonal trajectory. α_1 is an interval on a trajectory with its initial point and endpoint on the same, say the upper side of β, but no other point in common with β. α_2 is another such interval with its initial point and endpoint on the lower side of β and no other points in common with it. The interval β_1 (β_2) on β then joins the endpoint of α_1 (α_2) with the initial point of α_2 (α_1). We assume that the orientation of the two trajectory intervals α_1 and α_2 is chosen in such a way that the two intervals β_1 and β_2 are separated (Figure 2) except for the case where none of the two points belonging to α_2 lies between the two points belonging to α_1 and vice versa (Figure 3). γ is then again a Jordan curve except for the last case. We allow one of the orthogonal intervals β_1 or β_2 to be degenerate.

It is claimed that the length $|\tilde{\gamma}| = \int |\phi(z)|^{\frac{1}{2}} |dz|$ of any closed curve $\tilde{\gamma}$ on R which is freely homotopic to γ is

$$|\tilde{\gamma}| \geq |\alpha| \text{ resp.} \geq |\alpha_1| \text{ in cases (1) and (2),}$$

$$|\tilde{\gamma}| \geq |\alpha_1| + |\alpha_2| \text{ in case (3).}$$

In case (1), equality holds if and only if $\tilde{\gamma}$ is a closed trajectory which is homotopic to α, or possibly one of the two boundary curves of the ring domain swept out by these trajectories (closed curves composed of critical trajectories).

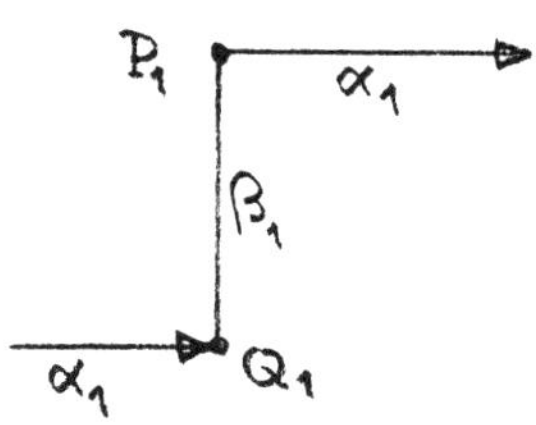

Fig. 1

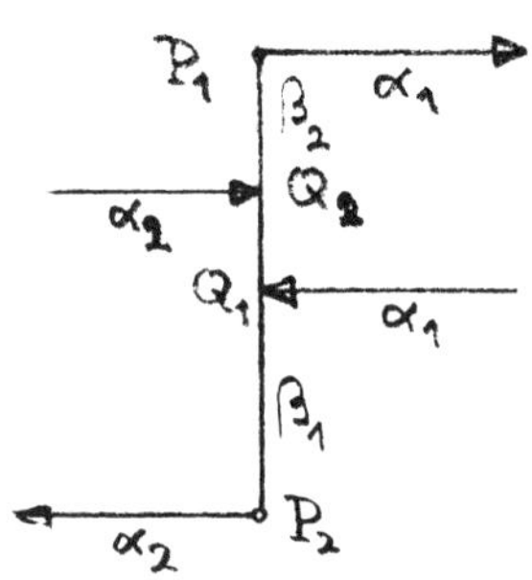

Fig. 2

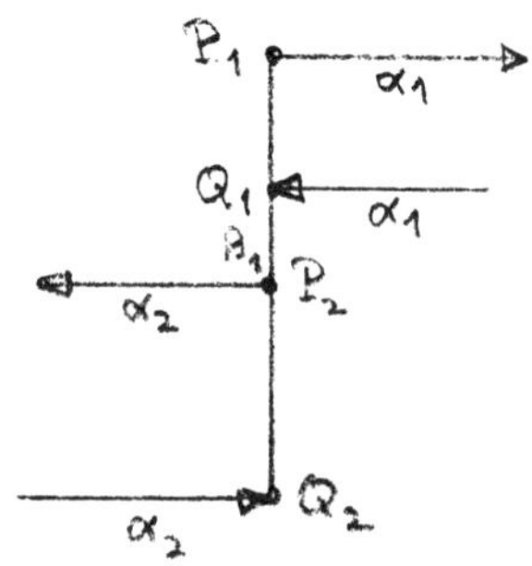

Fig. 3

For the proof we partially refer to [2], no. 17, where cases (1) and (2) are considered.[1] We first lift the curves γ and $\tilde{\gamma}$ as well as the quadratic differential ϕ on a properly chosen doubly connected covering surface R′. This surface can be represented conformally as a plane circular ring domain.

Let $P_1 \in \alpha$ in case (1); in cases (2) and (3) we choose P_1 to be the initial point of α_1. The closed curve γ generates a cyclic subgroup G of the fundamental group F of R with initial point P_1. Let R′ be the unlimited smooth covering surface of R the fundamental group F′ of which, with respect to a given point P_1' above P_1, projects into G.

The surface R′ is doubly connected. Let it be represented by the ring domain $0 \leq r_0 < |z| < r_1 \leq \infty$. The lift ϕ' of ϕ is a holomorphic quadratic differential on R′; the lift γ' of γ is a closed curve through P_1' which generates F′. Finally the curve $\tilde{\gamma}$ can be lifted into a closed curve $\tilde{\gamma}'$ which is freely homotopic to γ' on R′.

We claim that γ' is a Jordan curve containing the boundary component $|z| = r_0$ in its interior. This is clear in cases (1) and (2), as γ is a Jordan curve and the winding number of γ' around $z = 0$ must be ± 1.

To handle case (3) we only have to consider the following configuration: Let P_i, Q_i be the initial point respectively the endpoint of α_i, $i = 1, 2$. Then the order of the four points on β is $P_1\ Q_1\ P_2\ Q_2$, i.e., $\beta_1 \subset \beta_2$ (Figure 3). Now assume $\beta_1' \subset \beta_2'$. Then the closed curve γ' on R′ contains two Jordan curves, namely the two trajectory intervals α_1', α_2' together with the orthogonal intervals joining their endpoints (the two parts of $\beta_2' \setminus \beta_1'$). Both of these must, by the argument principle, contain the boundary component $|z| = r_0$ in their interior. So their winding number is ± 1. On the other hand, the whole curve γ' has winding number ± 1, which is a contradiction.

---

1 It is there tacitly assumed that the two endpoints of α lie on different sides of β.

For the rest of the proof we follow ([2], 17): We consider disjoint rectangles of orthogonal trajectories cutting γ'. Each of these contains exactly one interval of γ' and at least one interval of $\tilde{\gamma}'$. This proves the assertion.

§3. *Spirals*

6. *The cluster set of a spiral.* In this section ϕ and R are again arbitrary. Let α be a trajectory ($\phi(z)dz^2 > 0$ along α; a trajectory consists of regular points only) which is not closed. It is divided, by any point $P_0 \epsilon \alpha$, into two *rays*, say α^+ and α^-. We choose the branch of Φ along α in such a way that α^+ is mapped onto an interval $0 \leq u < u_\infty$ of the real axis, $\Phi(P_0) = 0$. A ray is called *critical*, if it tends towards a critical point; otherwise it is called *regular*. A critical trajectory contains at least one critical ray.

Let α^+ be a regular ray. If, for $u \to u_\infty$, the point $P = \Phi^{-1}(u) \epsilon \alpha^+$ tends to the boundary of R, α^+ is called a *boundary ray*, if not, it is called a spiral ray or a spiral. So if α^+ is a spiral there exists a sequence of real numbers $u_n \to u_\infty$ such that $P_n = \Phi^{-1}(u_n)$ tends to a point $P \epsilon R$. If P is regular and γ denotes the trajectory through P we can construct an adjacent rectangle (Figure 4) through which α^+ passes again and again so that necessarily $u_\infty = \infty$. If P is critical (Figure 5), one can do the same thing in one of the angles bounded by the trajectories emerging from P. So again $u_\infty = 0$. A spiral necessarily has infinite length.

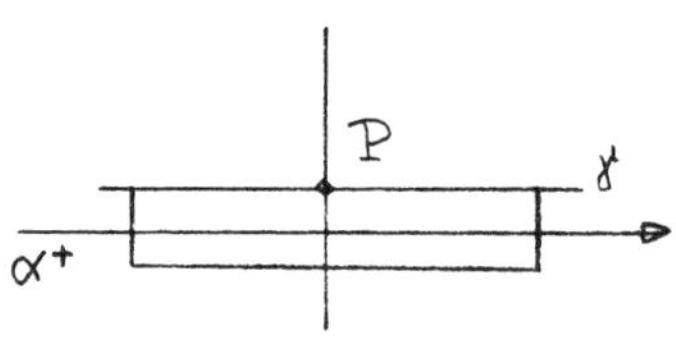

Fig. 4

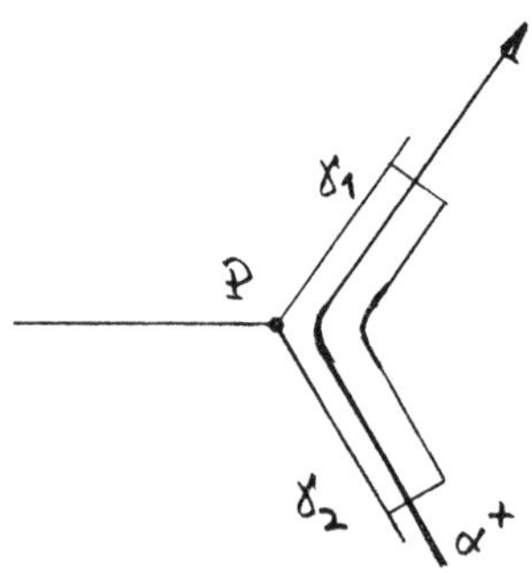

Fig. 5

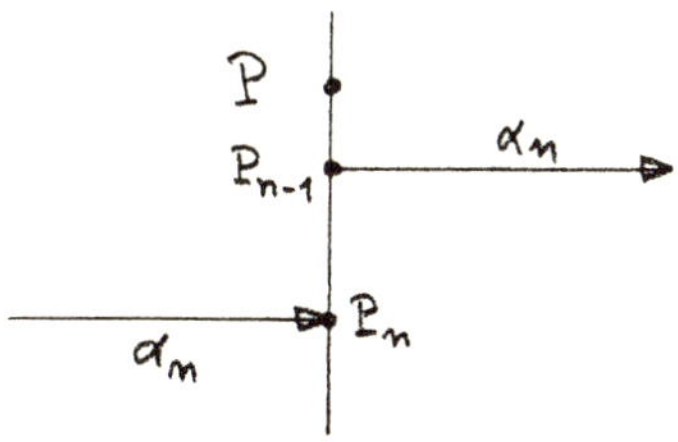

Fig. 6

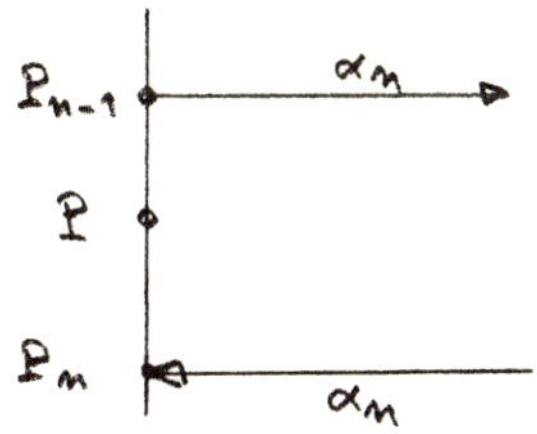

Fig. 7

Let A^+ be the *cluster set* of α^+; i.e., $P \in A^+$ if and only if there is a sequence of real numbers $u_n \to u_\infty$ such that $P_n = \Phi^{-1}(u_n) \to P$. For a regular ray A^+ is not empty if and only if α^+ is a spiral. We call it a spiral set. If $P \in A^+$, P regular, by the same reasoning as above (Figure 4) with arbitrarily long adjacent rectangles, we see that the whole trajectory γ through P lies in A^+. If P is critical this is true for at least two neighboring critical trajectories γ_1 and γ_2 emerging from P (Figure 5).

7. *Introduction of a symmetric surface*. A closed trajectory α determines a ring domain R_α. It is swept out by the closed trajectories of ϕ which are freely homotopic to α. They all have the same ϕ-length $a = |\alpha|$. In order to explore this fact for the spiral sets we are going to introduce an auxiliary symmetric surface.

Consider a regular point P in the cluster set A^+ of the spiral ray α^+. Let β be a closed interval on the orthogonal trajectory through P containing P in its interior. We can assume that the initial point P_0 of α^+ lies on β and that α^+ does not pass through the endpoints of β. We denote the successive intersections of α^+ with β by $P_0, P_1, P_2, \ldots$. Let α_n be the interval on α^+ between P_{n-1} and P_n. It is said to be of the first kind (Figure 6) if P_{n-1} and P_n lie on different borders of β (i.e., α^+ cuts β in the same direction at P_{n-1} and P_n), of the second kind in the opposite case (Figure 7).

We now cut the surface R along β (the surface with the cut is again called R) and glue it with its mirror image $\tilde{R}$ along the two borders β^- and β^+ of β. The new surface $S = R \cup \tilde{R}$ is symmetric with respect to $\beta^- \cup \beta^+$. In order to visualize it we take $\tilde{R}$ as an identical copy of R, put one above the other and glue them along the *same* borders of the cut (not crosswise). We can now continue ϕ by reflection to the whole surface S. Let $\tilde{a}_n$ be the mirror image of a_n. Then $a_n \cup \tilde{a}_n$ is a closed trajectory of ϕ on S. In our model a_n lies on the upper side of R and $\tilde{a}_n$ lies on the lower side of $\tilde{R}$, so if a point P runs through $a_n \cup \tilde{a}_n$ it goes from P_{n-1} to P_n on a_n and backwards above a_n to P_n.

The trajectories $a_n \cup \tilde{a}_n$ determine at most denumerably many ring domains on S, symmetric with respect to $\beta^- \cup \beta^+$. Their intersections with R are rectangles R_k with respect to the quadratic differential ϕ. These rectangles are swept out by trajectory intervals of ϕ. We again distinguish between rectangles R_k of the first and of the second kind.

The whole trajectory a^+ is contained in these rectangles, so

$$A^+ \subset \overline{a^+} \subset \overline{\cup R_k} .$$

This is true for arbitrarily small β.

8. *Let* a^+ *be a spiral with initial point* P_0. *If it cuts the orthogonal trajectory* β *through* P_0 *in every neighborhood of* P_0, *then it also does that in the direction of* a^+ *(which we call the positive direction).*

Proof. Let β_0 be an arbitrarily small closed interval on β which contains P_0 in its interior. Necessarily $P_0 \subset A^+$, as any finite interval on a^+ cuts β_0 only finitely many times, so $\Phi(P_n) = u_n \to \infty$ for the successive crossings P_n of a^+ with β_0. We consider the rectangles R_k induced by the intervals a_n as before. a_1 lies in the interior of one of them, say R_1. It is either of the first kind or of the second kind. For some large enough n $a_n \subset R_1$. If a_1 is of the first kind, a^+ cuts β_0

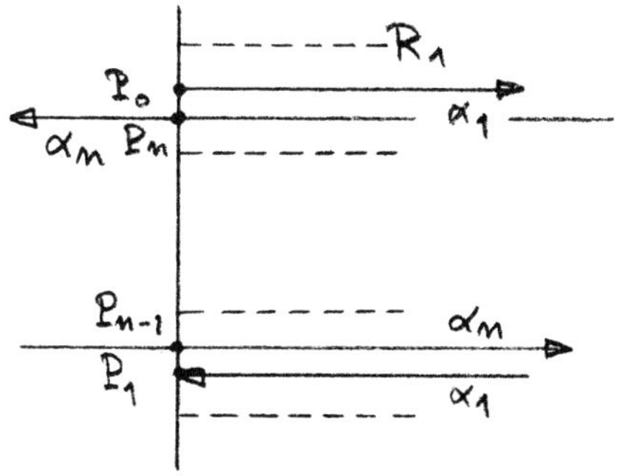

Fig. 8

in the positive direction at P_1. If it is of the second kind (Figure 8), the interval α_n cuts β_0 in the positive direction either at P_n or at P_{n-1} (Figure 8). This proves the assertion.

§4. *Spirals on finite Riemann surfaces*

9. In this section ϕ is an arbitrary holomorphic quadratic differential on a finite Riemann surface R. Then the symmetric surface S is finite and therefore the number of ring domains induced by the closed trajectories of any holomorphic quadratic differential is bounded independently of β. The same bound is valid for the number of rectangles R_k (see Section 7). Let their length be a_k. The spiral ray α^+ is subdivided, by the points P_n, into intervals α_n of lengths a_k. This has a number of consequences.

10. *Let* $P \in A^+$ *be the initial point of a trajectory ray* γ^+ *of infinite length. Then* γ^+ *is a spiral and its cluster set* Γ^+ *contains* γ. *Therefore* $\Gamma^+ = \overline{\gamma}$.

Proof. Let β be an arbitrarily small closed orthogonal interval containing P in its interior. We must consider β as a slit and P as lying on the upper border of it. It is either a boundary point or an interior point of the basis interval β_k of one of the rectangles R_k. But it cannot be a boundary point as γ^+ is regular and has infinite length, whereas R_k has length

$a_k < \infty$. So it must be interior (Figure 9). Let β'_k be the top interval of R_k. The ray γ^+ cuts β in a point P′ interior to β'_k. With $|\beta| \to 0$ the length of the interval on γ^+ between P and P′ must of course tend to infinity, so $P \in \Gamma^+$. The ray γ^+ is a spiral and its cluster set contains P. Q.E.D.

11. *Next we show that equality holds*: $\Gamma^+ = A^+$. *Consequently any spiral ray that is contained in* A^+ *generates* A^+.

Proof. Let $P' \in A^+$. We can of course assume, by a shift along a trajectory, if necessary, that P′ is a regular point. Let β' be a closed orthogonal interval with P′ its midpoint (Figure 10). By β'_0 we denote its contraction towards P′ by the factor $\frac{1}{2}$. Let the successive intersections of α^+ with β'_0 be denoted by $P'_0, P'_1, \ldots$. The lengths of the intervals α'_n are bounded, as there are only finitely many such lengths. Let a be a bound. A rectangle R_0 of length a can be constructed (Figure 11), which contains an interval $|P, Q|$ of length a on γ^+ in its interior. Let β_0 be its basis. We can assume that $|\beta_0| < \frac{|\beta'|}{4}$. The ray α^+ intersects β_0 after P'_0. Therefore γ^+ intersects β'. As, instead of P, we can take any point on γ^+ to start with, $P' \in \Gamma^+$. Therefore $A^+ \in \Gamma^+$, so $A^+ = \Gamma^+$.

If A_1^+ and A_2^+ are two spiral sets with a common point P which lies on a trajectory γ of infinite length, then $A_1^+ = A_2^+$. For one of the rays of γ with initial point P at least is a spiral which generates A_1^+ and A_2^+.

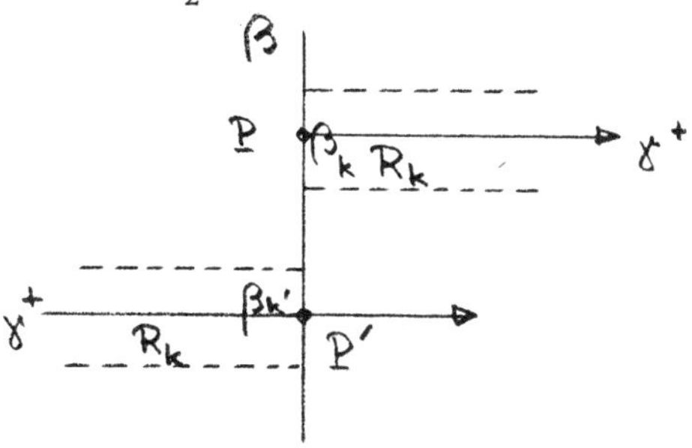
Fig. 9

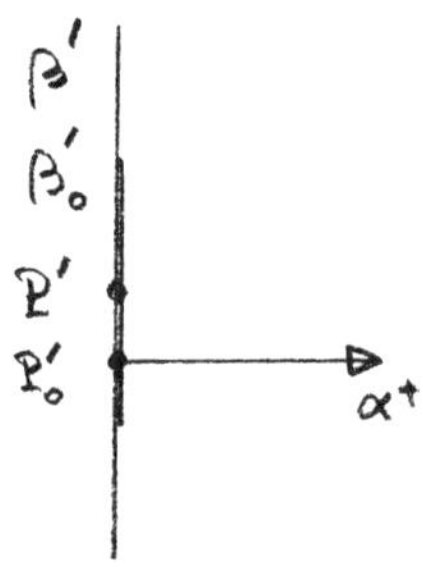
Fig. 10

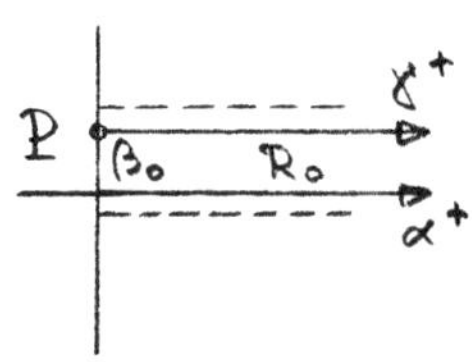
Fig. 11

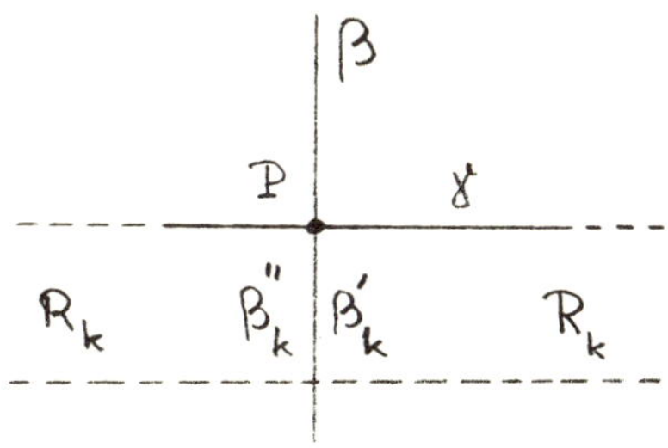

Fig. 12

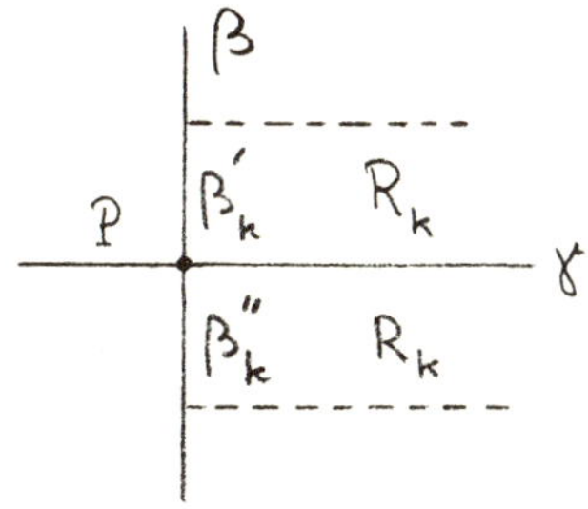

Fig. 13

12. *Let* β_0 *be an arbitrary open orthogonal interval. Then the set* $\overline{A^+ \cap \beta_0}$ *is perfect.*

To prove it we consider a point $P \in A^+ \cap \beta_0$ and an arbitrarily small closed orthogonal interval $\beta \subset \beta_0$ which contains P in its interior. As $P \in A^+$, we can construct the finitely many rectangles R_k defined by β and the ray α^+. The point P is either an interior point or an endpoint of the basis interval β'_k of a rectangle R_k. Assume first that it is an interior point. Then the trajectory γ through P can be continued within R_k and it cuts β at a point $Q \neq P$ (if $Q = P$, γ would be closed, which is impossible). But $\gamma \subset A^+$, so $Q \in A^+ \cap \beta$. Now let P be a boundary point of β'_k. If the corresponding endpoint of the upper side β''_k of R_k does not coincide with P we get another point $Q \in A^+ \cap \beta$ as before. If it coincides and if both β'_k and β''_k lie on different borders of β, then R_k is swept out by closed trajectories (Figure 12). So γ is on the boundary of a ring domain and therefore the ray α^+ must accumulate towards P from the other side. There is a rectangle R_ℓ with a basis interval β'_ℓ which has P as one of its endpoints. But now β''_ℓ cannot coincide with β'_ℓ so we get a point Q as before. The remaining case is the following: The two intervals β'_k and β''_k are on the same, say the upper border of β and have the common endpoint P (Figure 13). But now there is a rectangle R_ℓ with its basis interval β'_ℓ on the lower border of β, $P \in \overline{\beta'_\ell}$. If we would have the last situation again, namely that the top interval β''_ℓ of R_ℓ lies also on the lower border of β and that both β'_ℓ and β''_ℓ have P as common endpoint, then P would be surrounded by closed trajectories, an impossibility. To sum it up we always can find a point $Q \neq P$, $Q \in A^+ \cap \beta$. As β is arbitrary the theorem is proved.

REMARK. We will later show, for quadratic differentials with finite norm, that $\alpha^+ \subset A^+$. Then the above result is immediate, as P is accumulation point of the intersections of α^+ with β which are then points of $A^+ \cap \beta$.

COROLLARY. *As there are only denumerably many critical trajectories and therefore only denumerably many intersections of these trajectories with any orthogonal interval,* A^+ *must contain regular trajectories* (*i.e., no ray of which tends to a critical point*).

§4. *Quadratic differentials with finite norm*

13. In this section we consider quadratic differentials of finite norm on finite Riemann surfaces. *We first prove that* $\alpha^+ \subset A^+$. *It is enough to show that* α^+ *accumulates towards its initial point* P_0. *We claim that this is even true from both sides.* Assume the contrary: Then there is a closed orthogonal interval β_0 with P_0 as one of its endpoints which has no other point in common with α^+. From this follows that there is a closed subinterval $\beta_1 \subset \beta_0$ which contains P_0 such that no ray which emerges from β_1 in the positive direction (i.e., the one defined by α^+) meets β_0 again. If this were not so we would have a sequence of rays γ_n^+ with initial points $P_n \to P_0$ which would meet β_0 again. By an earlier argument they would have bounded length. Let a be a bound. But then an interval of length a on α^+ with initial point P_0 would meet β_0 again.

We now claim that there is an even smaller subinterval β_2 from which no boundary ray γ^+ emerges in the positive direction. Otherwise there would exist a sequence of rays γ_n^+, $n = 1, 2, \ldots$ with initial points

$P_n \to P_0$ such that γ_1^+ and γ_n^+ form, together with the subinterval of β_0 between P_1 and P_n, the boundary of a simply connected domain b_n of R. This follows from the fact that R is finite. We start off with a sequence of boundary rays γ_n^+ the initial points P_n of which tend to P_0. There is a subsequence of rays which tend to the same boundary component Γ of R. By letting Γ shrink to a point and blowing it up again we can assume that they converge towards different points on Γ. If we now cut R along β_0 and glue it with its mirror image along the two borders of β_0 and along Γ we get, by reflecting each γ_n^+, a sequence of disjoint Jordan curves of the form $\gamma_n^+ \cup \tilde{\gamma}_n^+$. Infinitely many of them must belong to the same free homotopy class and therefore bound ring domains. The intersections of those ring domains with R are the simply connected domains we were looking for.

Now let $P \in A^+$, $P \notin \gamma_1^+$. Such a point exists as A^+ does not consist of a single trajectory. Consider an orthogonal interval β through P. If P_n is close enough to P_0 the boundary ray γ_n^+ evidently will have at least three points in common with β, as α^+ does. But this is impossible as we would then find an interval of β in G_n with its two endpoints on γ_n^+.

The rays γ^+ starting from a sufficiently small interval β_2 in the positive direction must be spirals (except for the critical ones). For we have seen that they cannot tend to the boundary. Neither can they be closed, as they don't meet β_0 again. The set of points on R covered by these spirals is schlicht and evidently has infinite area in the ϕ-metric. This contradicts $\|\phi\| < \infty$.

COROLLARY. $A^+ \subset \overline{\alpha^+} \subset \overline{\alpha} \subset A^+$.

Therefore equality holds: *The cluster set of* α^+ *is equal to the closure of* α *which is equal to the closure of* α^+.

14. *Next we want to show that there are only finitely many spiral sets.* We make use of the fact that every spiral ray is contained in the spiral set it generates.

Let $A_1^+ \neq A_2^+$, where A_i^+ is the cluster set of α_i^+ with initial point P_i. Of course the intersection $A_1^+ \cap A_2^+$ does not have to be void. There is a closed orthogonal interval β_1 with P_1 as one of its endpoints such that $\beta_1 \cap A_2^+ = \emptyset$. Otherwise $P_1 \in A_2^+$ and so $A_1^+ = A_2^+$ by 11. It is easy (see [2], Section 11) to find a Jordan curve γ_1 composed by an interval on α_1^+ and the interval on β_1 between its two endpoints (Figures 14 and 15). If we do the same thing for A_2^+, we get a Jordan curve γ_2, and of course $\gamma_1 \cap \gamma_2 = \emptyset$ (Figure 16).

We want to show that γ_1 and γ_2 belong to different homotopy classes.

First, γ_1 is not homotope to a point. If this were the case it would bound a simply connected domain $G \subset R$ and by the argument principle G would have to contain a pole of ϕ.

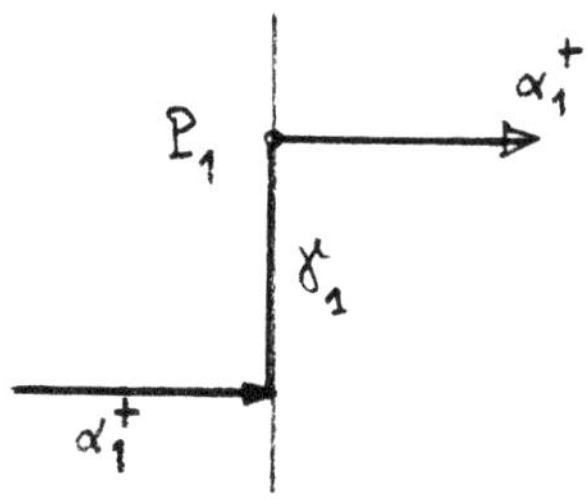

Fig. 14

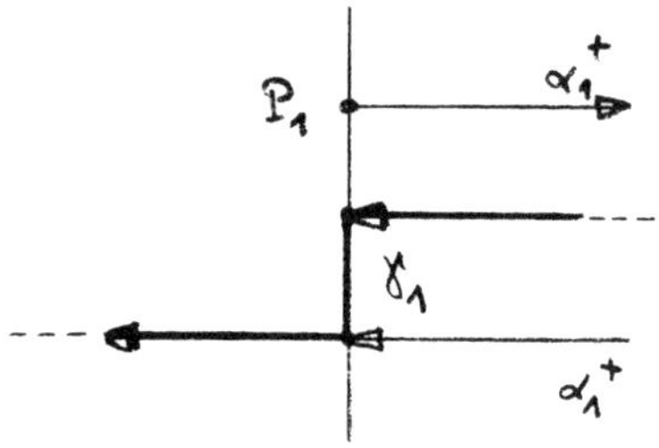

Fig. 15

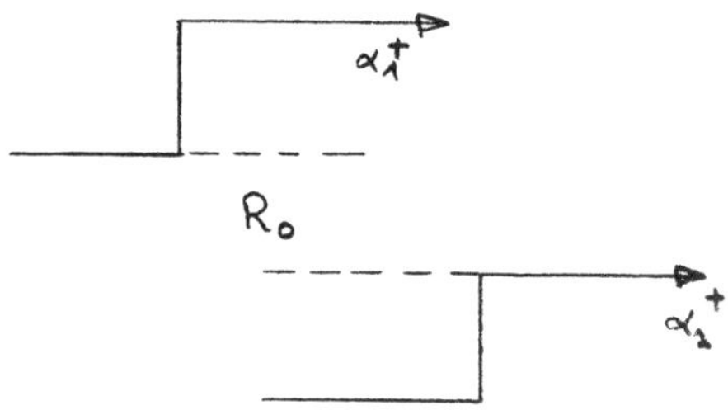

Fig. 16

Secondly, we see that γ_1 and γ_2 belong to different homotopy classes. For otherwise they would bound a ring domain $R_0 \subset R$. Again by the argument principle we would conclude that ϕ does not have a zero in R_0. But then it is easy to see that the continuation of a_1 inside R_0 must meet β_2. As $a_1 \subset A_1^+$, this is a contradiction.

So with each spiral set A_i^+ we can associate a Jordan curve γ_i. These curves are all disjoint and belong to different homotopy classes. As R is finite, the number of such curves is bounded. Actually the number of spiral sets plus the number of ring domains of closed trajectories is at most equal to the maximal number of disjoint Jordan curves which are not homotope and not homotope to zero.

15. *Let* a^+ *be a spiral with initial point* P_0. *Then there is a neighborhood of* P_0 *which has no point in common with a cross cut (trajectory, both rays of which tend to the boundary).*

Assume the contrary. Let β, as before, be a closed orthogonal interval with P_0 one of its endpoints. There is a sequence of points $P_n \in \beta$, $P_n \to P_0$ such that the trajectory γ_n through P_n is a cross cut. By a topological argument similar to the one before we can find a subsequence which we denote again by (γ_n) such that γ_1 and γ_n connect the same boundary components Γ_1 and Γ_2 of R, and, furthermore, bound a simply connected domain G_n of R. By the same argument as before (see Section 13) this would lead to a contradiction.

16. *Covering of a spiral set. Let* a^+ *be a spiral ray of a holomorphic quadratic differential of finite norm on a finite Riemann surface. Then every sufficiently small closed orthogonal interval* β *which contains the initial point* P_0 *of* a^+ *is the basis of a set of finitely many rectangles* R_k *such that the cluster set* A^+ *of* a^+ *is* $A^+ = \cup \overline{R_k}$. ($\beta$ is the basis of the rectangles R_k means that their bottom and top intervals β_k' resp. β_k'' cover both borders of β.)

Proof. First we choose β such that its intersection with every spiral set different from A^+ and with the closure of every ring domain of ϕ is void. This is possible because there are only finitely many spiral sets and ring domains and P_0 is outside the closure of any of them. Next we make β smaller, if needed, such that no cross cut meets β. Let $P \in \beta$, P not on a critical trajectory. Then at least one of the rays of the trajectory γ through P must be a spiral: Call it γ^+. The cluster set Γ^+ of γ^+ contains P and therefore is identical with A^+. Therefore $P \in A^+$. As the set of intersections of β with the critical trajectories is denumerable, this is true for every $P \in \beta$, so $\beta \subset A^+$. Now let $P \in \beta$ be arbitrary. By the construction of the rectangles R_k P is contained in the closure $\overline{\beta_k}$ of the side of a rectangle R_k on the upper edge of β and similarly in a $\overline{\beta_\ell}$ on the lower edge of β. (One simply has to notice that the ray α^+ passes arbitrarily close to P.) This proves the assertion.

§5. *Appendix*

17. We are now in the position of establishing inequality 8, p. 6 of [1], in the more general setting of Theorem 2, p. 7 for finite Riemann surfaces. Let f and $\tilde{f}$ be quasi-conformal mappings of a finite Riemann surface R onto another such surface R^* which are homotopic modulo the boundary of R. Then $f_1 \circ f$, with $f_1 = \tilde{f}^{-1}$, is a quasi-conformal self-mapping of R which is homotopic to the identity modulo the boundary. We denote the differential of f, expressed in local coordinates z of R and w of R^*, by

$$dw = p\,dz + q\,d\bar{z}, \qquad \kappa = \frac{q}{p}\ .$$

Then for every holomorphic quadratic differential ϕ on R with norm $\|\phi\| = 1$ the following *main inequality* holds:

$$1 \leq \iint_R |\phi(z)|\, \frac{\left|1-\kappa\dfrac{\phi}{|\phi|}\right|^2}{1-|\kappa|^2}\, \frac{\left|1-\kappa_1\dfrac{\bar{p}}{p}\dfrac{\phi}{|\phi|}\dfrac{1-\bar{\kappa}\dfrac{\bar{\phi}}{|\phi|}}{1-\kappa\dfrac{\phi}{|\phi|}}\right|^2}{1-|\kappa_1|^2}\, dx\,dy\ .$$

Here $\kappa_1 = \kappa_1(w)$ is the complex dilatation of the mapping f_1.

The analytic part of the proof is the same as it is for the disk (see [1], pp. 5-7). But the trajectory structure of ϕ (respectively the orthogonal trajectory structure) on a Riemann surface imposes three different kinds of parameter domains instead of just one on the disk.

(i) The closed orthogonal trajectories define finitely many ring domains R_i. They are radially cut by a trajectory interval and the Φ-images of these quadrilaterals are used as parameter domains. Any closed orthogonal trajectory is mapped by $f_1 \circ f$ onto a Jordan curve of at least the same length.

(ii) In each spiral set B_j of orthogonal trajectories we choose an interval α_j of a trajectory such that its image $f_1 \circ f(\alpha_j)$ has ϕ-length smaller than a given $\varepsilon > 0$. We construct the rectangles R_{jk} swept out by orthogonal trajectory intervals with their endpoints on α_j. The rectangles of the first kind are used as parameter domains. The image of each orthogonal trajectory interval β_{jk} in R_{jk} has length $\geq |\beta_{jk}| - \varepsilon$. The rectangles of the second kind are ordered in pairs of equal width, each pair consisting of a rectangle with its trajectory sides on the upper border of α_j and another one with its sides on the lower border of α_j. In order to make them have equal width it might be necessary to split them into parts by cutting them along orthogonal trajectories. The length inequality for a pair then holds up to 2ε.

(iii) The rest of the surface is covered by generalized strips of the same kind as for the disk.

We now integrate in the analogous way as in [1] and let ε (and therefore the α_j) tend to zero. This establishes the inequality.

18. The main inequality has the following *corollaries*:

(i) *Let* f_0 *be an extremal quasi-conformal mapping of* R *onto* R^*. *Then for every holomorphic quadratic differential* ϕ *of finite* *norm on* R *and every quasi-conformal mapping* f *of* R *onto* R^* *which is homotopic is* f_0 *modulo the boundary we have,*

$$\frac{1}{K_0} \leq \iint_R |\phi(z)| \frac{\left|1 - \kappa \frac{\phi}{|\phi|}\right|^2}{1 - |\kappa|^2} \, dx\, dy$$

with κ *the complex dilatation of* f *and* K_0 *the maximal dilatation of* f_0 *(i.e., the minimal maximal dilatation in the homotopy class determined by* f*).*

Proof. Let $\tilde{f}$ be an extremal mapping in its homotopy class. Then the maximal dilatation of $f_1 = \tilde{f}^{-1}$ is K_0. As an upper estimate of the third factor of the integrand we immediately get K_0.

(ii) *A Teichmüller mapping associated with a quadratic differential of finite norm is unique extremal.*

Proof. Let f have complex dilatation $\kappa = k \frac{\overline{\phi}}{|\phi|}$, where ϕ is a holomorphic quadratic differential of norm $\|\phi\| = 1$. Then, with $K = \frac{1+k}{1-k}$, we get

$$K \leq \iint_R |\phi(z)| \frac{\left|1 - \kappa_1(w) \frac{\overline{p}}{p} \frac{\phi}{|\phi|}\right|^2}{1 - |\kappa_1(w)|^2} \, dx\, dy .$$

An easy computation gives

$$\frac{\overline{p}(z)}{p(z)} \cdot \frac{\phi(z)}{|\phi(z)|} = \frac{\psi(w)}{|\psi(w)|} ,$$

where ψ is the quadratic differential on R^* induced by the Teichmüller mapping f (f maps the trajectories of ϕ onto those of ψ and $\|\psi\| = K$). So we have

$$K \leq \iint_R |\psi(z)| \frac{\left|1-\kappa_1(w)\frac{\psi(w)}{|\psi(w)|}\right|^2}{1-|\kappa_1(w)|^2} \, dx\, dy \leq K_1 \ .$$

Therefore f is extremal. Let $k = k_1$. Then

$$\kappa_1(w) = -k \frac{\psi(w)}{|\psi(w)|} \quad \text{a.e.,}$$

which shows that $\tilde{f} = f_1^{-1} = f$.

(iii) *Given a quasi-conformal mapping* $f: R \to R^*$. *Choose finitely many points on the boundary curves (if there are any) of* R. *Let* f_n *be the extremal quasi-conformal mapping of* R *onto* R^* *in the class of all mappings which are homotopic to* f *modulo the chosen boundary points. Then, with* $\kappa_n = k_n \frac{\bar{\phi}_n}{|\phi_n|}$, $\|\phi_n\| = 1$, *we get*

$$K_n \leq \iint |\phi_n(z)| \frac{\left|1+\kappa(z)\frac{\phi_n(z)}{|\phi_n(z)|}\right|^2}{1-|\kappa(z)|^2} \, dx\, dy \ .$$

Proof. We write the Teichmüller mapping f_n locally in the form

$$f_n = \Psi_n^{-1} \circ F_n \circ \Phi_n \ ,$$

where $\Phi_n(z) = \int \sqrt{\phi_n(z)}\, dz$, $F_n(\zeta) = K_n\xi + i\eta$, $\zeta = \xi + i\eta$. f_n induces a holomorphic quadratic differential $\psi_n = (\Psi_n')^2$ on R^* with norm $\|\psi_n\| = K_n$. Let $g_n = f_n^{-1}$. Then g_n is a Teichmüller mapping with complex dilatation $-k_n \frac{\bar{\psi}_n}{|\psi_n|}$. Let $\psi_n^* = \frac{\psi_n}{K_n}$. The main inequality now is applied to the surface R^* and the mappings g_n, f. We get

$$1 \leq \iint_{R^*} |\psi_n^*| \frac{(1-k_n)^2}{1-k_n^2} \cdot \frac{\left|1-\kappa \frac{\overline{p_n}}{p_n} \cdot \frac{\psi_n^*}{|\psi_n^*|}\right|^2}{1-|\kappa|^2} \, du\,dv \ ,$$

from which it follows that

$$K_n \leq \iint |\psi_n^*| \frac{\left|1-\kappa \frac{\overline{p_n}}{p_n} \frac{\psi_n^*}{|\psi_n^*|}\right|^2}{1-|\kappa|^2} \, du\,dv \ .$$

An easy computation shows that

$$\frac{\overline{p_n}}{p_n} \frac{\psi_n^*}{|\psi_n^*|} = - \frac{\phi_n}{|\phi_n|} \ ,$$

$$|\psi_n^*| \, du\,dv = |\phi_n| \, dx\,dy$$

which proves the theorem.

The inequalities (i) and (iii) can now, as in [1] be used to prove necessity and sufficiency of Hamilton's condition.

UNIVERSITY OF ZURICH

REFERENCES

[1]. Reich, E., and Strebel, K., Extremal quasi-conformal mappings with given boundary values. Forschungsinstitut für Mathematik, ETH, Zürich, 1972.

[2]. Strebel, K., Quadratische Differentiale mit divergierenden Trajektorien, Colloquium on Mathematical Analysis, Jyväskylä 1970, Springer lecture notes, 1973.

THE MAXIMAL INSCRIBED BALL OF A FUCHSIAN GROUP

Jacob Sturm and Meir Shinnar

The purpose of this paper is to prove the following:

THEOREM. *There exists a* $\delta > 0$ *with the following property: Let* Γ *be a Fuchsian group and let* N *be any normal polygon for* Γ. *Then* N *contains a non-Euclidean ball,* B_δ, *of radius* δ.

This result was first obtained by W. Harvey as a corollary to the Kazdan-Margulis Theorem. He also obtained a geometrical proof for finite sided polygons (verbal communication). Marden [3] has proven the theorem independently, but his proof is different from the one presented here. He also discusses the relationship with the Kazdan-Margulis theorem. Greenberg has pointed out that if one allows N to be a fundamental domain which is a convex polygon whose sides are conjugate in pairs, the same proof will work.

The research for this paper was done while the authors were participants in an NSF summer undergraduate program at SUNY in Stony Brook. Our adviser there was Professor Bernard Maskit to whom we extend our warm thanks for his helpful advice.

Throughout this proof we make use of the Poincaré model of plane non-Euclidean geometry in U, the closed unit disc in the complex plane centered at O. The prefix "H-" will mean "in the non-Euclidean sense" or, more simply, "non-Euclidean." If R is a measurable subset of U, $|R|$ will denote its H-area. The upper half plane, the unit circle and the lower half plane are denoted H^+, Q, and H^- respectively. We set $U^+ = U \cap H^+$ and $U^- = U \cap H^-$. We use the method of [4] to construct a normal polygon N for a Fuchsian group whose fixed circle is Q. Finally, $v \in \overline{N}$ is called an ordinary vertex if $v \in \mathring{U}$ and if v is the intersection of two sides of N which are not H-colinear. Hence the angle at an ordinary vertex is always strictly less than π.

It suffices to prove the theorem in the case where the center of N is O. If its center is at w, there exists a transformation T mapping U onto U and w to O, and TN is a normal polygon for $T\Gamma T^{-1}$ centered at O.

LEMMA 1. *Fix* $\varepsilon > 0$. *Let* $S_\varepsilon = \{$*H-triangles,* $T : |T| \geq \varepsilon\}$. *Then there exists* δ_ε *such that* B_{δ_ε} *is contained in each element of* S_ε.

Proof. Let $T \in S_\varepsilon$ and let a_T be the intersection of the angle bisectors of T and let B_T be the H-ball centered at a_T with radius r_T, tangent to the three sides of T. If the lemma is false, there exists a sequence $\{T_n\}_{n=1}^{\infty} \subset S$ such that $r_{T_n} \to 0$ as $n \to \infty$. Translate each T_n by a Mobius transformation so that a_{T_n} is mapped to O, and call the new triangle T_n again. Now as $n \to \infty$, $r_{T_n} \to 0$, the three sides of T_n pass arbitrarily close to O, and their curvature approaches O. Therefore the sum of the vertex angles of T_n approaches π and $|T_n| \to 0$, a contradiction.

LEMMA 2. *Let* N *be a normal polygon for any Fuchsian group, and let* $\mathcal{O}$ *be an ordinary cycle of vertices of* N, $\mathcal{O} = \{v_1, \ldots, v_r\}$. *Let* ω_i *be the angle at* v_i, $0 \leq \omega_i < \pi$. *Then* $\frac{1}{r} \cdot \sum_{i=1}^{r} \omega_i \leq \frac{2\pi}{3}$.

Proof. We know from [1] that $\sum_{i=1}^{r} \omega_i = \frac{2\pi}{\ell}$ where ℓ is the order of the stabilizer group of any v_i, $1 \leq \ell < \infty$. If $\ell = 1$, then $r \geq 3$ since $\omega_i < \pi$, so $\frac{1}{r}\sum_{i=r}^{r}\omega_i \leq \frac{1}{3}\cdot 2\pi$. If $\ell = 2$ then $r \geq 2$ and $\frac{1}{r}\sum_{i=1}^{r}\omega_i \leq \frac{1}{2}\pi < \frac{2\pi}{3}$. If $\ell \geq 3$ then $\frac{1}{r}\sum_{i=1}^{r}\omega_i = \frac{1}{r}\frac{2\pi}{\ell} \leq \frac{2\pi}{3}$ since $r \geq 1$. This lemma holds trivially for parabolic cycles since the angle at each vertex of a parabolic cycle is O.

COROLLARY. *If* N *has* n *sides, then* $|N| \geq \frac{\pi}{3}n - 2\pi$.

Proof. Let the vertices of N be $v_1, \ldots, v_n$ and let ω_i be the angle at v_i. Triangulate N into n triangles by drawing H-lines from the center of N to all the vertices. Then, by the Gauss-Bonet theorem, $|N| = \sum_{i=1}^{n}(\pi - \omega_i) - 2\pi$. But Lemma 2 $\Longrightarrow \sum_{i=1}^{n}\omega_i \leq n\frac{2\pi}{3}$. Hence $\sum_{i=1}^{n}(\pi - \omega_i) = n\pi - \sum_{i=1}^{n}\omega_i \geq n\pi - n\frac{2\pi}{3}$. Therefore $|N| \geq n\frac{\pi}{3} - 2\pi$.

Proof of theorem

Case 1. $|N| < \infty$, $n < 24$. Triangulate N from the center into triangles $T_1, \ldots, T_n$. Since $|N| = \sum_{i=1}^{n}|T_i|$ and since $|N| \geq \frac{\pi}{21}$ we must have $|T_i| \geq \frac{\pi}{21\cdot 24}$ for some i, $1 \leq i \leq n$. By Lemma 1, B_δ is contained in T_i and therefore in N, where $\delta = \delta_\varepsilon$, $\varepsilon = \frac{\pi}{21\cdot 24}$.

Case 2. $|N| < \infty$, $n \geq 24$. Triangulate again as in Case 1. By the corollary, $|N| \geq n\frac{\pi}{3} - 2\pi$ or $\sum_{i=1}^{n}|T_i| \geq \frac{\pi}{4}n$. Hence, for some i we must have $|T_i| \geq \frac{\pi}{4} \Longrightarrow B_\delta$ is contained in T_i and therefore in N.

Next we assume $|N| = \infty$. As remarked earlier, we may assume without loss of generality that the center of N is at O. If N has a free side, B_δ is certainly contained in N. Hence we may assume N has infinitely many sides and no free sides. The following cases remain:

Case 3. $|N| = \infty$, infinitely many vertices of N lie on Q. Assume infinitely many lie on $H^+ \cap Q$. Choose three such vertices, v_1, v_2, and v_3 such that $i < j \Longrightarrow \arg v_i < \arg v_j$. Draw the H-lines Ov_1, Ov_2, Ov_3, v_1v_2, and v_2v_3, forming triangles T_1 and T_2. Call the angle of T_i at O α_i. Now $\alpha_1 + \alpha_2 \leq \pi \Longrightarrow$ or some i, i = 1 or 2, $\alpha_i \leq \frac{\pi}{2}$. Therefore, $|T_i| = \pi - \alpha_i \geq \frac{\pi}{2}$ and B_δ is contained in T_i and therefore in N.

Case 4. $|N| = \infty$ and N has infinitely many ordinary vertices. Since each ordinary cycle is finite, Lemma 2 $\Longrightarrow S = \{$ordinary vertices v of N: the angle at v is $\leq \frac{2\pi}{3}\}$ is infinite. As before, assume $S \cap U^+$ is infinite since $S \cap [-1, 1]$ is finite (no three vertices of N are H-colinear). Fix $\varepsilon > 0$. Choose five elements of $S \cap U^+$, $v_1, \ldots, v_5$, such that $i < j \Longrightarrow \arg v_i < \arg v_j$ and such that $\arg \frac{v_{i+1}}{v_i} \leq \varepsilon$, $i = 1, 2, 3, 4$. This is possible since $S \cap U^+$ is infinite and since no three vertices of N are H-colinear. Triangulate as before: Since $\beta + \alpha_6 \leq \frac{2\pi}{3}$, we may assume $\alpha_6 \leq \frac{\pi}{3}$ (if $\beta \leq \frac{\pi}{3}$ a similar proof will work). Now

$$|T_1| + |T_2| = 2\pi - \sum_{i=1}^{6} \alpha_i \quad \text{but} \quad \sum_{i=1}^{6} \alpha_i < \varepsilon + \varepsilon + \frac{2\pi}{3} + \frac{2\pi}{3} + \frac{\pi}{3}$$

therefore $|T_1| + |T_2| > \frac{\pi}{3} - 2\varepsilon \Longrightarrow$ one of the T_i has area greater than $\frac{\pi}{6} - \varepsilon$. Since ε may be chosen arbitrarily small, B_δ is contained in T_i and the theorem is completed.

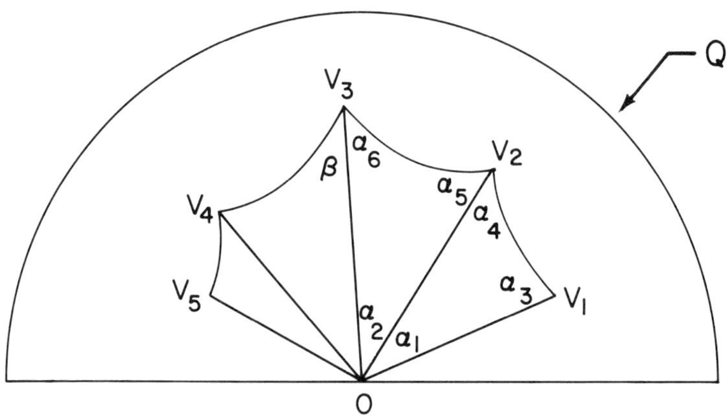

Using Lemma 1, one is able to calculate the δ of the theorem. A quick estimate shows that $\delta = \frac{1}{6} \cdot \frac{1}{24 \cdot 21}$ will work, although this is not sharp.

COLUMBIA UNIVERSITY

REFERENCES

[1]. Lehner, Joseph, *A Short Course in Automorphic Functions*, Holt, Rinehart and Winston, New York, 1966.

[2]. Lehner, Joseph, *Discontinuous Groups and Automorphic Functions*, American Mathematical Society Surveys, Providence, 1964.

[3]. Marden, A., *Universal Properties of Fuchsian Groups in the Poincaré Metric*, (these proceedings).

[4]. Siegel, C. L., *Topics in Complex Function Theory*, vol. II, Wiley-Interscience, New York, 1971.

Library of Congress Cataloging in Publication Data

Main entry under title:

Discontinuous groups and Riemann surfaces.

(Annals of mathematics studies, no. 79)
Includes bibliographies.
1. Riemann surfaces—Congresses. 2. Discontinuous groups—Congresses. I. Greenberg, Leon, ed.
II. Series.
QA333.D57 1974 512′.2 73-16783
ISBN 0-691-08138-7